FACILITY LOGISTICS

Series on Resource Management

Rightsizing Inventory
by Joseph L. Aiello
ISBN: 0-8493-8515-6

Integral Logistics Management: Operations and Supply Chain Management in Comprehensive Value-Added Networks, Third Edition
by Paul Schönsleben
ISBN: 1-4200-5194-6

Supply Chain Cost Control Using Activity-Based Management
Sameer Kumar and Matthew Zander
ISBN: 0-8493-8215-7

Financial Models and Tools for Managing Lean Manufacturing
Sameer Kumar and David Meade
ISBN: 0-8493-9185-7

RFID in the Supply Chain
Judith M. Myerson
ISBN: 0-8493-3018-1

Handbook of Supply Chain Management, Second Edition
by James B. Ayers
ISBN: 0-8493-3160-9

The Portal to Lean Production: Principles & Practices for Doing More With Less
by John Nicholas and Avi Soni
ISBN: 0-8493-5031-X

Supply Market Intelligence: A Managerial Handbook for Building Sourcing Strategies
by Robert Handfield
ISBN: 0-8493-2789-X

The Small Manufacturer's Toolkit: A Guide to Selecting the Techniques and Systems to Help You Win
by Steve Novak
ISBN: 0-8493-2883-7

Velocity Management in Logistics and Distribution: Lessons from the Military to Secure the Speed of Business
by Joseph L. Walden
ISBN: 0-8493-2859-4

Supply Chain for Liquids: Out of the Box Approaches to Liquid Logistics
by Wally Klatch
ISBN: 0-8493-2853-5

Supply Chain Architecture: A Blueprint for Networking the Flow of Material, Information, and Cash
by William T. Walker
ISBN: 1-57444-357-7

ERP: Tools, Techniques, and Applications for Integrating the Supply Chain
by Carol A. Ptak with Eli Schragenheim
ISBN: 1-57444-358-5

Introduction to e-Supply Chain Management: Engaging Technology to Build Market-Winning Business Partnerships
by David C. Ross
ISBN: 1-57444-324-0

Supply Chain Networks and Business Process Orientation
by Kevin P. McCormack and William C. Johnson with William T. Walker
ISBN: 1-57444-327-5

Collaborative Manufacturing: Using Real-Time Information to Support the Supply Chain
by Michael McClellan
ISBN: 1-57444-341-0

The Supply Chain Manager's Problem-Solver: Maximizing the Value of Collaboration and Technology
by Charles C. Poirier
ISBN: 1-57444-335-6

Lean Performance ERP Project Management: Implementing the Virtual Supply Chain
by Brian J. Carroll
ISBN: 1-57444-309-7

Integrated Learning for ERP Success: A Learning Requirements Planning Approach
by Karl M. Kapp, with William F. Latham and Hester N. Ford-Latham
ISBN: 1-57444-296-1

Basics of Supply Chain Management
by Lawrence D. Fredendall and Ed Hill
ISBN: 1-57444-120-5

Lean Manufacturing: Tools, Techniques, and How to Use Them
by William M. Feld
ISBN: 1-57444-297-X

Back to Basics: Your Guide to Manufacturing Excellence
by Steven A. Melnyk and R.T. Chris Christensen
ISBN: 1-57444-279-1

Enterprise Resource Planning and Beyond: Integrating Your Entire Organization
by Gary A. Langenwalter
ISBN: 1-57444-260-0
ISBN: 0-8493-8515-6

FACILITY LOGISTICS

Approaches and Solutions to Next Generation Challenges

MAHER LAHMAR

Auerbach Publications
Taylor & Francis Group
New York London

CRC Press is an imprint of the
Taylor & Francis Group, an **informa** business

Auerbach Publications
Taylor & Francis Group
6000 Broken Sound Parkway NW, Suite 300
Boca Raton, FL 33487-2742

Printed in the United States of America on acid-free paper
10 9 8 7 6 5 4 3 2 1

International Standard Book Number-13: 978-0-8493-8518-6 (Hardcover)

Library of Congress Cataloging-in-Publication Data

Facility logistics : approaches and solutions to next generation challenges / edited by Maher Lahmar.
p. cm. -- (Engineering management innovation series)
Includes bibliographical references and index.
ISBN-13: 978-0-8493-8518-6 (alk. paper)
ISBN-10: 0-8493-8518-0 (alk. paper)
1. Warehouses--Management. 2. Business logistics. 3. Warehouses--Management--Mathematical models. 4. Warehouses--Management--Data processing. 5. Physical distribution of goods--Management--Data processing. I. Lahmar, Maher.

HF5485.F33 2008
658.7'85--dc22 2007027061

Visit the Taylor & Francis Web site at
http://www.taylorandfrancis.com

and the Auerbach Web site at
http://www.auerbach-publications.com

Contents

Preface

Since the early days of the industrial revolution, facility logistics have played a central role in the operations of enterprises. With the advances in automation, computation hardware and software, and the emergence of online retailing, facility logistics practices have evolved to parallel the changes in the scope and scale of manufacturing and service enterprises. In today's competitive environment, firms are expected to rely on modern management techniques and technological advancements to offer customers a wide range of products at lowest possible prices, while ensuring the efficiency of their supply chains and the profitability of their operations. The design of facility layouts, warehouses, and material-handling systems is often an important decision that can significantly impact the success of industrial projects. Equally important is the efficient management of logistics operations within facilities that can reduce handling times, increase inventory turnover, and improve accounting accuracy, which allows supply chains to be more flexible and responsive.

The trends toward more globalization and outsourcing and the emphasis on facility sustainability and security may have provided solutions to some of firms' problems but also raised further logistics challenges. The material in this book is mainly driven by the recent trends in technology, industrial practices, and business environments. In Chapter 1 the main trends impacting facility logistics operations—visibility, security, flexibility, labor, globalization, and sustainability—are discussed in detail. This chapter provides valuable guidelines to managers on the ongoing and future changes that will impact the design and management of facility logistics. It also offers insights on where future research in the facility logistics area is heading. Chapter 2 reviews the basic functionalities of warehouse management systems (WMS) and examines their capabilities. It also addresses the current shortcomings of WMS and lists potential improvement opportunities. A warehouse manager can use this reference to understand the limitations of current WMS, evaluate his or her objectives, and assess whether investing in such systems will meet these objectives. Chapter 3 outlines a comprehensive yet simple method that allows a quick assessment of warehouse performance. This method can help managers and students visiting the facility to quickly identify the weaknesses and strengths of the warehouse operations.

Chapters 4 through 10 present a set of solution approaches to emerging challenges in facility logistics. The design and planning algorithms included in these chapters intend to provide any interested person having a minimal background in operations research and mathematical modeling, but with a basic understanding of facility logistics, to explore new techniques and managerial insights. Specifically, Chapters 4 and 5 address two issues related to the integration of different facility logistics decisions—the concurrent design of facility layouts and material-handling systems and the insertion of cross-aisles in warehouses. Chapter 6 discusses the importance of the stochastic modeling of facility layout problems and presents models that can capture the effects of variability in layout and logistical decisions. Chapter 7 uses stochastic modeling to compare different protocols for staging pallets in crossdocking facilities. Chapters 8 through 10 address management issues related to different types of material-handling systems. In particular, Chapter 8 provides insights on the impact of different automated storage/retrieval systems (AS/RS) dispatching rules on the throughput performance. Chapter 9 reviews a wide range of applications of carousel systems, existing analytical models, and discusses challenging open carousel-related problems. Chapter 10 offers a unifying and comprehensive treatment of policies that are effective, and efficient logical control solutions to avoid deadlocks in a variety of problems. To fully grasp the material presented in this chapter, it is preferable that the reader be familiar with logical control. Practitioners may use the findings of these chapters as guidelines to conduct changes to the design of their facilities. These chapters may help academicians interested in the facility logistics area identify future research topics or create teaching material packages.

The teaching and learning processes have been significantly metamorphosed with the boom in Internet usage, the proliferation of online content, and the commercialization of user-friendly developing tools. Facility logistics instructors have to draw from these advances to continuously update their educational materials and tools. Chapter 11 provides a review of educational resources available to instructors of facility logistics-related courses and training workshops. It also presents examples of how multimedia tools can be used to develop new teaching material.

This book is unique in the sense that it combines a presentation of the challenges facing facility logisticians, research advances in the area, and modern teaching tools and resources. While many professionals, students, and educators may enjoy reading this book, the intended audience for this book is the large and heterogeneous group of researchers, instructors, and practitioners who must, due to the nature of their work, deal with the modeling, simulation, design, management, and analysis of complex facility logistics systems. Researchers can use this book to familiarize themselves with the latest trends that are shaping the industry and driving research and current achievements in the area as well as future topics and research directions. Instructors will find innovative educational material that will make teaching facility logistics more effective and learning a more enjoyable experience for students. Students at the senior and graduate level can use this book as a concise yet comprehensive reference that presents the latest trends and covers the essentials about layout design, warehousing, and material-handling systems. Practitioners who are interested in the cutting-edge management and design tools will benefit from the algorithms, heuristics, innovative practices, and the variety of decision-support systems presented in this book. A set of references are available at the end of each chapter to satisfy the needs of those who are interested in deeper knowledge on the subject.

The preparation of this book would not have been possible without the support and help of several colleagues and partners. I would like to extend my thanks to Dr. Omar Al-Araidah, Mehmet Can Arslan, Dr. John Bartholdi, Dr. James H. Bookbinder, Dr. Yavuz Bozer, Dr. Gürdal Ertek, Dr. Kevin Gue, Dr. Elkafi Hassini, Dr. Sunderesh S. Heragu, Dr. James K. Higginson, Bilge Incel, Dr. Keebom Kang, Dr. M.B.M. de Koster, Dr. Ananth Krishnamurthy, Dr. Charles J. Malmborg, Dr. Gang Meng, Dr. Mike Ogle, Dr. Spyros Reveliotis, and Dr. Henk Zijm for accepting my invitation to contribute to this book. Auerbach Publications has been extraordinarily supportive and patient with the process: a thousand thanks to Raymond O'Connell, and Dr. Vinithan Sethumadhavan.

I am also very grateful to Dr. Michael Ogle, the Vice President of the Technical and Engineering Services at the Material Handling Industry of America, for his assistance and guidance through the different phases of the book preparation. I am also thankful to Dr. Kevin Gue from Auburn University for his numerous suggestions that significantly improved the content and the format of this book. Finally, I am deeply indebted to Dr. Hamid Parsaei, the chair of the Department of Industrial Engineering at the University of Houston, for his help and encouragements during the preparation phase of the book proposal.

Maher Lahmar PhD

Editor

Maher Lahmar received the BS and MS degrees in industrial engineering from Bilkent University, Ankara, Turkey, and a PhD in industrial engineering from the University of Minnesota, Twin Cities, United States. He is currently an assistant professor in the Department of Industrial Engineering at the University of Houston and the director of the Enterprise Logistics Laboratory. His research interests are in the areas of manufacturing systems, facility logistics, production and inventory control, and supply chain management. His articles have appeared in a variety of journals including *IIE Transactions, IEEE Transactions in Robotics and Automation*, and *International Journal of Manufacturing Systems*. He was a finalist in the 2000 MSOM Student Best Paper Award competition and he received the third place in the 2006 Outstanding Material Handling and Logistics Research Paper Award competition. Dr Lahmar serves on the Texas Department of Transportation (TxDOT) Technical Assistance Panel (TAP) for the Research Management Committee (RMC2). He is also a member of The Institute for Operations Research and the Management Sciences (INFORMS), the Institute of Industrial Engineers (IIE), and the Society of Manufacturing Engineers (SME).

Contributors

Omar Al-Araidah
Department of Industrial Engineering
Jordan University of Science and Technology
Irbid, Jordan

Mehmet Can Arslan
Faculty of Engineering and Natural Sciences
Sabancı University
Istanbul, Turkey

John J. Bartholdi, III
School of Industrial and Systems Engineering
Georgia Institute of Technology
Atlanta, GA

James H. Bookbinder
School of Industrial and Systems Engineering
Department of Management Sciences
University of Waterloo
Waterloo, Ontario, Canada

Yavuz A. Bozer
Department of Industrial and Operations Engineering
University of Michigan
Ann Arbor, MI

Gürdal Ertek
Faculty of Engineering and Natural Sciences
Sabancı University
Istanbul, Turkey

Kevin R. Gue
Department of Industrial and Systems Engineering
Auburn University
Auburn, AL

Elkafi Hassini
DeGroote School of Business
McMaster University
Hamilton, Ontario, Canada

Sunderesh S. Heragu
Department of Industrial Engineering
University of Louisville
Louisville, KY

James K. Higginson
Odette School of Business
University of Windsor
Windsor, Ontario, Canada

Bilge Incel
Roketsan Inc.
Ankara, Turkey

Keebom Kang
Graduate School of Business and Public Polidy
Naval Postgraduate School
Monterey, CA

M.B.M. de Koster
Department of Decision and Information Sciences
RSM Erasmus University
Rotterdam, The Netherlands

Ananth Krishnamurthy
Department of Decision Sciences and Engineering Systems
Rensselaer Polytechnic Institute
Troy, NY

Charles J. Malmborg
Department of Decision Sciences and Engineering Systems
Rensselaer Polytechnic Institute
Troy, NY

Gang Meng
Oracle, USA, Inc.
Cambridge, MA

Mike Ogle
Industry Council on Material Handling Education
Material Handling Industry of America
Charlotte, NC

Spyros Reveliotis
School of Industrial and Systems Engineering
Georgia Institute of Technology
Atlanta, GA

Henk Zijm
Faculty of Electrical Engineering, Mathematics and Computer Sciences
University of Twente
Enschede, The Netherlands

Chapter 1

Facility Logistics: Forces of Change

Mike Ogle

CONTENTS

Abstract Once considered simply a cost center concentrating on productivity and quality, facility logistics must now concern itself with increased challenges related to visibility, security, flexibility, labor, globalization, and sustainability to act as a key enabling link in the global supply chain. This chapter addresses these challenges and their likely impact. It then provides an overview of the technologies and techniques that have been developed and are being developed to address those challenges.

1.1 Challenges

Facilities are traditionally designed to perform value-added processes as one link in a value chain. They manage space, resources (including people), and time to facilitate the transformation and flow of materials and information. It must be emphasized that facility design and management does not end at the four walls of the facility. Many forces "outside the box" affect how facilities should be designed and operated. Local optimization of processes, work cells, departments, and facilities may work against the strategic and tactical objectives of the organization or the value chain of suppliers and customers.

End users (consumers and businesses) want everything "now" in an ever-increasing variety with 24/7/365 visibility into inventory status and expected delivery dates and times. They also want to change their mind and their tastes at any time and expect instant response on the part of their suppliers. Waste is expected to be at a minimum to get the job done. At the same time, energy, labor, and transportation costs rise and all links in the chain seem to be exposed to increasing uncertainty due to tight multimodal capacity and security concerns. Along with the rise in costs and risk, service-level expectations continue to increase (e-commerce, next- or same-day delivery, always-in-stock guarantees, specific delivery windows with penalties, etc.), even as facilities become leaner. All these forces of change impact how the future facilities must be designed to provide solutions that provide desired customer service levels at the lowest cost per item, box, pallet, or shipment while hedging bets against future product mixes.

This chapter focuses on today's most influential forces of change defined in the list below:

- Visibility: Total knowledge of the status and plans for every resource and item that may be flowing into, within, or out of the facility.
- Flexibility: Flexible facilities prepare for and adjust to change by utilizing plans, resources, people, and products that are adaptable and scalable.
- Labor and Management: Facility labor and management are driven by a culture of responsibility and self-improvement to support the goals of the facility and its customers.
- Green and Sustainability: Green and sustainable facilities are energy efficient, support reuse, and minimize the waste stream.
- Security: Secure facilities protect the physical and intellectual assets of the facility while supporting their productive capability.
- Globalization: Global facilities apply resources to minimize differences in language, regulation, culture, and methods to support the multimodal, global flow of goods and services.

You may notice significant overlap in the definitions of the forces. Unfortunately, the forces identified cannot easily be separated, divided among groups of people, and addressed using stand-alone initiatives. Savvy managers of facility logistics must learn how to apply a balance of responses to the forces to be successful. This balance is going to be different for each company and potentially for each facility within a company's network of facilities. The greatest overriding theme and the one that has the greatest affect on the above-mentioned challenges is uncertainty. Uncertainty is the enemy of lean, clouds visibility, impedes flexibility, and stresses both management and labor. Globalization extends planning horizons and adds several new uncertainties to supply chains in the form of culture and procedures. If one finds ways to reduce uncertainty, improvements in many measures will also be found.

Although the forces identified have a great influence on facility logistics, they are not the only forces of change. The six forces were chosen due to their clear impact, the types of tools that can be utilized to deal with them, and the relative newness of their importance. The greatest force with the greatest potential impact is to provide the information necessary to plan or react to the remainder of the forces. For that reason, visibility is the first force to be addressed.

1.1.1 Visibility

The greatest influence on facility logistics comes from the need for visibility. The flow of information has become as important as the flow of goods and services. Visibility in its purest facility logistics definition means that you have total knowledge of the status and plans for every resource and item that may be flowing into, within, or out of the facility. Obtaining that kind of knowledge requires a great deal of information, using both automatic identification technologies and human input.

There are facilities where the size, volume, timing, value, and variety of transactions are such that they may be effectively managed using written paper records. The great majority of facilities, however, require software applications using computers, automatic identification, and wired/wireless communications to have any chance of keeping up with customer demand. Handling the volume of transactions productively is of great concern, but accuracy is an even greater concern. People make mistakes when they try to hear or read information, then attempt to say, type, or write that information. Automated identification technologies are applied to assist or replace human identification and information processing. Wireless communications provide the ability to share that information from and to any two locations within or outside of facilities. The Internet and private networks provide the ability to securely control access and sharing of information among trading partners. Although the emphasis here appears to concentrate on technologies, the greatest challenge for visibility continues to be the human element. Individuals, departments, and other facilities within the company must cooperate and choose the data flows that really make a difference in adding value. Trading partners must learn how to trust each other and share the data that helps everyone plan and execute those plans better.

The following are a few examples of the tools and techniques that enhance visibility.

1.1.1.1 Metrics

Operations tend to manage what is expected, measured, and rewarded. Applying metrics such as defect rates, work-in-process levels, order-fill rates, end-to-end production time, or hundreds of other metrics require information visibility. Many customers and suppliers now demand that information be provided that has typically never left the facility, let alone ventured outside of the company.

1.1.1.2 Dashboards

Once the problem of gathering all needed information from its point of generation has been solved, one must find a way to put it into the hands of the right people, at the right time, in the right format so that they may make the right decisions. Management dashboards provide a presentation interface to the massive amounts of raw data that may be available to decision makers.

1.1.1.3 Radio-Frequency Identification and Automatic Identification

Far too often, radio-frequency identification (RFID) has been touted as a replacement for all other automatic identification (autoID) technologies. RFID is a very powerful complementary solution that helps provide some unique new capabilities when compared with other autoID technologies. The least expensive, read-only passive tags (those that rely on a reader or interrogator to send them a signal so they may react) typically contain only a number with little difference from the numbers printed as bar codes. The more expensive tags have greater functionality that allows them to actively make their presence known, plus they may add the additional capability of being rewritten with a great deal of information. Add some additional processing or sensory capability to record aspects of the tag's supply chain journey and one can add significantly to the ability to track, trace, and adapt to supply chain events.

1.1.1.4 M2M

Machine-to-machine (M2M) (also machine-to-man) technologies allow devices and equipment to communicate with each other and possibly make decisions without human intervention. They also aid the decision-making processes of people. If resources and items in the facility such as pallet loads of goods, machine tools, industrial trucks, and conveyor systems provide better information to people and control systems, great improvements can be made in the response to problems or opportunities. Machines now have the ability to provide alerts to off-site maintenance personnel by phone, pager, or e-mail. Service technicians and engineers can respond remotely as well, providing a Man-to-Machine closing of the loop. This off-site monitoring and response provides an opportunity to outsource maintenance either to the supplier of the system or a third party that specializes in the equipment or processes.

1.1.1.5 Mobile Labeling/Tagging

Much like the rental car companies that realized that on-the-spot printing greatly speeded up returns of cars, many transactions in the facility can happen on the spot with mobile printing solutions. Mobile printers/taggers could be on forklifts, on people's belts, or at the receiving/shipping dock. Cross-docking, receiving, putaway, shipping, etc. can all take advantage of mobile labeling and any other tools that separate a person or other resources from a single physical location for retrieval and application of information that needs to flow with a product.

1.1.2 Flexibility

Flexibility is needed because of rapid change and the inability to always effectively see that change coming. Visibility can help lessen the impact of change by providing more precise information regarding upcoming changes. Consumer tastes change; supply sources change; regulations are created or changed. Flexibility for facility logistics is defined as the ability to easily adjust to changes in demand and supply of materials and resources.

Demand change is primarily driven by changes in customer needs and may either come directly from customers in the form of changes in buying patterns, or may come in the form of internal changes in production requirements. Products that are not selling need to be flushed out of the system. Products that are selling may need additional space that was not planned to be provided. Customers demand more custom labeling, application of RFID tags, special packaging, etc. that cause changes to pick, pack, and ship procedures.

Supply change is primarily driven by the ability to source materials or resources that are needed to satisfy the end customer. Materials may need to be changed due to supply pricing pressures or shortages. Technology may need to be added to deal with unexpected increases in the cost of labor. New procedures may be needed to decrease the turnaround time in the receiving dock. These are all forms of supply change.

Flexibility is increasingly a competitive strategy for survival in developed industrial countries. Companies that find it increasingly difficult to compete by producing commodity items have developed flexible facilities that can produce low volumes of custom products. High-volume production lines disappear in favor of something that resembles a job shop organized according to expected order flow over a period of a few months to a few years depending on the projected volatility of the industry.

The initial cost of designing for flexibility may be higher, but changeovers will be less expensive both in terms of the physical assets and the human cost of change. Advances are still needed to enable easier movements of machine tools and other physical resources (controls, pneumatic and water supply lines, etc.) to make layout changes faster and more cost effective. Too many companies have found out the hard way that facilities designed to handle large orders cannot handle the same volume of smaller, more frequently changing orders.

The following are a few examples of the tools and techniques that enhance flexibility.

1.1.2.1 Product Life Cycle Management

Product life cycle management (PLM) tools capture all the product and process information associated with the life cycle of manufacturing products, and attempt to store them in a way that will make it easier to profitably produce a portfolio of products by taking advantage of similarity. Facilities are designed to quickly change between an ever-changing mix of products that use similar components. Commitment to a final product assembly may even be delayed beyond the facility to allow for customization at the point of purchase.

1.1.2.2 Stock-Keeping Unit Rationalization

Although not a primary function but related to PLM, products can be analyzed for profitable activity to help understand when they should be taken out of service. Significant space and time is wasted on products that have outlived their usefulness and are only carried "because we've always offered them." The proper term for this analysis is SKU rationalization. Eliminating those products increases capacity and flexibility.

1.1.2.3 Automation

Automation is flexible? Automation that has a single goal of maximizing productive capability for a specific product is not flexible. Automation that is built to support the ability to quickly changeover production or distribution is flexible.

1.1.2.4 Postponement, Modularity, and Interchangeability

Designing products and systems to delay commitment supports the desire to reduce waste in the form of products or work-in-process that limits the ability to adjust to changes in demand. Modular subproducts that support the great variety of end products have been emphasized for years in the form of group technology concepts. It is advised to take advantage of similarity to hedge bets regarding supply and demand.

1.1.2.5 Cross Training

People can be one of the major roadblocks to achieving process flexibility. Standard job definitions with prescriptive task instructions built for productivity alone create jobs that will burn workers out and disrupt the high-productivity goals intended to be met in the first place. Cross-trained individuals can perform multiple tasks/jobs and adjust easier to change. They are happier and tend to have better long-term productivity, even if their productivity in a small time window is not as high.

1.1.2.6 Smaller/Shared Facilities Closer to Customers

Many facilities are shrinking in size and locating closer to customers to provide quick response with a smaller subset of capabilities. Outgoing transportation costs are reduced and customer response is enhanced. In some cases, facilities may even be shared with customers when they have downsized and have created unused space within their facility. What was once a factory within a factory may have indeed become a factory within a customer's factory, taking partnering to another level.

1.1.2.7 Outsourcing

Facilities operating at a high but manageable utilization may have difficulty dealing with the impact of growth, seasonality, and unusual, nonrepetitive spikes in order volume. The challenge can either be dealt with by letting business opportunities pass by or they may be addressed by finding additional capacity. Outsourcing is one way to reach out and find additional capacity offered by other companies. That capacity should be found and contracted well before there is a need. Plans should be in place dealing with the preferred design and operation of contracted space so that when the need arises, the transfer of goods is efficient and effective. Plans must also be in place to ramp down outsourced operations and bring them back into a company's own facilities.

1.1.2.8 Virtual Facilities

There is growing pressure to structure contracts such that equipment may be leased from suppliers or third parties that integrate equipment from many vendors. This helps reduce the amount of financial commitment made to a particular technology, but may increase costs in an attempt to hedge bets. In addition, companies that specialize in performing certain kinds of operations may establish a virtual center in a center and take responsibility for classes of value-adding operations. These operations may be at the beginning, the end, or in the middle of operations performed by the customer.

1.1.2.9 Lean

Lean facilities are driven by a culture of improvement and change that applies only the best mix of resources needed to produce value for its customers. Waste and commitment to an excess of resources and materials stifles flexibility by either forcing a commitment to utilize those resources rather than switching to more valuable alternatives, or by requiring the company to sell or dispose of those resources at a discount.

1.1.3 Labor

Although relatively cheap labor may still be found throughout the world, many developed industrialized countries find that labor costs continue to grow. The number of people available to perform manufacturing and warehousing jobs will continue to decline, resulting in higher labor costs and the inability to fill positions. The baby boom generation is nearing retirement age. The education, skill, and desire levels of the labor force create challenges. Employee loyalty continues to decrease due to the direct influence of decreasing employer loyalty. Additional costs such as medical care, workmen's compensation, and repeated cycles of hiring and training will add to the burden. In such conditions, automation becomes more desirable and generally results in a labor cost decrease, but is not the only answer.

The following are a few examples of the tools and techniques that enhance labor effectiveness.

1.1.3.1 Labor Management Systems

Software systems can help plan task management so people can be more productive. Time studies can act as guidelines to help software perform planning, execution, and management of labor resources. Providing on-the-spot direction and measurement of labor activities can help minimize empty, unproductive trips. Task interleaving built into the software can help reduce unproductive walking time by providing other tasks to be performed (carrying loads, verifying inventory counts, maintenance, etc.) while returning from or moving to another task.

1.1.3.2 Safety and Ergonomics

Facility safety and ergonomics minimize the risk of injury while supporting the productive capability of the facility. Injuries may be quick and dramatic, they may be slow and cumulative, or they may be the great majority somewhere in-between. Ergonomic devices can reduce the strain on people to help them stay on the job longer, and lower or eliminate lost day injuries. Designing facilities and the accessory devices in work areas to present items in the golden zone (mid-chest to waist level) helps reduce the bending and strain that results when repetitive picks or manufacturing operations take place. Safety training, proper signage, protective barriers, guarding and automatic interlocks all contribute to a facility that can protect the long-term, productive well-being of its employees.

1.1.3.3 Item-to-People Technologies

If maximizing the productivity of the workers is desired, they should not spend time walking. A number of devices have been created to bring items to the person for picking, and then take them away after the pick is complete. The worker can concentrate on the picking task while the electromechanical devices return product back to the proper staging or storage location.

1.1.3.4 Voice Systems

Voice systems allow labor resources to utilize both hands while performing and reporting on a task. Real-time task assignment and verbal verification help keep labor more productive and focused on the task.

1.1.3.5 Solutions, Not Components

Companies are increasingly challenged to build solutions for their facilities as they continue to hire fewer facility support personnel. Instead, they are outsourcing the design, installation, and maintenance of their operations. The vendor community has responded individually or by teaming with other companies to provide complete solutions rather than offering individual components. Quickly disappearing are the days when component suppliers would deal directly with facility managers and expect them to piece together a solution.

1.1.4 Green and Sustainability

Green and sustainable facilities minimize the waste stream and encourage reuse. Green facilities use less energy and materials (chemicals, water, packaging, etc.). Green facilities generate fewer waste products, and find ways to creatively reuse those it must produce. Pressure is building around the world to use green thinking in every step of product design, manufacturing, and distribution. Currently, incentives are provided in some states and localities to pursue green thinking, but regulations are likely on the way that will require the use of many of the products and techniques described in the following sections.

The following are a few examples of the tools and techniques that help create green facilities.

1.1.4.1 Green Facility Design

Green facilities make use of natural light and solar power to decrease energy use. Rain collection systems have been built by some facilities to capture and store water for use in toilets or for other nonpotable water usage situations. Green facilities are constructed or renovated using a high percentage of recycled content (steel, crushed concrete, plastic, etc.). Lighting systems are controlled to apply light only when and where it is needed by sensing the presence of people or movement. Temperature, humidity, and ventilation controls also are applied only where people will be present or when materials require specific conditions. Bike racks are provided along with priority parking locations for those carpooling. Creative facility designers have responded to programs such as Leadership in Energy and Environmental Design (LEED) point system by the U.S. Green Building Council (www.usgbc.org).

1.1.4.2 Reusable Pallets and Containers

Although the majority of boxes and pallets are made from renewable resource materials, they do create a waste stream that requires additional handling and disposal along with increased sourcing activity. Plastic, metal, composite, and other longer-lasting materials are used to construct pallets and containers that may make many more trips prior to disposal.

1.1.4.3 Fuel Cells for Industrial Vehicles

One way to decrease the localized output of exhaust gases is to switch from internal combustion vehicles to electric vehicles or to fuel cell vehicles. Initial purchase costs are higher for electric vehicles, but they do have a lower cost per hour of operation. Fuel cell vehicles are just starting to be adopted, but will likely take some time to become a significant percentage of the total number of vehicles in facilities. Rather than a complete redesign of the vehicles to accommodate fuel cell systems, some companies have created fuel cell packs that are the same size as the battery packs that they replace. This somewhat standardized replacement approach will increase adoption rates. The packs are quickly refueled and installed with standard electrical connectors, plus they run 2–3 times longer than the batteries they replace.

1.1.4.4 Motors and Controls

According to the U.S. Department of Energy, motor-driven equipment accounts for nearly two-thirds of the electricity consumed in U.S. industry. New low-energy motors with controls help provide only the power that is needed to perform a task and know when to turn themselves off or put themselves into a low-power readiness state.

1.1.4.5 Fumes to Fuel

Some facilities are capturing their volatile organic compounds (VOCs, primarily paint fumes) and converting them to energy—providing both an additional energy source and decreasing a potential source of air pollutant.

1.1.5 Security

Secure facilities protect the physical and intellectual assets of the facility while supporting the productive capability of the facility. Security of the supply chain has become a much greater concern in recent years due to terror threats. Where once the greatest concern was with the protection of goods from theft, now there is great concern for contamination or the addition of explosives or chemical weapons to shipments. There are many national and international regulations and guidelines that help provide a more secure, and visible, supply chain. Savvy facility managers will find ways to provide double benefits to security requirements. For example, motion sensors to detect entry into or exit from sensitive areas may also serve to turn on and off lighting and heating, ventilation, and air-conditioning (HVAC) to save on energy usage. RFID tags may provide a nonvisible way to detect unauthorized movement, but may also be used to enhance visibility.

Facility managers must also learn how to develop lists of prioritized threats, learn how to detect them, and minimize their impact by training everyone how to respond and how to recover. They also need to periodically test the systems to see if they actually detect and respond to threats as planned.

The following are a few examples of the tools and techniques that enhance security.

1.1.5.1 Tracking Technologies

Global and local positioning systems not only enhance visibility, but they also help a facility understand the location of resources, goods, and people as they flow throughout the facility. Food, pharmaceuticals, luxury, and other high-value, highly regulated items must be closely tracked to maintain their integrity. People must be tracked to aid in planning and make sure that they have authorized access to various points in the facility.

1.1.5.2 Physical and Electronic Seals

Once only used for sensitive shipments in large freight containers, various forms of physical and electronic seals may be used to help insure that a shipment has been sealed throughout its journey. Facilities have had to add capabilities to apply and to read these seals as they become more prevalent.

1.1.5.3 Access Technologies

Entry and exit within the facility and various sensitive areas require proper identification. People generally need to provide two out of the three following forms of information to prove their identity. One, they may provide something they know, such as a password. Two, they may provide something they have, such as an access card with a unique identity. Three, they may provide something they are, such as a fingerprint or an iris scan. Facilities wishing to tighten security must have fewer entry and exit points, or they must provide ways to increase security through automated means (motion detection sensors, automatic checks of identity, etc.) or by increasing the security staff.

1.1.6 Globalization

Since mankind first realized that it was possible to rely on others to provide specialized goods and services in exchange for his or her own goods and services, there has been a pursuit of better communications and transportation capabilities to support the flow of information and materials. Tablets, papers, horses, carts, ships, trains, planes, the Internet, and many other tools have now combined to create a more interconnected, more interdependent world. Work flows to where it may be done best to create a proper balance of cost and customer service. Cost reductions have been a strong contributor to the flow of work between countries as many manufacturers have chosen better providers of goods or they may simply be chasing cheap labor. Although global shipments can be arranged at a mouse click, they may take months to complete their journey into the hands of the final consumer. Cost reduction amounts may simply move into other cost categories and cause unforeseen increases in cost, control, and response time.

If facilities wish to be global players, they must apply the proper resources to minimize differences in language, regulation, culture, and methods to support the multimodal, global flow of goods and services. Global facilities must also learn what they do best and utilize outsourcing to those who know how to do it better. At a local or international level, virtual facilities may actually consist of several facilities with different corporate parents collaborating to offer a set of goods or services that were once provided by a single company. The variety of goods and services flowing within each physical facility continues to grow and stretches the capabilities and capacities of each company to deal with the variety of components, subassemblies, labeling, and packaging. Adopting standardized communication, packaging, and shipping methods helps companies deal with the variety and increase the flow of goods and information. Flexible automation helps provide shorter changeover times and allows companies to provide smaller shipments on a more frequent basis. Reusable containers—from large intermodal containers to small totes—help reduce waste at both ends of an international journey, but are found much more often in closed-loop systems where trading partners agree to their use. Regional, national, and international package pooling companies are growing to provide greater standardization and an increase in reusables.

1.2 Summary

Facility logistics no longer fits neatly into a box called a plant or a warehouse. Nor is facility logistics merely a cramped engineering exercise that can be justified solely on the amount of human labor that was replaced by machines. Today's facility logistics

> integrates management decision-making, information processing, automatic identification/data collection, and a variety of handling equipment into broad strategies like supply chain management.

This quote substitutes "facility logistics" for "material handling." The quote appeared at the beginning of the article "New Directions in Material Handling" from the December 1997 issue of *Material Handling Engineering*. Most of it sounds like it could have been written today. Has there really been so little change in facility logistics in ten years' time? Yes...and no. The goal has not changed from the quote, but most of the supporting tools and techniques for dealing with the challenges have undergone extensive change. Information technology changes have been extensive. The reliability of equipment and availability of interchangeable spare parts have been greatly increased. Senior management understands the value of logistics, even if they do not really understand what is happening inside the box. Globalization and multinational sourcing of products and components are much easier and more prevalent by far than ten years ago. Transportation infrastructure is being challenged to keep up with the growth in product volume and variety. Security issues threaten to disrupt product flow either from actual acts of terrorism and sabotage or from the time-consuming processes put in place to prevent such acts or events.

Lean, flexible, green, and global facilities are becoming more transparent and automated as visibility is emphasized and labor resources continue to become more expensive and scarce. Various conflicting measurement systems for facilities, departments and individuals may not support higher level goals. It takes a savvy facility logistics designer and manager to develop a good balance of resources and processes that match corporate goals and customer needs. That balancing act becomes even more difficult as the underlying factors continue to shift during and shortly after facilities are designed, laid out, and operations begin. Customer expectations and interests shift from one set of features and functions to another. Building and land use limitations shift. Labor availability shifts. Equipment and information technology product developments happen in shorter cycles. The only constant is uncertainty and change. Those who can reduce sources of uncertainty, plus plan for, and adjust to, change will be the only ones with longer-term success.

Chapter 2

Warehouse Management Systems and Product Flow: Reconciling the Two

James K. Higginson and James H. Bookbinder

CONTENTS

Abstract When a shipment does not move directly from its origin to destination, a number of flow-related tasks (such as sorting, merging, diverting, and crossdocking) must occur. Although modern facilities are designed to encourage flow activities, warehouse management systems (WMS) historically have focused on *storage*, and often did not adequately support product movement. With increased demands for improved supply-chain integration, that has changed.

This chapter discusses WMS, with an emphasis on the management of product flow. We begin by reviewing the basic functionalities of WMS, and then examine their capabilities in areas related to product movement. Integration with other logistics software (e.g., enterprise resource planning [ERP], systems for transportation management, or yard management) is also discussed, followed by conclusions and suggestions for further research.

2.1 Introduction

Prior to the mid-1990s, warehousing was seen as a necessary evil of logistics, with numerous practitioners, consultants, and academics calling for reduction in the number of warehouses used in a supply chain. By the end of that decade, however, demands for faster response and improved customer service had forced a change in thinking, with the trend to fewer warehouses slowing and, in some cases, reversing (Copacino, 1997).

Warehousing is now seen as an important competitive component of a supply chain. Consequently, WMS have, in most applications, become necessary to efficiently achieve the increasing levels of warehouse performance required (Faber et al., 2002). WMS have frequently not kept up with changes inside and outside the warehouse, nor reflected applicable developments in Operations Research (OR).

When a shipment does not move directly from its origin to destination, a number of flow-related activities—including disaggregating, sorting, merging, diverting, and crossdocking—may occur. These activities might be performed on a vehicle, but more commonly they will take place in a facility such as a warehouse or distribution center. (Theoretically, a distribution center is a warehouse in which product storage is not done.)

Warehouses can play various roles in a supply chain (Higginson and Bookbinder, 2005). These include acting as product storage centers, make-bulk/break-bulk consolidation centers, crossdock facilities, transshipment points, assembly facilities, product-fulfillment centers, depots for returned goods or for trucks or drivers, and bases for local customer sales and support. Many of these roles focus on the movement of product, rather than its storage. Our objective is to examine WMS by emphasizing the coordination of item flow. We take an academic view of this subject; hence, the chapter will not discuss considerations common to the practitioner literature (such as WMS implementation and vendor-specific systems). The chapter begins by reviewing the basic functionalities of WMS, and then examines their capabilities in areas related to product flow. Issues in the integration of WMS with other software systems are briefly noted. We close with a summary of our findings and suggestions for additional research.

2.2 Warehouse Management Systems: Overview

A WMS is a computer software package that collects, analyses, and reports the information necessary to move goods through a warehouse or distribution center. From this information, a WMS is able to instruct employees on the best ways in which to perform warehouse activities (e.g., where to put or where to find an item), and, as such, governs the flow and storage of products through the facility.

Like ERP systems, WMS software is supplied as a set of modules from which the organization in question must choose. Typical modules include:

- Inbound shipment control: Assigns docks and time slots to deliveries; records data on incoming products and their characteristics; and assigns storage locations to those items.
- Stock locator system: Manages all inventory locations (including pallets and forklift trucks) and provides information for item put-away and picking operations.
- Inventory control: Maintains information about the status of each item, cycle counting, product shrinkage and spoilage, and damaged goods.
- Order fulfillment: Develops and prints pick lists, bar codes, and other documents, and estimates the requirements for labor and material handling in order fulfillment.
- Outbound shipment control: Generates packing lists, manifests, and bills of lading; plans the packing and consolidation of materials and the loading of the truck; and records information on the shipment and vehicle.

Common to all modules is the ability to summarize and report activity information, such as inbound and outbound movements, product activity, and facility performance.

A WMS should support electronic communication inside the facility, such as via radio frequency and voice-activated technology. More advanced

WMS enable communication with external parties through electronic data interchange (EDI) or advance shipment notification (ASN). The general role of WMS has also expanded, with some systems offering capabilities related to trailer and yard management, light manufacturing, returned-goods management, labor planning, coordination of multiple warehouses, import–export control, full accounting systems, and the ability to link to advanced planning and other software systems.

Potential benefits of implementing a WMS have been reported by authors such as Boggs (1995), Alex (2000), and Faber et al. (2002). Positive effects include increased asset productivity; more accurate records; improved utilization of labor, equipment, and space; the ability to track product movement in real time; lower cycle times, and enhance customer service. Many WMS vendors also claim the potential to decrease inventory. Piasecki (2005), however, notes that the size of this reduction in comparison to total inventory often is minimal because "the predominant factors that control inventory levels are lot sizing, lead times, and demand variability. It is unlikely that a WMS will have a significant impact on any of these factors."

A WMS is considered a "logistics execution system." It is, thus, intended for planning day-to-day activities. Actually a WMS plans at a fairly low level, typically not extending beyond the facility walls. The primary function of a WMS is not optimization; instead, it performs a supervisory function concerned with the activities within the warehouse with the goal of making the best use of the resources (capital and human) (Ballard, 1996). (There is, of course, the question of if or how much optimization a WMS should perform: Optimizing the activities within a warehouse may suboptimize operations in other areas, such as transportation.)

Planning at a level higher than that of the WMS requires software dedicated to analysis of one particular area. Advanced planning and scheduling (APS) systems, for example, provide OR capabilities in optimization that are lacking in WMS and other supply-chain planning systems (Green, 2001). APS software may or may not use information from the WMS. A categorization of computer packages according to supply-chain application and OR techniques is given by Aksoy and Derbez (2003).

WMS often developed independently from, and hence have a unique relationship with, enterprise resource planning (ERP) systems. Most ERP systems were designed by firms with little experience or expertise in logistics. Their goal, nevertheless, was to integrate, via a "suite" of applications, all of the company's functions (including warehousing). As Frazelle (2002) notes, in ERP systems, the warehousing module "is typically an afterthought application for these [software] providers, and the full-suite providers typically have very little expertise in warehousing." As a result, most ERP systems lack the functionality required to adequately support warehousing, transportation, etc. (Frazelle, 2002; Handfield and Nichols,

2002; Spiegel, 2003). Organizations wishing to include warehouse management functionality as part of their ERP have had to undertake expensive software integration projects, or purchase an ERP system that was probably less than desired. Lacefield (2005) notes that several major ERP vendors now claim their systems to be capable of supporting supply-chain processes. Research in ERP systems is discussed by Esteves and Pastor (2001), Al-Mashari (2002), and Pairat and Jungthirapanich (2005).

Because the market for WMS and many other logistics execution systems is mature, it is hard to differentiate warehouse software based only on processes "inside" the facility. This has led WMS developers to give much greater attention to product flow. Applications have been added to better coordinate warehousing with other activities in the supply chain (McCormick, 2003). As discussed later in this chapter, the integration of WMS with other logistics execution systems has not always been easy or successful. Nevertheless, product flow is now a crucial part of WMS.

2.3 Fundamental Principle of WMS

Practitioners and researchers must keep in mind a fundamental principle: WMS operate according to a set of predefined policies.* As a result, a WMS "imposes its own logic on a warehouse's operations and organization. Implementing a standard WMS . . . [involves] making compromises between the way a warehouse wants to work and the way the system allows [it] to work" (Faber et al., 2002). If adding a WMS to a facility requires major compromises in the way that the organization wishes to manage warehouse activities, the implementation can significantly affect the performance of the facility (Faber et al., 2003).

Lindgaard (2004) provides an example of this in a grocery distribution center. Here, the WMS calculated the time to pick and assemble each order, and workers were required to perform within 5 percent of the estimated time. The WMS did not, however, consider the product weight or volume when designing order-picking routes. Thus, an order picker who wanted to meet the time standard often had to either deviate from the pick list or hope that the products picked did not exceed the capacity of handling equipment and were compatible (based on their weight and size) with each other. Ultimately, this forced vehicle loaders to repack orders, and when the

* A second important principle is that even in automated facilities almost all the activities controlled by a WMS are performed by "people." When an unusual event occurs, people will change their normal behavior or activities and try to find a way around the system constraints in an attempt to meet the schedule (Banks and Gibson, 1997). This fact should not be understated when modeling facility operations.

loading docks became congested, order pickers "were unable to drop their loads in the locations specified by the WMS, leaving them wherever they could in the rush to report back to their supervisors within the allocated time frame."

A WMS requires extensive work before implementation. This has in some cases encouraged the purchase of simpler packages—a potential error because such packages may be overgeneralized. Footlik (2005), for example, notes that many WMS algorithms for determining storage locations focus on dollars sold or quantity sold, which creates problems when the facility handles both expensive and cheap products or when sales units differ dramatically. Examples of necessary product data include the cubic dimensions and weight of each stock-keeping unit (SKU) in every size in which the item is stocked (e.g., cases and pallets), whether it is rackable, its maximum stacking height, maximum quantity per location, hazard classifications, and whether a finished good or a raw material (Piasecki, 2005). Similar details are necessary for each storage location in the facility. Lindgaard (2004) illustrates the importance of accurate data with the example of one WMS program that recorded the outside dimensions of trucks, resulting in many orders being too tall to fit inside the vehicle. This led to increased labor as loads had to be repacked to fit, increased need for pallets and packing materials, and inaccurate WMS scheduling and route calculations.

Given the potential complexity of a WMS implementation, a major question is whether the buyer of a WMS should purchase a standard (off-the-shelf) package or a customized version. With a standard product, "the WMS is leading and the planning and control structure follows. With a tailor-made WMS... the planning and control structure is leading and the tailor-made WMS follows" (Faber et al., 2002). In general, the more complex the warehouse (measured, e.g., by the number and variety of items to be handled each day, and the nature and variety of processes and technology), the more customized a WMS package should be. Standard, off-the-shelf packages are usually sufficient for simpler warehouses (Faber et al., 2002).

2.4 WMS and Product Flow

We now turn our attention to the interaction of WMS and product flow. Traditional flow activities in a warehouse include product receiving, putaway, order picking and assembly, and shipping. Each of these categories includes several more specific tasks such as breaking bulk, order consolidation, and crossdocking. A thorough discussion of product storage, flow activities, and decisions in a warehouse is given by Gu et al. (2007). We will now discuss the relationship between WMS and these decisions and activities.

2.4.1 Product Receiving

In a standard WMS environment, the receipt of goods begins with assignment of dock to arriving vehicle, followed by the manual reconciliation of what was actually received with what was expected according to the ASN. Adjustments to the product database are made from the dock, quality checks are done, and the updated information is used by the WMS to plan the items' putaway.

Product receipt may require one or more of the following:

Breaking bulk, the disaggregation of large-quantity shipments, is basic to the movement of materials in a warehouse or distribution center. *Transshipment* concerns the unloading of one item or a group of items from one vehicle and reloading it onto another. If goods are not added or removed from the shipment, the process sometimes is referred to as *transloading*.

A *crossdock* denotes a facility from which items are dispatched within a short period (48 hours is often mentioned) after their arrival, without putting it into storage. Before shipping from a crossdock, some sorting and consolidation of items may occur, as in the case of transshipment. There are technical differences: Crossdocking is a customer-focused strategy that attempts to move a product through a facility as quickly as possible, while transshipment is a carrier strategy that aims to improve truck utilization by better matching the characteristics of loads and vehicles. A discussion of transshipment in logistics networks is given by Beuthe and Kreutzberger (2001). Crossdocks are discussed by, for example, Apte and Viswanathan (2000), Napolitano (2001), Gümüş and Bookbinder (2004), and Higginson and Bookbinder (2005).

Effective scheduling and coordination of inbound and outbound shipments from the warehouse require timely and accurate flows of information between supply-chain members (Bookbinder and Barkhouse, 1993). This may be accomplished via, for example, ASN, EDI, and automatic identification (autoID) technologies such as bar codes and radio-frequency tags. Because planning for crossdocking goes beyond the warehouse to include inbound and outbound transportation, crossdocking functionality was virtually nonexistent in WMS before 2003. Improvements in this area have appeared only recently, but capabilities remain weak.

Arguably, the most important function of a WMS in a crossdocking environment is determination of those docks to which incoming and outgoing trucks should be assigned. The overall goal is to avoid congestion, for example, due to high product accumulation around an outgoing vehicle (rule of thumb: do not assign busy trucks to corner docks); too many forklifts (or other shipment-handling equipment) traveling long distances and thus blocking other movements (remedy: do not assign busy vehicles to docks in the middle of the facility); and imbalances between where empty shipment-handling equipment is and where it is needed (cure: intersperse incoming and outgoing trucks) (Bartholdi and Hackman, 2006). (Note that these three causes of congestion ignore the timing of incoming and outgoing vehicles entirely.) Developing an optimal assignment of docks is beyond the capabilities of many WMS; an add-on optimization package seems to be needed.

2.4.2 Item Putaway

Item putaway refers to "the act of placing merchandise in storage" (Frazelle, 2002). This includes the identification and determination of storage locations, as well as physically moving and placing product. Historically, managing item putaway is one of the primary functions of a warehouse and of a WMS; the other function is order picking (discussed in Section 2.4.3). Both activities benefit from a huge body of academic research.

The storage location assignment problem (SLAP), for example, attempts to determine an optimal assignment of products to storage sites, often with the goal of minimizing the average workload of employees. There are three common approaches to the choice of storage location: dedicated (always assign a particular item to the same fixed location), random (assign an SKU arbitrarily to any empty location), and group (item locations are based on product characteristics) (e.g., Hausman et al., 1976). Although these assignment methods generally are available in WMS, their application is not always problem-free. Kosfield and Quinn (1999), for example, describe a WMS that used random storage but lacked algorithms to minimize vehicle time or to efficiently route vehicles. As a result, large incoming shipments often were placed into multiple locations spread throughout the facility, rather than in a few contiguous bins; similarly, orders were retrieved from many dispersed locations. Changes to the WMS coding resulted in an increase in facility throughput of 110 percent.

As mentioned previously, and as seen in the above-mentioned example, WMS typically provide good, but often not optimal, solutions. A major reason relates to the quantity of data required to create a truly realistic (i.e., "geometric") model of the warehouse in the WMS. Most WMS do not have an adequate geometric representation of the facility or complete knowledge of distances between all the pairs of storage locations—data required by most optimization techniques. "Such detailed information would not only be time-consuming to gather but would have to be specialized to every site and updated after any change in physical layout" (Bartholdi and Hackman, 2006).

Some literature has incorporated geographic information systems (GIS) into warehouse-related research, although at levels of decision-making higher than order picking. Johnson et al. (1999) discuss a GIS model to integrate multiple-facility warehousing and production, while Min and Melachrinoudis (2001) use GIS in a decision support system for warehouse restructuring. Min and Zhou (2002) treat information technology (IT)-driven models within supply-chain modeling.

A true geometric model of a warehouse and more detailed data would allow OR to be better incorporated in WMS. Some simple OR models for warehousing are commonly included in WMS, but are not necessarily problem-free or practical. Cube logic as exhibited by the well-known cube-per-order-index (COI) (e.g., Malmborg, 1995) is an illustration.

The COI is constructed as follows. For each SKU, one calculates the ratio of the item's volume (space required per unit) to its turnover (annual demand). SKUs with smallest values of COI should be located nearest to the shipping area (i.e., the point of pickup or delivery).

The COI has been proven to be optimal (to minimize average workload) under particular conditions. These include single command systems, dual command systems, or a zoned warehouse (Malmborg, 1995), assuming in each case that the cost of moving any item is constant and proportional to the distance traveled. Caron et al. (1998) integrate routing policies with storage based upon COI.

Order picking, however, may also depend on *human factors*, for example, the impact of load weights, in addition to distance. Petersen et al. (2005) conducted a simulation study that showed the effectiveness of COI in conjunction with "golden zone" picking, whereby SKUs with high demand are slotted at a height between the picker's waist and shoulders. Hwang et al. (2003) consider a low-level picker-to-part warehouse system. Relatively heavy items are stored and usually retrieved by an "out-and-back" method. These authors, too, emphasize the importance of *weights* in manual material handling operations. Hwang et al. (2003) propose a rule that, although yields somewhat less throughput than COI, still performs considerably better in terms of human safety.

Although cube logic is included in most WMS, it is rarely used: If the items are capable of being stacked into the location in a manner that fills every cubic inch of space in the location, cube logic will work. As this rarely happens in the real world, cube logic tends to be impractical (Piasecki, 2005).

Academic research has also studied methods for determining how much space in a fast-pick area needs to be allocated to a particular item (e.g., Gu et al., 2007). Allocation of this space is generally done in the context of overall warehouse layout. Roodbergen and Vis (2006) design the order-picking area so as to minimize the average distance which the pickers must travel. They obtain a formula for the length of a typical route; this is then the objective function in a nonlinear programming model. Their results relate the number of aisles in an order-picking area to the required storage space.

In spite of the above, WMS use fairly simple methods for determining where in the facility to store products and how much space each should be given. Bartholdi and Hackman (2006), for example, note that most WMS permit only two methods: Equal space allocation (the same amount of space is apportioned to each SKU) or equal time allocation (assign sufficient space to make the time interval between restocking equal for all SKUs). Birkholz (2004) describes one WMS that determined storage locations simply by looking for empty bins or locations containing the identical item. The WMS also did not apply physical coordinates to locations to optimize travel time or improve order-picking efficiency. As a result, employees overrode 72 percent of the WMS suggested storage locations, usually because of inadequate space or inefficient travel route. Mark (2006) suggests that large facilities handling many A-type items consider linking slot optimization software to the WMS to improve productivity in picking orders. Conversely, Macro and Salmi (2002) describe a warehouse simulation model developed for analyzing storage capacity of warehouses of various sizes. Their results with fairly simple decision algorithms were outperformed by those of a WMS of moderate complexity.

2.4.3 Order Picking and Assembly

Order picking is "the process of removing items from storage to meet a specific demand" (Frazelle, 2002). *Order assembly* refers to combining items picked by different employees, or collected from different locations of the warehouse, into a single load to be delivered to the buyer. This includes checking the shipment for completeness; packing in appropriate containers; weighing the load; and printing packing slips, address labels, and shipping documents.

Order picking has been studied extensively; many optimization algorithms have been developed (see, e.g., Cormier, 2005; de Koster et al., 2006). Petersen (1997) provides an evaluation of routing policies for order picking, while de Koster and van der Poort (1998) compare optimal and heuristic methods for developing order-picking routes.

Gue et al. (2006) consider issues related to the organization of workers into an order-picking system, and the effects of pick density (frequency that workers stop to make picks). Analytical as well as simulation models enable those authors to study congestion due to narrow aisles and to the preceding factors. Hsieh and Tsai (2006) are also concerned with order picking efficiency and pick density. The paper develops a methodology and series of tables and graphs that can be used to evaluate the design of a warehouse under alternative combinations of cross-aisle layout, storage assignment policy, and order picking approach.

Cormier (2005) notes that order-picking policies fall into two major types, namely, policies for routing and for batching. Let us first discuss routing policies.

The order-picking algorithms most commonly found in WMS are (Piasecki, 2005):

- Location sequencing: All possible locations are numbered. A flow route is identified independently of the WMS, which develops picking routes by following the numerical sequence of locations.
- Pick-to-clear: Picking is done first from that location with the smallest quantity on hand.
- Fewest locations: Pick from the least number of bins that can fill the order.
- Nearest location: Pick item $(i + 1)$ from that site closest to where item i was picked.
- LIFO, FIFO, lot sequence: The picking progress is based upon location of the most recently arrived item (last in, first out—LIFO), the oldest arrival (first in, first out—FIFO), or a particular lot number or lot date (lot sequence).
- Quantity of measure: Orders for more than a prespecified quantity are picked from a location different than are smaller orders.

The conclusion for OR from this list is that WMS tend to use fairly simple algorithms for determining order-picking sequence. As a result, the efficiency of order picking is primarily dependent on the location(s) selected when putting away the product. Implementers of basic WMS should thus pay particular attention to the WMS order putaway logic.

Many order-picking algorithms are special cases of the traveling salesman problem (TSP) where travel is constrained by aisles (e.g., Ratliff and Rosenthal, 1983). Several other warehouse flow activities, such as item putaway, also may be modeled as a TSP. Thus, optimal solutions often can be found. However, despite claims by WMS developers, most WMS do not optimize order-picking routes.

The previous section noted the difficulty in optimizing product putaway if the WMS does not have a true geometric model of the facility. This problem also exists when attempting to optimize order picking, because optimization algorithms require knowledge of the shortest distance between all pairs of storage locations and the route that yields this distance. Thus, as Bartholdi and Hackman (2006) observe, "no warehouse management systems that we know of manage an explicit geometric model of the layout of the warehouse. Therefore, true pick-path optimization is not currently done." They also note that pick lists developed by WMS typically give only the sequence of pick locations, leaving the picker to determine the shortest path from location to location. This may be difficult if the order picker has limited information. As a result, picking from a paper list can be more effective than from a radio-frequency device, because a paper list allows the picker to glance ahead and plan the most efficient route. This look-ahead is not possible with most radio-frequency devices, which display only the next location to be visited.

More sophisticated WMS allow batch picking, zone picking, and wave picking. Batch picking combines several customer orders into a batch or set. An order picker then uses a consolidated list to retrieve all orders in the batch during a single trip through the facility. Zone picking divides the storage area into zones, picks one zone at a time, and passes an order from one zone to the next when picking in the previous zone is completed ("pick-and-pass"). Wave picking is a variation of batch picking where all zones are picked simultaneously, and then the components of each order from all zones are later sorted and assembled into the actual individual orders. Methods for deciding which orders to assign to which batches are discussed in van den Berg (1999) and Gu et al. (2007).

Some WMS also allow "task interleaving," the concurrent performance of two dissimilar tasks such as putaway and picking. Task interleaving is most common in full-pallet operations (Piasecki, 2005). Graves et al. (1977) were one of the first to discuss storage–retrieval interleaving in automated warehousing systems; Gu et al. (2007) provide a classification of algorithms since then.

2.4.4 Freight Consolidation and Shipping

Freight consolidation (or shipment consolidation) refers to determining which shipments, destined to different consignees, should be transported on the same vehicle. Technically, a WMS views shipment consolidation as part of the order-assembly process: The WMS must decide where (to which loading dock) to move those shipments which will be tendered to a specific carrier. Freight consolidation has been studied, for example, by Hall (1987), Higginson and Bookbinder (1994), and Daganzo (1999). Gray et al. (1992) discuss the design and operation of an order-consolidation warehouse.

2.4.5 Managing Product Returns

Warehouses play an important role in the management of returned products. The handling of such goods is often very labor-intensive: Any returned item may require repair, repackaging, refurbishing, remanufacture, dismantling of parts, recycling, or temporary storage, as well as reshipping. This is, however, one area in which most WMS are weak, with successful application requiring customization of the software (Bowman, 2004). As Bostel et al. (2005) note, "A major difficulty in adequately handling reverse-logistics activities concerns the uncertainty of the reverse flows themselves."

Returned products entering a warehouse may arrive in several different ways. For example, a consumer might send the item, without prior notification, to a warehouse that has been designated as a "returns depot"; the customer may be instructed to send it to a warehouse only after being issued a returns authorization; or trucks returning from a retail store or other distribution centers may be transporting a pallet-load or more of returned goods. In the first two cases, returned items arrive one-at-a-time, rather than as a group; and also in the second case and perhaps the third, the warehouse has information before their arrival about products sent back.

Arriving items must be recorded, and then routed either to work stations for inspection or to loading docks for transfer to outgoing vehicles. In the former case, employees must determine if the product is re-saleable in its present condition, or if it requires repackaging, repairs, return to supplier, or scrap. Goods that can be resold reenter the forward logistics channel at this point. Clearly, managing returned products involves the interaction of both the reverse and forward channels. Bostel et al. (2005) provide a comprehensive review of OR literature on reverse-logistics systems, emphasising the transportation plan implied by this interaction.

A number of characteristics unique to the management of returns suggest important ingredients of a WMS reverse-logistics module. These include:

- The creation, at time of arrival, of a distinct identifier for each item
- The recording of details of the customer who returned that item
- The ability to link a returned product to its return authorization (if used)
- The user-input code (revised as the status of the item changes) denoting the condition of the item, which would determine its routing (e.g., inspect—repair—test—putaway)
- The automatic in-facility routing to appropriate docks or areas for return to suppliers when they next deliver
- The ability to handle seasonality, which may create fluctuations in resource requirements
- The linkages to other computerized systems, which include customer service (e.g., an outside call center where the return was initiated or approved) and finance (refunds, credits)

The effective management of returned goods requires the ability to link warehouse activities with transportation movements. It is well known that transportation costs can be reduced significantly if delivery vehicles pick up returned items on their trips back to the warehouse. Planning for this may be difficult, because the vehicles often will not be under the control of the organization operating the warehouse. Thus, a WMS returned-goods module should be able to access information about future deliveries to coordinate the collection and shipping of these items. If the supplier agrees to collect returned products when delivering new ones, the WMS must pass all relevant information to the supplier's vehicle-scheduling system.

"Returned goods" may also include pallets, containers, and reusable packing material. A tracking system will be required to record the type and number of containers or material held by each customer. When these are returned by customers to the warehouse, records must be updated to accurately reflect each customer's balance. A method is also needed to uniquely identify pallets and containers, for example, to allow specific pallets to be used by the WMS as storage or consolidation locations. We add that as customers increasingly demand compliance with their inventory-handling requirements (e.g., the use of standard totes, containers, and pallets), a WMS will have to ensure that facility personnel know exactly which types of containers and packing materials to employ in each case.

2.4.6 Performance Measurement

We close this section by briefly addressing the measurement of warehouse performance. This has been extensively discussed in the academic literature (e.g., Gu et al., 2005; Higginson and Bookbinder, 2005).

A WMS is of course an information system; hence it has the obvious ability to compile and report a large number of summary statistics. These include inbound and outbound movements, product activity, and facility and employee achievement. Some considerations must however be kept in mind. First, performance information reported by a WMS is "historical." WMS do not determine, for example, how long it should take a person to complete an activity. Also, standard performance measures (typically those most frequently reported and used by warehouse managers) can be very inaccurate assessments of output: "An experienced worker will grab orders requiring the least travel time, or orders with the most case quantities and fewest reaches, to make his numbers look good" (Drickhamer, 2005).

Lastly, traditional measures of worker productivity have become misleading as the responsibilities of distribution center employees have expanded beyond product putaway, retrieval, and order assembly to include, e.g., preparing floor-ready merchandise, assembling displays, and relabeling products. Bowman (2003) notes that increased time spent by warehouse workers performing tasks other than item receiving, putaway, and picking, "has led some executives, judging strictly on the basis of dollars shipped per man-hour, to wrongly conclude that warehouse productivity is on the decline." Labor management systems are discussed later.

2.5 WMS Integration with Other Systems

A major question in WMS implementation is the extent to which a WMS can be integrated with other software systems (most commonly ERP and logistics execution systems, discussed later). WMS developers cite lower costs, operational advantages, and improved efficiency from system integration, while attempting to offer it to remain competitive. Although common practice in the 1990s was to implement ERP packages with built-in logistics capabilities (of varying quality), the recent trend has been to purchase the most suitable logistics execution program regardless of its developer, and hope that it can be linked easily to other software packages. Kim and Narasimhan (2002) discuss strategies for information systems utilization in initiatives to integrate supply-chain activities, while Sahay and Gupta (2003) discuss models and criterion for selecting supply-chain software, including WMS.

The term "WMS integration" usually refers to links with ERP or logistics execution systems. Before discussing issues related to integration, we briefly define the functionalities of the most common logistics execution systems.

- Order management systems (OMS): An OMS is a software package that automates customer-order handling activities. Common functionalities include order receiving, entry, routing, and tracking; real-time inquiries related to product prices and specifications, inventory availability, and customer accounts; customer communication; and order file updating.
- Transportation management systems (TMS): TMS plan activities related to transportation operations, such as load building and consolidation; transportation documentation; shipment scheduling, routing, and tracking; carrier selection; bill auditing and payment; and claims management. Some of these functionalities (particularly those related to shipment routing, scheduling, tracking, and cost) appear in more advanced WMS. Mason et al. (2003) discuss the benefits of integrating the warehousing and transportation functions of a supply chain.
- Yard management systems (YMS): These systems focus on the management of warehousing and transportation activities in the areas

surrounding, but outside, the facility. YMS coordinate product movement as goods arrive in the yard, enter the warehouse, and later depart the property. Functionality includes managing inbound and outbound trucks, tractor and trailer locations, and inventory in trailers, as well as assigning docks, monitoring vehicles owned by other parties, and measuring and reporting dock utilization and vehicle wait times.

- Labor management systems (LMS): LMS focus on work standards to manage labor at the activity level, and to compile productivity information at the individual and activity levels. An LMS calculates target times, based on engineering standards, for the labor component of specific tasks. From this, the required number of employees, the assignment of tasks to workers, and how long it should take these people to complete those tasks can be determined.

WMS must often interface with other less-mentioned systems, including those controlling data collection devices (such as radio-frequency terminals and radio-frequency identification [RFID] readers) and automated material handling equipment. The characteristics of these other systems may impose restrictions on the WMS. For example, because the cost of automated material handling equipment may exceed the cost of a WMS, the software that controls the former will often dictate the choice of WMS. In addition, control systems for automated handling equipment may require the WMS to deal with inventory in locations served by this equipment differently from that elsewhere in the facility.

A major task in linking software packages is mapping the information flow necessary for communication between programs and modules. For example, when the warehouse receives goods, the WMS must know to which other modules and systems (e.g., purchasing, inventory control, and accounts payable) the accompanying information should go (Saxena, 2000). Most integrated systems apply "static integration," where one software package makes a decision and then communicates the result to the next system (Jeroen van den Berg Consulting, 2004). Commonly, orders are accepted by the order management system (OMS) based on inventory availability, then passed to the TMS for delivery planning. The TMS transmits the resulting delivery information to the WMS, which determines order picking and assembly. After shipping, the WMS sends confirmation of the completed order to the OMS.

The alternative is "dynamic integration," under which each software package contributes to decision-making. For example, when deciding whether to accept an incoming order, the OMS will consider not only inventory availability, but also transportation and warehouse costs and capacities (as provided by the TMS and WMS) (Jeroen van den Berg Consulting, 2004).

Areas of WMS and TMS collaboration could include dock planning, load building, and documentation.

A second problem is the actual translation link or interface through which data will flow. Most package developers have created standard interfaces to guarantee that their various software products can communicate with each other, and often establish certification programs to ensure that external suppliers of subsystems (including add-on WMS modules) are incorporating that interface correctly.

The problem, however, is the interface between systems purchased from different vendors. By 2006, a standardized interface did not exist, and unless the WMS developer has created a specific interface to work with systems developed by an "approved business partner," time-consuming and expensive programming often is necessary. Some early integration projects were reported to cost over a million dollars, almost as much as the cost of software itself (Cooke, 1998). Detailed discussions of how data is interfaced and how data flows between systems are given by numerous authors. Helo and Szekely (2005) provide an overview of enterprise application integration software, while Chandra and Kumar (2001) examine the architectural requirements for supply-chain integration.

Lastly, the integration of computer packages may create problems with data integrity. For example, upgrading a WMS package may render computer code written to link the previous version of the WMS to another system inoperative or inaccurate. Moreover, software vendors will support only those interfaces sold with their package and not customized code for system integration (Trunk, 1999).

2.6 Conclusions and Suggestions for Further Research

This chapter discussed WMS and their interaction with product flow activities. We have related various issues on the functionality of WMS, that is, what the systems have been intended to do, to what they should "ideally" do. Differences between those capabilities typically concern product flow tasks, such as sorting, merging, and crossdocking. These activities indicate that the facility is functioning as a distribution center, whereas the very name, WMS, emphasizes storage.

Three major implications follow from this chapter. First, academic models of warehouse activities must recognize the widespread use of WMS and the barriers (and opportunities) that these systems create. Researchers should ask if algorithms are compatible with current WMS logic and functionality, and if not, what modifications or information would be required to make them so.

Second, it is evident that new warehouse-planning algorithms will not be readily adopted by WMS developers. A better route for researchers wishing to commercialize their models will be through "add-on" systems, rather than through WMS. Lastly, the market for WMS is seeking greater breadth over greater depth. Future WMS will require inclusion of multiple supply-chain functions, rather than more exact methods for planning activities inside the warehouse. Thus, algorithms for warehouse activity planning that can logically incorporate aspects of the broader supply chain are called for.

We are nevertheless optimistic, from our overview of WMS, and their fundamental principles presented here, that research is being done to remedy those discrepancies. We have discussed a number of models which link two or more aspects of warehousing. Additional studies of this type are desirable. Just as APS systems have been developed as "add-ons" to the transaction-based ERP systems, the WMS of the near future will need to incorporate its own add-ons. WMS implementation will then become more difficult, but the resulting system will be considerably more applicable in industry.

References

Aksoy, Y. and Derbez, A. 2003. 2003 software survey: Supply chain management. *OR/MS Today*, 30 (3), 34–41.

Alex, M. 2000. Where's the warehouse? *Manufacturing Engineer*, 74 (5), 214–215.

Al-Mashari, M. 2002. Enterprise resource planning (ERP) systems: A research agenda. *Industrial Management and Data Systems*, 102 (3), 165–170.

Apte, U.M. and Viswanathan, S. 2000. Effective cross docking for improving distribution efficiencies. *International Journal of Logistics: Research and Applications*, 3 (3), 291–302.

Ballard, R.L. 1996. Methods of inventory monitoring and measurement. Presented at the National Materials Handling Centre Conference on Warehouse Accuracy—Approaches to Error Reduction. Cranfield, UK: Cranfield University. February 20.

Banks, J. and Gibson, R.R. 1997. Don't simulate when . . . ten rules for determining when simulation is not appropriate. *IIE Solutions*, 29 (9), 30–32.

Bartholdi, J.J. and Hackman, S.T. 2006. *Warehouse & Distribution Science.* http://www.warehouse-science.com

Beuthe, M. and Kreutzberger, E. 2001. Consolidation and trans-shipment, In Brewer, A.M., Button, K.J., and Hensher, D.A. (eds.), *Handbook of Logistics and Supply-Chain Management.* Oxford: Pergamon, pp. 239–252.

Birkholz, A.L. 2004. *Optimization of Material Handling in a Manufacturing Warehouse Facility*, Master's thesis. Menomonie, WI: University of Wisconsin–Stout.

Boggs, K.C. 1995. Warehouse management systems. *Industrial Engineering*, February, 27 (2), 30–32.

Bookbinder, J.H. and Barkhouse, C.I. 1993. An information system for simultaneous consolidation of inbound and outbound shipments. *Transportation Journal*, 32 (4), 5–20.

Bostel, N., Dejax, P., and Lu, Z. 2005. The design, planning, and optimization of reverse logistics networks, In Langevin, A. and Riopel, D. (eds.), *Logistics Systems: Design and Optimization*. Dordretch, The Netherlands: Kluwer Academic, pp. 171–212.

Bowman, R. 2003. Racking up gains in productivity. *Global Logistics & Supply Chain Strategies*, August 1. Available at www.supplychainbrain.com. Last accessed September 7, 2007.

Bowman, R. 2004. Seeking excellence within the four walls of the warehouse. *Global Logistics & Supply Chain Strategies*, April 1. Available at www.supplychainbrain.com. Last accessed September 7, 2007.

Caron, F., Marchet, G., and Perego, A. 1998. Routing policies and COI-based storage policies in picker-to-part systems. *International Journal of Production Research*, 36, 713–732.

Chandra, C. and Kumar, S. 2001. Enterprise architectural framework for supply chain integration. *Industrial Management & Data Systems*, 101 (6), 290–303.

Cooke, J.A. 1998. Crossing the great software divide. *Logistics Management and Distribution Report*, 37 (6), 72–74.

Copacino, W.C. 1997. *Supply Chain Management: The Basics and Beyond*. Boca Raton, FL: St. Lucie Press.

Cormier, G. 2005. Operational research methods for efficient warehousing, In Langevin, A. and Riopel, D. (eds.), *Logistics Systems: Design and Optimization*. Dordretch, The Netherlands: Kluwer Academic, pp. 93–122.

Daganzo, C.F. 1999. *Logistics Systems Analysis*, 3rd edn., Heidelberg, Germany: Springer-Verlag.

de Koster, M.B.M., Le-Duc, T., and Roodbergen, K.J. 2006. Design and control of warehouse order picking. A review, ERIM Report Series Research in Management report #ERS-2006–005-LIS. Rotterdam, The Netherlands: Erasmus Research Institute of Management.

de Koster, R. and van der Poort, E. 1998. Routing order pickers in a warehouse: A comparison between optimal and heuristic solutions. *IIE Transactions*, 30, 469–480.

Drickhamer, D. 2005. Labor management software: The final profit frontier. *Material Handling Management*, 60 (9), 26–29.

Esteves, J. and Pastor, J. 2001. Enterprise Resource Planning Systems research: An annotated bibliography. *Communications of the AIS*, 7 (8), 1–52.

Faber, N., de Koster, R.B.M., and van de Velde, S.L. 2002. Linking warehouse complexity to warehouse planning and control structure: An exploratory study of the use of warehouse management information systems. *International Journal of Physical Distribution and Logistics Management*, 32 (5), 381–395.

Footlik, R.B. 2005. Optimizing WMS performance. *Material Handling Management*, 60 (1), 30.

Frazelle, E.H. 2002. *World-Class Warehousing and Materials Handling*. New York: McGraw-Hill.

Graves, S.C., Hausman, W.H., and Schwarz, L.B. 1977. Storage–retrieval interleaving in automatic warehousing systems. *Management Science*, 23 (9), 935–945.

Gray, A.E., Karmarker, U.S., and Seidmann, A. 1992. Design and operation of an order-consolidation warehouse: Models and application. *European Journal of Operational Research*, 58, 14–36.

Green, F.B. 2001. Managing the unmanageable: Integrating the supply chain with new developments in software. *Supply Chain Management*, 6 (5), 208–211.

Gu, J., Goetschalckx, M., and McGinnis, L.F. 2005. Warehouse design and performance evaluation: a comprehensive review, Working paper. Atlanta, GA: Virtual Factory Laboratory, Georgia Institute of Technology.

Gu, J., Goetschalckx, M., and McGinnis, L.F. 2007. Research on warehouse operations: A comprehensive review. *European Journal of Operational Research*, 177, 1–21.

Gue, K.R., Meller, R.D., and Skufca, J.D. 2006. The effects of pick density on order picking areas with narrow aisles. *IIE Transactions*, 38 (10), 859–868.

Gümüş, M. and Bookbinder, J.H. 2004. Cross-docking and its implications in location-distribution systems. *Journal of Business Logistics*, 25 (2), 199–228.

Hall, R.W. 1987. Consolidation strategy: Inventory, vehicles and terminals. *Journal of Business Logistics*, 8 (2), 57–73.

Handfield, R.B. and Nichols, E.L., Jr. 2002. *Supply Chain Redesign*. Upper Saddle River, NJ: Financial Times Prentice-Hall.

Hausman, W.H., Schwarz, L.B., and Graves, S.C. 1976. Optimal storage assignment in automated warehouses. *Management Science*, 22 (6), 629–638.

Helo, P. and Szekely, B. 2005. Logistics information systems: An analysis of software solutions for supply chain coordination. *Industrial Management & Data Systems*, 1051, 5–18.

Higginson, J.K. and Bookbinder, J.H. 1994. Policy recommendations for a shipment consolidation program. *Journal of Business Logistics*, 15 (1), 87–112.

Higginson, J.K. and Bookbinder, J.H. 2005. Distribution centres in supply chain operations, In Langevin, A. and Riopel, D. (eds.), *Logistics Systems: Design and Optimization*. Dordretch, The Netherlands: Kluwer Academic, pp. 67–91.

Hsieh, L.F. and Tsai, L.H. 2006. The optimum design of a warehouse system on order picking efficiency. *International Journal of Advanced Manufacturing Technology*, 28 (5–6), 626–637.

Hwang, H., Oh, Y.H. and Cha, C.N. 2003. A stock location rule for a low level picker-to-part system. *Engineering Optimization*, 35 (3), 285–295.

Jeroen van den Berg Consulting, 2004. *WMS Market 2004*. Utrecht, The Netherlands: Jeroen van den Berg Consulting.

Johnson, D.A., Taylor, G.D, and Visweswaramurthy, G. 1999. Highly constrained multi-facility warehouse management system using a GIS system. *Integrated Manufacturing Systems*, 10 (4), 221–232.

Kim, S.W. and Narasimhan, R. 2002. Information system utilization in supply chain integration. *International Journal of Production Research*, 40 (18), 4585–4609.

Kosfield, M.A. and Quinn, T.D. 1999. Use of dynamic simulation to analyze storage and retrieval strategies. *Proceedings of the 1999 Winter Simulation Conference*, 376–741.

Lacefield, S. 2005. Overwhelmed by your WMS choices?. *Logistics Management*, 44 (5), 61–63.

Lindgaard, G. 2004. Business: Making business our business—one path to value-added HCI. *Interactions*, 11 (3), 12–17.

Macro, J.G. and Salmi, R.E. 2002. A simulation tool to determine warehouse efficiencies and storage locations, In Yücesan, E., Chen, C.-H., Snowdon, J.L., and Charnes, J.M. (eds.), *Proceedings of the 2002 Winter Simulation Conference*, San Diego, CA: Institute of Electrical and Electronics Engineers, pp. 1274–1281.

Malmborg, C.J. 1995. Optimization of cube-per-order index warehouse layouts with zoning constraints. *International Journal of Production Research*, 33, 465–482.

Mark, K. 2006. Exceeding expectations: Why WMS is looking more and more like an affordable way to boost warehouse productivity. *Canadian Transportation & Logistics*, 109 (1), 14–19.

Mason, S.J., Ribera, P.M., Faris, J.A., and Kirk, R.G. 2003. Integrating the warehousing and transportation functions of the supply chain. *Transportation Research E*, 39 (2), 141–159.

McCormick, C. 2003. WMS: Expect more functionality in 2003 as providers strive to stand out in a mature market. *Canadian Transportation & Logistics*, 106 (1).

Min, H. and Melachrinoudis, E. 2001. Restructuring a warehouse network: Strategies and models, In Salvendy, G. (ed.), *Handbook of Industrial Engineering* (3rd ed), New York, Wiley, pp. 2070–2082.

Min, H. and Zhou, G. 2002. Supply chain modeling: Past, present and future. *Computers and Industrial Engineering*, 43, 231–249.

Napolitano, M. 2001. *Making the Move to Cross-Docking*. Oxford, OH: Warehousing Education and Research Council, pp. 308–320.

Pairat, R. and Jungthirapanich, C. 2005. A chronological review of ERP research: An analysis of ERP inception, evolution, and direction. *Proceedings of the IEEE 2005 Engineering Management Conference*. Austin, TX: University of Texas, pp. 288–292.

Petersen, C.G. 1997. An evaluation of order picking routing policies. *International Journal of Operations and Production Management*, 17, 1098–1111.

Petersen, C.G., Siu, C. and Heiser, D.R. 2005. Improving order picking performance utilizing slotting and golden zone storage. *International Journal of Operations & Production Management*, 25 (9–10), 997–1012.

Piasecki, D. 2005. Warehouse Management Systems (WMS), Inventory Operations Consulting LLC, www.inventoryops.com/warehouse_management_systems.htm. Last accessed September 7, 2007.

Ratliff, H.D. and Rosenthal, A.S. 1983. Order picking in a rectangular warehouse: A solvable case of the traveling salesman problem. *Operations Research*, 31 (3), 507–521.

Roodbergen, K.J. and Vis, I.F.A. 2006. A model for warehouse layout. *IIE Transactions*, 38 (10), 799–811.

Sahay, B.S. and Gupta, A.K. 2003. Development of software selection criteria for supply chain solutions. *Industrial Management and Data Systems*, 103 (2), 97–110.

Saxena, A. 2000. *Linking Enterprise Resource Planning Systems and Warehouse Management Systems*. Waterloo, Ontario, Canada: Department of Management Sciences, University of Waterloo.

Spiegel, R. 2003. ERP vendors muscle into logistics. *Logistics Management*, 6 (4), 45–48.

Trunk, C. 1999. Building bridges between WMS & ERP. *Transportation and Distribution*, 40 (2), SCF6–8.

van den Berg, J.P. 1999. A literature survey on planning and control of warehousing systems. *IIE Transactions*, 31, 751–762.

Chapter 3

Warehouse Assessment in a Single Tour

M.B.M. De Koster

CONTENTS

Abstract This paper presents an assessment method for warehouses based on a single facility tour and some questions and answers (Q&A). The method helps managers and students who visit a facility to get more information from tour visits through a simple and rapid assessment form. Since its inception, it has been applied to a number of cases, successfully identifying weak and strong points of the operations.

3.1 Introduction

Over the past decades, many companies have offshored manufacturing activities to Asia-Pacific and Eastern Europe. Because the consuming markets have not moved, this has put an increasing burden on the distribution operations of such companies. Companies have centralized warehouse operations in few, but often large facilities responsible for distributing products over a large region. Managing efficiency and effectiveness (service) is a great challenge for managers of such facilities. As a result, they feel a great need to benchmark warehouse operations, not only their own, but also their competitors'. However, assessing the performance of a distribution facility is a tricky business. Even after having visited a large number of them, it is still difficult to tell after a visit whether this was a best-in-class operation, just above average, or even relatively poor performing. Nevertheless, even short-tour visits can reveal a lot of information to the trained eye.

This paper proposes a method to help managers get more information from tour visits, through a simple and rapid assessment form. The form should be filled out immediately after the visit. The evaluation has been inspired by the ideas of Gene Goodson in *Harvard Business Review* on rapid plant assessment (Goodson, 2002). Since its development, the method has been successfully applied in several visits, with different groups of managers (with and without warehouse experience) and students.

The major functions of a warehouse are to store products to make an assortment for customers, to assemble customer orders, sometimes to add value to the orders by customization activities, organize transport for the customers, and ship orders timely, in the way desired by the customer. Warehouse performance, therefore, has multiple dimensions. Often, performance is measured in terms of ratios of output and input factors. Output factors include production (shipped orders, lines, and units); quality (e.g., order completeness, and error-free and on-time delivery); flexibility (possibility to cope with changes in customer demand); agility (process adaptation to changed environment); and innovativeness (use of new supply-chain concepts yielding competitive advantage). Inputs are the resources used to achieve the outputs. These include the number of full-time equivalents (work hours used per year), investment in systems, buildings, and information technology (IT) infrastructure, process organization (i.e., the management), or the assortment carried.

Some researchers have tried to develop benchmark tools for warehouses (Hackman et al., 2001; McGinnis et al., 2002; De Koster and Balk, 2007). One such tool is data envelopment analysis (DEA), which expresses the warehouse efficiency as a ratio of weighed output and weighed input factors, normalized on a 0 to 1 scale. Although DEA is a powerful tool, it is usually difficult to obtain the necessary data at the required accuracy level. Also, for every factor that is included in the efficiency analysis, more cases are needed to have statistically meaningful results. Furthermore, the warehouses should be comparable, which in practice may be difficult to realize. It is also difficult to compare warehouses in different countries, even when they operate in the same industry branch (think of cultural differences, or just of the number of working hours per full-time employee). Finally, it is difficult to include factors in DEA that are not measured on interval scales, or more subjective assessments (like teamwork, motivation, safety, and cleanliness).

As an alternative, or addition, to more quantitative analyses, this tool is based on a single facility tour and can be carried out in a few hours, including some questions and answers (Q&A). It is not necessary to have deep insight in the operations. The main objectives of the tool are to discern the warehouse's strengths and weaknesses—after some elementary training on how to use the tool. The tool can also be used to evaluate operations of logistics' service providers, operating public or dedicated warehouses. This is not to say that the tool can be a substitute for due diligence and care when analyzing company performance. In particular, financial performance is not part of the tool. However, all too often managers ignore visual signals that can be easily acquired in favor of seemingly objective data, like quantities processed, inventory turns, or company profits (which are rarely directly attributable to a warehouse).

3.2 Assessment Method

The tool is based on a factor-rating method (see, e.g., Heizer and Render, 2004) and consists of 11 areas that have to be assessed, each on a six-category scale (see Exhibit 1). Seven areas (1 to 5, 8, and 10) are more or less generally applicable to industrial facilities and have been adapted to warehouse environments from Goodson (2002). Areas 6 and 7 (storage and order-picking systems) form the heart of any warehouse (Tompkins et al., 2003) and must, therefore, be included in an assessment. Areas 9 (level and use of IT) and 11 (managing efficiency and flexibility) are equally important in an assessment. To aid filling out the assessment form, a number of "yes or no" questions have been formulated (Exhibit 2), which serve the purpose of conveying the opinions on the area and aiding area scoring. A score is measured on a six-category ordinal scale and ranges from poor (1 point) to excellent (9 points) with an additional category "best-in-class" (11 points). Best-in-class means that there is no better. We first discuss the areas in more detail and then discuss results as well as further validation of the method.

3.2.1 Area 1: Customer Satisfaction

Customer satisfaction is difficult to rate in a facility visit. However, all the people in the facility—and particularly workers—should clearly know who the customers are, both internal and external. Management can take care of this by explicitly showing external quality performance indicators to the workers. Signboards with picking or shipping errors, customer complaints, and returns over time, quality guidelines for workers, and so on indicate sensitivity to wishes of customers and quality assurance. Try asking an order picker, packer, or dispatcher: "What is the impact for customers when you make an error?" When this person answers that it will result in a complaint (or return, or a customer credit note), it should lead to a higher score than when the employee has no idea at all, or when she or he deems there are no clear consequences.

Even when products are picked by article (batched over multiple customer orders), the person should have an idea of the customers' wishes, whether there are deadlines for the (batch) order to be shipped (many large warehouses work with fixed departure schedules to reach their customers timely), and what the consequences are for not finishing the work in a timely manner.

Questions 1, 4, 14, and 21 are related to this area.

Exhibit 1: Warehouse Rating Sheet

Warehouse:
Date visit:
Group:

	Area	*Related Questions*	*Poor (1)*	*Below Average (3)*	*Average (5)*	*Above Average (7)*	*Excellent (9)*	*Best-in-Class (11)*	*Total*
1	Customer satisfaction	1, 14, 21							
2	Cleanliness, environment, ergonomics, safety, hygiene	2a, 2b, 3, 17, 21							
3	Use of space, condition of building, and technical installations	5a, 5b, 6a, 6b, 15, 21							
4	Condition and maintenance of material-handling equipment	16							
5	Teamwork, management, and motivation	1, 12, 21							
6	Storage systems and strategies, inventory management	7a, 7b, 8, 9a, 9b, 19							
7	Order-picking systems and strategies	10, 11a, 11b, 20							
8	Supply-chain coordination	19							
9	Level and use of IT	20							
10	Commitment to quality	4, 11a, 11b, 12, 13, 14, 17, 20							
11	Managing efficiency and flexibility	18							
Total									

Exhibit 2: Questionnaire

Warehouse:
Date visit:
Group:

		Yes	No
1	Are visitors welcomed and given information about warehouse operation, customers, and products?	☐	☐
2a	Is the facility clean, safe, orderly, and well lit? Is the air quality good and noise level low?	☐	☐
2b	Is the environment attractive to work in?	☐	☐
3	Are the work processes ergonomically well-thought over?	☐	☐
4	Do the employees appear committed to quality?	☐	☐
5a	Is the warehouse laid out in a U-shape rather than an I-shape?	☐	☐
5b	Does the layout prevent major crossing flows?	☐	☐
6a	Is material moved over the shortest/best-possible distances?	☐	☐
6b	Is double handling prevented and are appropriate product carriers used?	☐	☐
7a	Are products stored on their right locations? Do storage strategies lead to operational efficiency?	☐	☐
7b	Are the locations used dynamically?	☐	☐
8	Is the number of different storage systems (with different racks, material-handling systems, and storage logic) justified?	☐	☐
9a	Is appropriate (non-)splitting of inventory in bulk- and forward-pick stock applied?	☐	☐
9b	Is there an effective process management for introducing new products, getting rid of nonmovers, and internal relocations?	☐	☐
10	Is the organization of the picking process well designed without obvious improvement possibilities?	☐	☐
11a	Are storage and receiving processes monitored and controlled online?	☐	☐
11b	Is the response to mistakes and errors immediate?	☐	☐
12	Are work teams trained, empowered, and involved in problem-solving and ongoing improvements?	☐	☐

(*continued*)

Exhibit 2: Questionnaire (continued)

Warehouse:
Date visit:
Group:

		Yes	*No*
13	Are up-to-date operational goals and performance measures for those goals prominently posted?	☐	☐
14	Are ratings for customer satisfaction and shipping errors displayed?	☐	☐
15	Are the buildings, floors, and technical installations in good quality and well maintained?	☐	☐
16	Are the material-handling systems used, the racks and the product carriers in good operating condition and well maintained?	☐	☐
17	Are inventories accurate?	☐	☐
18	Has a right balance been struck between order customization, process flexibility, and efficiency?	☐	☐
19	Are receiving and shipping processes, and inventory levels tuned with suppliers and customers?	☐	☐
20	Is the level of IT, picking, and storage technologies adequate for the operation?	☐	☐
21	Is this a warehouse you would like to work in?	☐	☐

Total: Yes or No

The total number of "yeses" on this questionnaire is an indicator of the warehouse's overall performance. The more yeses, the better the performance. A question should be answered a "yes" only if the warehouse obviously adheres to the principle implied by the question. In case of doubt, answer "no."

3.2.2 Area 2: Cleanliness, Environment, Ergonomics, Safety, and Hygiene

This is an area that is relatively easy to assess. If a facility is clean, it usually indicates that management organizes the processes well. In clean facilities, items do not get lost, inventory accuracy is higher (as well as order fulfillment accuracy), and there is an overall sensitivity to orderliness. Order-picking warehouses (where case and item picking occur) typically generate much waste (pallets have to be unwrapped, boxes have to be opened) and workers have to be able to get rid of it in an easy way. In well-run warehouses, one can find waste baskets in front of the racks, where waste can be separated immediately at the source by type (which is compulsory in the European Union [EU]). In a well-run facility, the air is clean, noise levels are low, and it is well lit. In short, it is comfortable to work in. All location codes are easily readable (also from a distance) and bar coded, such that there is no confusion as to which code refers to which location (particularly for the lower beams in a pallet rack, or in a shelf area where location sizes are often tiny).

Worker positions should have been designed with attention for ergonomics. As much of the work is repetitive or strenuous, ill-designed work positions lead to high absence rates and labor turnover.

In many warehouses, pickers do not have fixed work positions because they drive trucks or walk with pick carts. Even in such cases ergonomics pays off. The use of tiny screens and buttons on mobile terminals leads to low productivity and even to errors (reduction of which often was the main reason for the use of such terminals). In the European warehouse of a large Japanese manufacturer of consumer electronics, pickers use mobile terminals to receive pick instructions and confirm the picks. When they were asked about the contents of their work, it appeared that for a single order (of a few units) about 20 entries had to be made to confirm this. If 20 cases of the same product had to be picked from a pallet, and then labeled, scanned, and put on a conveyor belt, it might take minutes to confirm this via the radio frequency (RF)-terminal/scanner in the information system. Workers obviously find workarounds (do first and confirm when convenient), which may compromise the system integrity.

Safety is of utmost importance in many warehouses, especially where heavy-pallet lifting or order-picking trucks or cranes are used. Order-pick and forklift trucks may weigh up to several tons and can drive at considerable speeds. Warehouses should have safe travel paths for pedestrians and safety collision protection devices. Workers on foot should not work in narrow aisles together with heavy order-pick trucks. Unsafe working conditions can be discerned from the amount of damage at the racks, at the trucks, or signboards indicating the number of accidents, or if people smoke in a battery-charging area. Unsafe working conditions should lead to a low score on this criterion.

Hygiene (based on hazard analysis and critical control points [HACCP]) is of particular importance for warehouses which process (pet) foods, drugs, or raw materials for such products. If deep-frozen products wait for a considerable time in an insufficiently conditioned receiving or shipping area, the condition of the product may deteriorate. Questions 2a, 2b, 3, 17, and 21 relate to this area.

3.2.3 Area 3: Use of Space, Condition of Building, and Technical Installations

Although (particularly in distribution warehouses) labor is the most important ingredient of operational cost (in particular the order pickers, see Tompkins et al., 2003), facility cost (including technical installations) is a close second. Whether buildings and technical installations are owned, rented, or leased is irrelevant. Therefore, space should not be wasted. Excessively large warehouses do not only lead to high costs, but often also to inefficient processes, due to long travel times for storage, order picking, or crossdock. In case of storage of large numbers of loads of slow-moving products, high-bay stacking is preferred. There is, of course, a difference between countries in the costs of land and labor. If labor and land are relatively cheap (United States), buildings are usually lower. If land is expensive (Japan), buildings are higher.

On the other hand, insufficient space may prevent a process from being executed effectively and efficiently. If products have to be dropped at temporary locations because of lack of space in the proper area, if products have to be dug up because they are stored at wrong locations, or if much waiting and delays occur because maneuvering spaces are used by other workers, this area receives a low score. It may be necessary that multiple persons work in the same area (e.g., order pickers and replenishers in a pallet-storage area); nevertheless, blocking and congestion should be avoided. This can be enforced by having one-directional traffic or distribution of fast-moving articles over multiple storage zones.

Many facilities have undergone natural expansion: gradually more and more buildings and systems have been added. In many cases, this leads to suboptimal logistic processes. Warehouses spread over multiple locations lead to necessary transport movements between the parts. How is this process organized? Can inventory get lost while in transport? If not handled properly, it should lead to a low score for this area.

The technical state of buildings, doors, floors, dock levelers, dock shelters, sprinkler installation, and heating and cooling installations is fairly easy to assess during a visit. The quality of floors (i.e., flatness and absence of pits and ramps) is particularly important if forklifts, reach trucks, and high-bay trucks are used for discrete transport.

The basic facility layout is important for achieving top performance. U-shaped layouts, where dock doors are mainly located along one façade, usually lead to better performance (greater expansion possibilities, more flexible use of dock doors and receiving/shipping personnel, less crossing flows, shorter average travel distances) than layouts with dock doors on opposite sides of the buildings (I-shaped layout).

Questions 5a, 5b, 6a, 6b, 15, and 21 support the assessment of this area.

3.2.4 Area 4: Condition and Technical State of Material-Handling Equipment

Although it may seem wise at first sight to use a special truck for every different type of work, multiple brands of material-handling equipment lead to less flexibility, higher risk of unavailability, and higher maintenance cost. Material-handling equipment that breaks down frequently or batteries that do not charge sufficiently may lead to an inefficient operation and missed deadlines. Even old trucks can work properly if well maintained. You might try to ask a driver whether he or she experiences any problems with the trucks. While asking this in a warehouse of a Serbian food retailer, it appeared that the batteries of one of the narrow-aisle pallet trucks charged insufficiently. This made the truck unavailable for a substantial part of the day, leading to orders that could not be filled completely on time.

Proper working material-handling equipment shows from maintenance recorded on the equipment, the looks of the equipment, and few failure records or performance obstructions in the operation. Question 16 supports this area.

3.2.5 Area 5: Teamwork, Management, and Motivation

As Bartholdi and Eisenstein (1996) and Bartholdi et al. (2000) showed, bucket brigades (a teamwork order-picking concept) can lead to substantial performance (particularly throughput) improvements in picker-to-parts order-picking systems. Although the bucket-brigade concept is only applicable under special circumstances, people working as a team will perform better than as individuals. This is particularly true in order picking, receiving, and shipping. If people are multiskilled and rotate in different areas of the warehouse, this might be an indicator of team spirit. If people are proud of their work and the company, this is a positive indicator. One might try to discern this factor by asking questions to the employees and management.

Questions 1, 12, and 21 support this area.

3.2.6 Areas 6 and 7: Storage and Order-Picking Methods

Storage and order picking form the heart of most warehouse operations. Warehouse efficiency depends to a large extent on the methods used for storing products and picking the orders. The question is whether the appropriate methods are used. This is probably difficult to assess, particularly for inexperienced visitors. Also, great varieties of storage and picking technologies are available on the market. The choice of these also depends on the volume to be picked, the variety in the assortment, and quantity to be stored and the labor cost rate. Higher labor costs and larger throughput volumes justify more automated storage and picking systems, and a higher level of order-picking aids, like scanners, mobile terminals, or voice-recognition equipment. In low-volume warehouses, that is, with few orders, the preferred way is picking by order. Although multiple workers can work on the same order, the order is kept intact; it does not have to be split and sorted, but can, after possible order assembly, immediately be packed for shipping. In very high throughput volume warehouses, picking by order is impossible. Instead, orders are picked by article (in batch) after which the items are sorted and grouped by order.

3.2.7 Area 6: Storage Systems and Strategies and Inventory Management

To assess the methods used, the visitor might pay attention to the following elements.

- Are products stored at their appropriate locations? This includes storage based on physical properties (conditioning, dimensions, weight, and theft-proneness) and turnover speed: fast-moving items should be located on easily accessible locations at short distances from the dispatch position (Question 7a).
- Are the locations used dynamically? In many warehouses, fixed locations are used, from which products are picked. Even when products are initially assigned to these locations on turnover frequency (to reduce travel time), such an assignment will be far from optimal if not regularly maintained (like reassignment every month). Few companies do this. Companies that use dynamic locations, taking into account dynamic turnover frequency, score better than companies with fixed locations and little reassignment (Question 7b).
- Is the number of different storage systems (with different racks, material-handling systems, and storage logic) justified? Warehouses often store large numbers of products. The idea is to create the highest throughput efficiency possible, with the fewest systems used. These are often contradictory requirements, but a balance between the two should be struck. In case many different storage

systems are used, consideration should be given to merging two of them, without decreasing order-picking efficiency; or where few storage systems are used part of the assortment could be taken from a system and stored separately to increase efficiency and homogeneity of handling (Questions 7a, 7b, and 8).

- Is the inventory of certain products split into bulk storage and forward-pick storage? If items are picked in a condensed forward-storage area, the order-picking lead times are reduced considerably and storage activities can be decoupled from order picking. Such systems can be designed for box picking (bulk stored on pallets, lower pallet locations used for picking the boxes), or item picking (bulk stored on boxes on pallets, shelves used for item picking). Particularly if bulk quantities tend to be large and order-pick quantities are small, splitting inventory pays off and outweighs the replenishment efforts (Question 9a).
- Is family grouping applied in storage with the objective of making processes efficient? Many forms exist, such as grouping items that are frequently ordered together. Grouping methods that do not immediately lead to higher efficiency, such as products of the same supplier together, or products of the same owner together, score lower (Questions 7a and 7b).
- Is inventory managed appropriately? Are inventory levels appropriate? It may be difficult to answer these questions, but clear visible signals should not be ignored. For example, in a company with short product life cycles, there should be an explicit program to get rid of "old" products. Look for a corner in the warehouse where seemingly nonmovers are stored. These can be recognized by little pick activity, great product inhomogeneity, and sometimes small quantities stored per product. Inventory levels (ask for inventory turnover rate) depend on product properties, where suppliers are located and on the degree of supply-chain cooperation. If suppliers are located further and products are cheaper, higher inventory levels are justified. Expensive products with short life cycles should have low inventory levels (Questions 9b and 19).

3.2.8 Area 7: Order-Picking Systems and Strategies

Before making an assessment, the order-picking methods used (often more than one) should be classified. A typical classification and explanation of methods can be found in Exhibit 3. Have the weak points of the order-picking systems used been addressed adequately and sufficiently? Every order-picking system has strengths and weaknesses. The strengths are

Exhibit 3: Order-Picking Methods

The below figure shows different order-picking methods that can be found in warehouses (for a description of some of these methods, see Tompkins et al., 2003). In many warehouses, multiple methods are used. The large majority employs humans for order picking. Among those, the picker-to-parts system, where the picker walks or rides along the items, is most common. Parts-to-picker systems include automated storage and retrieval systems (AS/RS), mostly using aisle-bound cranes that retrieve one or more unit loads (bins: mini-load system, or pallets) and bring it to a pick position. At this position, the picker takes the number of pieces required by the customer order, after which remaining load is stored again. Other systems use vertical lift modules (VLM), or carousels that also offer unit loads to the picker, who is responsible for taking the right quantity. Put systems are positioned between the picker-to-parts and parts-to-picker systems, because they often combine the two principles. First, inventory has to be retrieved, which can be done in a parts-to-picker or picker-to-parts manner. Second, the carrier (usually a bin) with these parts is offered to a picker who distributes the items over customer orders. Put systems are particularly popular in case a large number of customer-order lines have to be picked in a short time window (e.g., at the Amazon German warehouse) and can result in about 500 picks: on average per picker hour (for small packages) in well-managed systems.

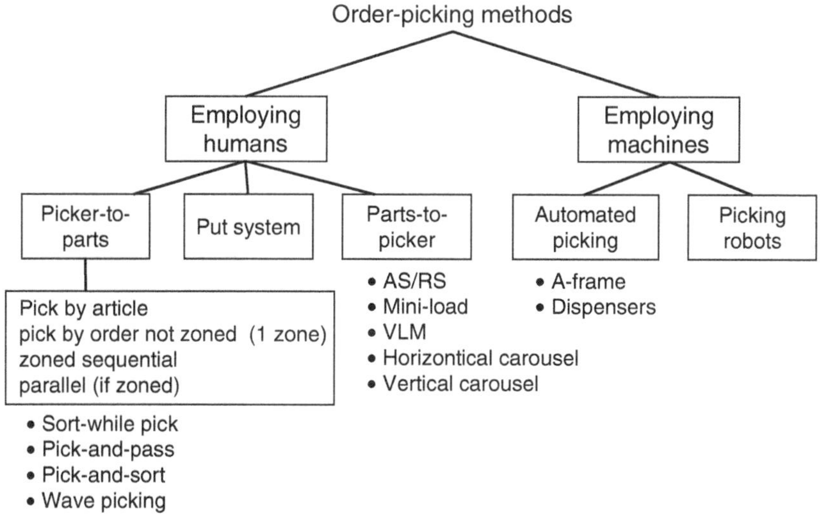

(*continued*)

Exhibit 3: Order-Picking Methods (continued)

Picker-to-part systems are the most common. The basic variants include picking by article (sometimes called batch picking) or pick by order. In the case of article picking, multiple customer orders (the "batch") are picked simultaneously by a picker. Many in-between variants exist: picking multiple orders followed by immediate sorting (on the pick cart) by the picker ("sort-while-pick"), or "pick-and-sort" in which case the sorting takes place after the pick process has finished. Another basic variant is zoning, which means that a logical storage area (this might be a pallet-storage area, but also the entire warehouse) is split in multiple parts, each with different pickers. The pickers can work sequentially, traveling along the locations in their zone, and pass the product carrier with pick instruction to pickers in the next zone, or they can work in parallel, and work on the same orders. If this is the case, the order parts have to be assembled before they can be packed and shipped. Parallel and batch picking speed up the picking process, at the cost of additional sorting and order assembly work. The term "wave picking" is used if orders for a common destination (e.g., departure at a fixed time with a certain carrier) are released simultaneously for picking in multiple warehouse areas. Usually it is combined with batch picking.

usually immediately visible in a visit (apparently, the system works); weaknesses are more difficult to discern. Batch picking, followed by sorting on an automated sorter, requires that all items (including the last items, which usually are missing) are picked in time for the sorter to start. Is this handled adequately? Order throughput times in picker-to-parts systems can sometimes be very long. Is this controlled sufficiently? For example, in Océ's parts warehouse (Océ is a manufacturer of professional copiers and printers), which supplies parts overnight directly to technicians in Western Europe, orders are picked in batches (of orders for technicians in the same country) of about 60–120 order lines per order picker. The throughput time can be very long and is difficult to predict. Also, pickers can decide themselves on the number of lines they want to work on. This makes it difficult to guarantee that the fixed departure times of the trucks can be realized, requiring extra control effort (regular progress checking and emergency help) to guarantee this. The European warehouse of Yamaha Motor Parts uses a zoned pick-by-order system. A conveyor passes the order bins between the zones. As there are many zones (about 60), and orders can sometimes be large, orders queue before every zone, making order throughput times close to unpredictable at busy moments. Yet, Yamaha has a fixed truck departure schedule for all customer destinations. The problem was solved by batching multiple small orders into the same order bin, thereby strongly reducing queuing. In C-Market's warehouse (a supermarket chain), pickers on order-pick trucks travel long distances in a large pallet warehouse to pick orders for a single supermarket. In competitors' warehouses, pickers on long-fork trucks pick two or three stores simultaneously in roll containers in one warehouse zone only, which leads to a large increase in productivity.

The following questions (see also Questions 10, 11a, 11b, and 20) might guide the evaluation of the order-picking process:

- Are throughput times sufficiently controlled?
- Does avoidable double handling occur?
- Are obvious improvements possible in the picking process? You might think of some improvements and ask the pickers for their evaluation.
- How is the progress of the order-picking process monitored and controlled?
- Are the used picking aids (order lists, labels, RF terminals, scanners, picking carts) well designed and of help to increase quality and efficiency?
- Have measures been taken to make the picking process sufficiently ergonomic?

3.2.9 Area 8: Supply-Chain Coordination

The degree of supply-chain coordination is visible at the shop floor in several areas. At the yard, inbound trucks may be waiting to be allocated to a dock door, due to inability to properly coordinate arrival times. In the receiving area, trailers and containers must be unloaded and goods must be processed. Is this a rapid, well-organized process, or very time consuming because the product carriers are wrong and products have to be restacked, information cannot be found or is incomplete, boxes of the same product are spread over multiple pallets or over the entire container? In case much paperwork is necessary to check incoming shipments, this is also not a sign for well-tuned processes. You might also ask what happens in case of wrong, under, or over receipts. Does this happen often? Does it delay the process? Attention also has to be paid to the frequency of supply and the drop size. Drop size might be identified at a visit, frequency not without asking. If you see small drop sizes, ask the receivers the frequency of supply of these suppliers. At some warehouses, powerful customers try to reduce their inventories by just-in-time (JIT) policies: frequently ordering small quantities. Although this leads to inventory reduction at the customer's facility, it leads to high handling and transportation cost for the supplier, which might retaliate against the customer.

In an extreme case, we asked a U.S. wholesaler where the customer returns were handled. In response to that question we were taken to a warehouse at the other side of the street, where an endless heap of mostly damaged boxes were waiting to be processed. These were the returns of mainly one customer, who returned "suddenly" a few dozen truckloads of excess stock. This was representative for the company's entire receiving process.

Even if products are loosely stacked in sea containers, it is still possible to have an efficient receiving process if adequate agreements have been made with suppliers. In the warehouse of Zeeman, a textile hard-discounting retailer mainly receiving products in sea containers from East-Asian suppliers, the boxes are grouped by product in the container, and box sizes are standardized. This allows rapid manual unloading of the containers using extendible conveyors, after which the boxes are automatically counted, labeled, and palletized. Conversely Schuitema, a franchise retail organization, has to restack all of Unilever's pallets (a main supplier), because they do not fit into the storage slots.

The level of supply-chain coordination is also visible in the shipping area. An abundance of paperwork needed to ship products is an indicator, as well as the carriers on which products are shipped. If products are shipped on product carriers that return (e.g., pool pallets or closed-loop bins), this often indicates an efficient distribution and collection process, coordinated with the recipients. It saves one-way packaging materials which, particularly in Europe, are expensive, not only because of material cost, but also because fees have to be paid to green-dot systems in different countries to organize proper recycling of these materials. If products are shipped in sea containers on slipsheets (loads on flat carton "pallets" that can be pushed into the container by "push–pull" trucks), this saves space in the container and it suggests advanced coordination with the receiving customer (who also needs such a truck).

Question 19 refers to this area.

3.2.10 Area 9: Level and Use of IT

Nowadays, warehouses do not run without a sufficient level of information systems. Best-in-class warehouses use systems for electronic-information exchange with suppliers, customers, carriers, customs authorities, and brokers in the supply chain. They use a warehouse management system

(WMS) for managing the warehouse processes and they use appropriate tools and aids to support important warehouse processes. WMS comes in a great variety, varying from simple spreadsheet applications, to standard modules of ERP software packages, specialized WMS packages, or tailor-made applications. In general, the more complex the operation (mainly measured in number of order lines, assortment size, different processes, and uncertainty in demand and supply, see Faber et al., 2002), the more justified or even necessary specific or tailor-made software becomes. A WMS is necessary to find the best location where an incoming load can be stored, the best location from which an order line can be picked, the right person to pick an order line (in the right sequence, minimizing travel time), the regular update of article-to-location assignments (based on turn-over frequency) to internally move products to make sure that articles are cycle counted regularly without disturbing the main work flows, and so on.

Tools that can be used to speed up processes and reduce errors include pick-to-light and put-to-light systems and use of the right communication means with drivers and pickers to guarantee real-time monitoring of work progress. Bakker, a mail-order company which specializes in flower bulbs, uses a put-to-light system for distributing bulbs that have been pre-picked over the right customer-order bins. A graphical screen helps the picker, as it shows visually which bins have to be addressed. These aids increase productivity significantly.

Question 20 reflects this area.

3.2.11 Area 10: Commitment to Quality

Commitment to quality can be derived from a number of factors in a facility. First, from the design itself—at which points is it easily possible to make errors? If an operator can determine where to store an incoming load and later provide confirmation, this is an obvious source for errors. Storage errors are very serious, as they potentially impact multiple customer orders. The same is true for picking: can an operator easily pick the wrong item or the wrong quantity? Best-in-class operations do not ensure quality by building in additional checks of the picked orders. Instead, they take measures that prevent people from making obvious errors ('poka-yoke,' or fool-proofness principle). In the previously mentioned warehouse of Yamaha, pickers at a mini-load workstation have to pick a unit from a compartmented bin, containing multiple products. To prevent errors, the computer screen is divided in the same way as the bin, with the proper part illuminated. In addition, a battery of spotlights illuminates exactly the right compartment of the bin.

Second, is continuous process improvement actively propagated in the facility? Are workers stimulated to improve their processes and can proof be found for this? Indicators for this can be an idea box, implementation of six-sigma improvement projects or the number of master black-belts, or the number of process improvements recently realized. You might try and ask about this. In a recent tour of the European distribution center of a U.S. manufacturer, we were told that people could be promoted to management level only if they at least owned a six-sigma green belt.

This area is addressed with Questions 4, 11a, 11b, 12, 13, 14, 17, and 20.

3.2.12 Area 11: Managing Efficiency and Flexibility, as a Function of Volume, Assortment, and Variety

It is very difficult—if not impossible—to manage a large number of orders, together with a large assortment and a variety of customer wishes efficiently, in a manner that is flexible enough to accommodate late changes. Process automation and mechanization with multiple solutions for different storage areas can help for efficiency, but usually bring down flexibility. Logistics service providers with public warehouses and short-term contracts usually opt primarily for flexibility and sacrifice efficiency to some extent. Flexibility is expressed as the ease to which different customer-order patterns (large versus small orders), different customer wishes (product and order customization) can be accommodated, the processes expanded or shrunk, and assortment changes handled. During a visit, attention can be paid to what extent any of these principles have been sacrificed. If processes seem very efficient, you might ask whether the above-mentioned flexibility features can be accommodated. In case an operation seems very flexible, it is interesting to estimate whether customers are really willing to pay for the inefficiency. If a right balance seems to have been struck, a company scores higher than when there are obvious flaws. This is addressed with Question 18.

3.3 Results and Validation

The assessment has been carried out with several groups of managers and students. Within a group, the areas are divided over different group members. Immediately after the visit, each group filled out the warehouse rating sheet as a team effort. Exhibit 4 shows the outcomes of some assessments carried out in 2004 and 2005 with different groups of international people (in total 96 persons from 22 countries participated, about 30–40 people per visit, with and without warehousing experience). For every facility, the maximum score is 121. The results show a clear distinction between high- and low-ranking facilities. Low-ranked facilities nearly always score 'NO' on

Exhibit 4: Some Examples of the Tool's Results

N = number of groups

Warehouse	*Description*	*Ave. Total Rating (N)*	*Std. Deviation**	*Average (max) SD Area*	*DEA Efficiency Score (Percent)*
A	Multinational interior-decoration retailer	65.9 (8)	10.8	1.6 (2.5)	58.8
B	Automotive manufacturer, spare parts	82.5 (8)	8.9	1.7 (2.6)	95.5
C	National wholesaler supermarket products	76.3 (6)	3.5	1.4 (2.0)	100
D	National food retailer	59.2 (9)	7.2	1.5 (2.1)	—
E	Multinational hard-discounting non-food retail chain	64.0 (6)	10.0	1.6 (2.5)	66.2
F	Multinational fashion products manufacturer/ wholesaler/retailer	73.0 (6)	3.1	1.4 (2.2)	44.2

*In calculation, we interpreted the scores as measured on an interval scale.

Question 21; high-ranked facilities 'YES.' The outcomes of area ratings are quite varied as well, although "Customer satisfaction" (area 1) obviously scores fairly high in general.

To validate the method, basically three different methods were used. First, we independently benchmarked the warehouses using DEA, based on a database of 71 warehouses. Second, we compared the standard deviations of area and total scores among groups. If these standard deviations are moderate, we can at least say that the scoring is reliable. Third, we asked the warehouse managers for feedback on the scores per area (the method was mailed to them prior to the visit).

To benchmark the warehouses with DEA, we asked the warehouse or logistics manager to fill out a questionnaire, addressing performance in the areas of shipment quality, production (volume and variety), and flexibility (for a full description of the method see De Koster and Balk, 2007). The resulting efficiency scores (the maximum efficiency to be obtained is 100 percent) can be found in Exhibit 4. Although the factor-rating and benchmarking methods look at different indicators, the correlation between the two scores is quite high: 64 percent for the companies listed in Exhibit 4, indicating that the assessment method is a good forecaster of performance (albeit the number of included warehouses is still small).

Exhibit 4 also displays the standard deviation of total and area scores. The maximum standard deviation of the total score is within 16 percent of the average. For individual area scores, the average standard deviation varies between 1.4 and 1.7 (less than 25 percent of the average area score). Usually there are one or two areas of some disagreement between groups, with standard deviations up to 2.6. No areas consistently showed a higher standard deviation in the scoring. The score reliability improves when the assessment is done with more experienced people: having seen more facilities obviously helps in calibrating one's judgment. However, it should be emphasized that all facilities were also visited by such inexperienced people, leading to the above-mentioned moderate standard deviations of scores.

After every visit, the warehouse manager was confronted with the area scores. In all cases, they agreed with the relative ranking of their scores. Obviously, warehouse management is often aware of weak points, but it is not always easy to improve. For example, a weak layout cannot easily be changed by the management; such a conclusion should serve as input for the company's facility development staff.

3.4 Conclusion

The method presented in this paper may help managers and students to rapidly assess warehouse facilities. The method serves as an addition to more quantitative methods, like financial analysis. We have validated the method with DEA benchmarking. Although the number of warehouses benchmarked with both methods is still small, first results indicate that indeed the method shows some value in an assessment. Total and area scores are reasonably homogeneous among the different groups (although every warehouse so far shows one or two areas with standard deviations higher than 2, which may be as much as 40 percent of the average area score). It is helpful, in this respect, that the assessors have applied the method more than once.

In conclusion, if a warehouse appears to score well, based on the visual information and Q&A, it usually is. If it scores poorly, there definitely is room for improvement, particularly in the low-ranked areas.

References

Bartholdi, J.J. and Eisenstein, D.D. 1996, A production line that balances itself, *Operations Research* 44(1), 21–34.

Bartholdi, J.J., Bunimovich, L.A., and Eisenstein, D.D. 2000, Dynamics of two- and three-worker "bucket brigade" production lines, *Operations Research* 47(3), 488–491.

De Koster, M.B.M. and Balk, B. 2007, Benchmarking American, Asian and third party international warehouse operations in Europe with DEA, in *Production and Operations Management*, forthcoming.

Faber, N., de Koster, M.B.M., and van de Velde, S.L. 2002, Linking warehouse complexity to warehouse planning and control structure, *International Journal of Physical Distribution & Logistics Management* 32, 381–395.

Goodson, R.E. 2002. Read a plant—fast, *Harvard Business Review* 80(5), 3–11.

Hackman, S.T., Frazelle, E.H., Griffin, P.M., Griffin, S.O., and Vlatsa, D.A. 2001, Benchmarking warehousing and distribution operations: An input–output approach, *Journal of Productivity Analysis* 16, 79–100.

Heizer, J. and Render, B. 2004, *Operations Management*. Upper Saddle River, NJ: Pearson-Prentice-Hall.

McGinnis, L.F., Chen, W.-C., Grifffin, P.M., Sharp, G.P., Govindaraj, T., and Bodner, D.A. 2002, Benchmarking warehouse performance: Initial Results for Internet-Based Data Envelopment Analysis for Warehousing (iDEAs-W 1.0), *Progress in Material Handling Research*. Charlotte, NC: Material Handling Institute, pp. 205–224.

Tompkins, J.A., White, J.A., Bozer, Y.A., and Tanchoco, J.M.A. 2003, *Facilities Planning*. New York: Wiley.

Chapter 4

Simultaneous Facility Layout and Materials-Handling System Design

Omar Al-Araidah, Ananth Krishnamurthy, and Charles J. Malmborg

CONTENTS

Abstract This chapter presents a new cost-based concurrent layout and materials-handling system design technique that takes into account flexibility in department shapes, manufacturer guidelines on equipment usage, and limitations on ergonomic deployment of material-handling operators. The methodology is aimed at providing management with the "optimal" layout, material-handling equipment portfolio, and the number and skill mix for material-handling operators. A cost-based, discrete-plane layout algorithm adapting material-handling system design in layout development is discussed. The cost function accounts for machine availability, machine capacity, investment and operational costs of equipment, manufacturer recommendations for usage of equipment, operator availability, and ergonomic usage of operators. The computational advantages of a cluster-based procedure that attempts to exploit the potential correlation between material-flow volume distance and material-handling cost are investigated. Because the cluster-based approach has certain limitations, a two-stage greedy simulated annealing (SA) algorithm adapting the characteristics of the best-available solutions in the estimation of the SA computational parameters is proposed. The algorithm is aimed at decreasing the computational times of classical single-stage SA algorithms while maintaining high-quality solutions. The algorithm benefits from the fast convergence of greedy algorithms and the high-quality results of SA procedures attained though exploring alternatives with inferior objective function values. The algorithm takes the properties of the noise range of the SA algorithms into consideration to obtain a proper melting temperature for the initial solution attained through the greedy heuristic thereby accelerating the SA stage. The algorithm is tested in layout environments where palletized products are transported between departments using variable path manual and powered industrial trucks.

4.1 Introduction

Typically, the layout of a facility is evaluated, based on its ability to guarantee the flow relationships between departments required to meet demand requirements, while minimizing the costs associated with storing and handling materials within the facility. Consequently, facility design involves designing both the layout and the associated material-handling system. The classical block-layout problem determines the optimal location of individual departments by considering the space requirements of individual departments, as well as the interactions (or flow) between departments. The design of the material-handling system supporting the layout involves the selection of the appropriate mix of handling equipment, determining the number of replicates of each handling equipment type, and assigning material-handling equipment to individual material moves between departments. In a deterministic setting, the design of the facility layout and associated material-handling system leads to a mixed-integer programming problem. The computational complexity of this problem has led to several studies aimed at improving the efficiency of the design process.

A common approach is to assume a constant material-handling cost per unit distance, and therefore restricting the layout design procedure to minimize material-flow volume distance or cost. Much of the layout literature focuses even further on complexity reduction through the use of procedures such as clustering (Scriabin and Vergin, 1985) and space-filling curves (SFCs) (Meller, 1992) to reduce the block-layout problem to a quadratic assignment form that is readily addressable through efficient heuristics such as simulated annealing (SA) and genetic algorithms. However, many material-flow-based block-layout applications are characterized by the use of multiple-handling equipment types with different movement costs per unit distance. Solution techniques designed around these simplified material-handling cost assumptions are not always appropriate because, for any given block-layout alternative, subsequent implementation is likely to involve assignment of equipment types to individual movements based on distance traveled. Some researchers estimate that material-handling costs could contribute up to 20–50 percent of the total costs in a facility (Tomkins and White, 1984). It is therefore important to accurately estimate the material-handling costs of a particular facility layout. Although, there have been effective methods proposed for using realistic material-handling cost criteria directly in the search for a solution, they tend to rely on knowledge-based approaches such as expert systems for codifying expert judgment in rule form. These approaches do not generally scale up well for large problems (Malmborg et al., 1989; Chu et al., 1995; Kim and Kim, 1997).

This chapter summarizes observations from two studies (Al-Araidah et al., 2006, 2007) that are both aimed at efficiently generating high-quality solutions for moderate- to large-scale block-layout problems based on realistic measures of material-handling costs. Both studies treat the block-layout problem as a quadratic assignment problem with a specialized cost objective. The scope of these studies is restricted to facilities with discrete material-handling equipment such as forklift trucks, pallet trucks, and lift trucks. This equipment profile is typical of discrete part operations where flexibility under relatively moderate traffic loads assumes significant importance. The two studies differ in terms of the approach used to improve the efficiency of the search procedure for determining the layout solution. First, a cluster-based procedure that attempts to exploit the potential correlation between material-flow volume distance and material-handling cost is investigated. A simplified "surrogate" objective based on material-flow volume distance is used to develop partial solutions. Subsequently, perturbations of these partial solutions are evaluated to determine the layout that minimizes the cost objective function. It is anticipated that clustering based on the simplified objective would significantly accelerate the search procedure without compromising solution quality. The second study investigates the use of specialized SA procedures as an alternative means to improve computational efficiency. Based on experiments conducted using the original cost function that incorporates material-handling costs, a two-stage optimization procedure is designed. Greedy search is applied in the first stage to determine a starting temperature that is used as the starting point of the second SA stage of the procedure. Numerical studies are conducted on several examples for both procedures and insights are summarized.

The rest of the chapter is organized as follows. Section 4.2 provides background information on the complexity-reduction techniques used for block-layout problems. Sections 4.3 and 4.4 provide the details and performance results of the cluster-based approach. Sections 4.5 and 4.6 provide the details and performance results of the specialized SA procedure. Section 4.7 summarizes the main observations and conclusions.

4.2 Background Discussion

Quantitative approaches for solving material-flow-based block-layout problems have been investigated extensively (Liggett, 2000). Approaches to the layout problem can be classified as construction heuristics, improvement procedures, and knowledge-based approaches (Meller, 1992). Under this classification system, construction heuristics and improvement procedures typically focus on volume distance minimization or maximization of activity relationships while knowledge-based approaches focus on material-handling cost. In most applications, work-center shape exerts a significant influence on the quality of alternative layouts. Explicit consideration of work-center shape requires first representing the shape requirements using discrete unit areas or continuous planar space, and then subsequently allocating available space to the different work centers. The particular method chosen often impacts the solution method and computational effort (Liggett, 2000). The use of discrete unit areas is attractive as it provides flexibility in the determination of work-center shapes subject to contiguity constraints. Continuous planar space representations usually involve fixed work-center shapes or selection among alternative shapes yielding a variation of the discrete bin-packing problem. In general, there is no known model based on a continuous planar representation that, when solved to optimality, yields the global optimal solution to the block-layout problem (Bozer et al., 1994; Bozer and Meller, 1997).

The use of SFCs is a strategy for controlling the dimensionality of the layout problem by maintaining work-center contiguity while reducing block layout to a sequencing problem (Meller, 1992). In the application of SFCs, a layout is initially represented as a matrix where each element of the matrix represents a unit area, and each work center assumes an integer number of unit areas based on its space allocation. The technique assures that work centers retain contiguity by assigning unit areas to neighboring grid spaces within the layout (Bozer et al., 1994). In improvement phases of the solution process, SFCs facilitate work-center position interchanges regardless of size or adjacency thereby facilitating the easy application of search techniques such as SA (Meller, 1992; Bozer et al., 1994; Balakrishnan et al., 2003) and genetic algorithms (Kochhar et al., 1998; Balakrishnan et al., 2003).

Clustering approaches, such as the facility layout by analysis of clusters (FLAC) model (Scriabin and Vergin, 1985), can also be used to simplify large-scale block-layout problems by decomposing them into a series of smaller problems. Many clustering techniques originated in cellular manufacturing system design (Askin and Standridge, 1993). They have been found to perform well relative to CRAFT-based perturbation procedures defined on individual work centers. A three-phase hierarchical clustering approach involving cluster formation for initial layout followed by refinement steps has been shown to be effective in solving problems with five to thirty work centers while incorporating a wide range of geometric constraints (Tam and Li, 1991).

Most layout development algorithms assume a fixed cost per unit distance of material movement even though appropriate material-handling equipment types may depend on the length of specific material-flow paths in a facility (Meller and Gau, 1996; Zollinger, 1996). This suggests the need for simultaneous layout development and material-handling system design. This type of concurrent approach is a key advantage of knowledge-based approaches to layout problems (Muther, 1973; Hassan et al., 1985; Malmborg et al., 1989; Matson et al., 1992; Chu et al., 1995; Kim and Kim, 1997). It is an advantage that comes at the cost of a complexity level that limits the extent to which the strategy can be applied in investigation of the extensive solution space associated with large-scale problems. Applications of practical interest can easily involve 40 or more work centers yielding a solution space of significant size. Perturbations of candidate solutions require recomputation of material-flow paths and possible reassignment of equipment types prior to material-handling cost evaluation. These necessitate the application of simplifying measures such as clustering based on simpler surrogate objectives or designing specialized search procedures, tailored for the particular class of problems under investigation.

The cluster-based approach investigates the possibility for enhancing the search process through the use of hybrid strategies integrating simplified volume distance measures as well as material-handling costs. The hypothesis is that although these criteria are not necessarily consistent, they may be sufficiently correlated to enable the effective application of a material-flow volume distance-based strategy such as clustering within a material-handling, cost-driven procedure. The main idea is that by imbedding volume distance efficiencies in initial solutions, overall search efficiency may be improved if material-flow volume distance reduction tends to support material-handling cost performance. To investigate this possibility, Section 4.3 first introduces a realistic cost model applicable to a restricted class of palletized material-handling systems. The scope of application includes systems where palletized loads are moved by four categories of handling equipment appropriate for progressively longer flow paths. Using a rule base derived from ergonomic guidelines for assigning equipment types to material moves, the model computes material-handling costs for layout alternatives based on the direct costs of equipment and labor. Subsequently, an SA algorithm is proposed for using the cost model to search for an optimal block-layout solution using a cluster-based approach. In the cluster-based procedure, initial solutions are conditioned to reduce material-flow volume distance through the application of a clustering step. The results from this approach are compared to those obtained with the single-phase approach where the algorithm uses the material-handling cost exclusively in the search for a solution.

4.3 Cluster-Based Approach for Facility Layout and Material-Handling System Design

As explained earlier, the cluster-based solution strategies attempt to exploit correlation between material-flow volume distance and material-handling cost subject to variation due to differing fixed and variable costs for alternative equipment types. For example, costs and travel speeds associated with a hand-pallet truck are significantly different from those associated with a conventional-rider lift truck. The distribution of move distances appropriate for these equipment types also varies although the material-handling costs associated with both will exhibit conditional correlation with volume distance. The appropriate allocation of material moves to equipment types changes with revisions to a layout solution. Therefore, a key question for the designer is whether the degree of correlation is still sufficient to justify the use of complexity-reduction tools to accelerate the search for a material-handling-cost-based solution. Building on this idea, the solution approach uses a preliminary clustering of work centers based on simple material-flow volume distance to enhance the search for a material-handling cost minimizing solution to the block-layout problem. This concept is summarized in Figure 4.1. The cluster-based approach involves volume distance-based cluster development followed by SFC selection, "cluster sequencing along the SFC" prior to work-center sequencing within clusters based on the material-handling cost objective. The intent is to compare this approach with a single-phased procedure that does not include cluster formation for complexity reduction. Specifically, it involves SFC selection followed by "sequencing of work centers along the SFC" based on material-handling cost. In both cases, calculation of the material-handling cost involves computation of material-flow distances between work centers, assignment of equipment types based on guidelines developed by the Occupational Safety and Health Administration of Canada, and calculation of the expected annualized cost of material-handling using manufacturer-supplied data for alternative equipment types. The following sections describe these steps in greater detail.

4.3.1 Cluster Development

The clustering step reduces complexity by partitioning individual work centers into clusters based on flow intensity. Once clusters are formed and located in the layout using an SFC, the sequence of work centers within clusters is determined to minimize volume distance while controlling for work-center shape. The following notation is used to describe the clustering step:

- M: The number of work centers
- F_{ij}: Flow volume per unit time between work centers i and j
- F_i: Outflow from work center i
- C_{ij}: Closeness coefficient for work centers i and j for $i,j = 1,\ldots,M$
- V: A minimum threshold value for combining clusters

Following a procedure for machine grouping in cellular manufacturing systems (Askin and Standridge, 1993), the closeness coefficients for pairs of work centers are computed using

$$C_{ij} = \max\left(F_{ij}/F_i, F_{ji}/F_j\right) \quad \text{for} \quad i,j = 1,\ldots,M$$

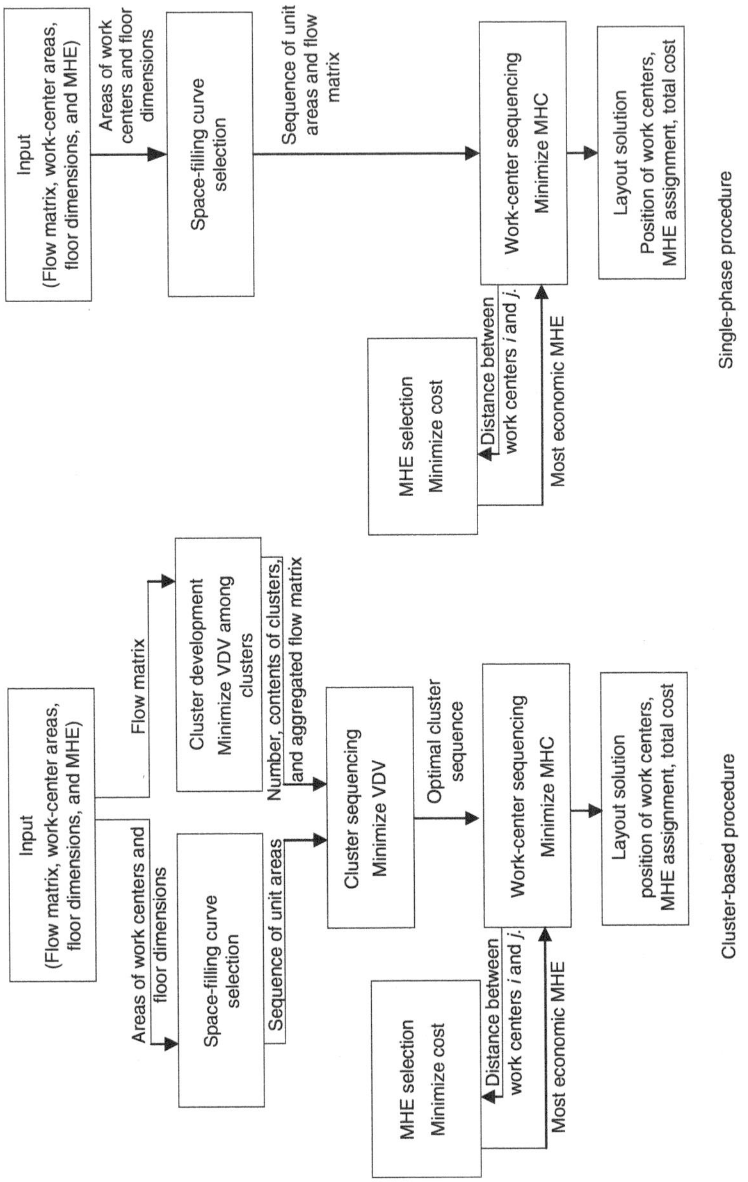

Figure 4.1 Cluster-based and single-phase procedures.

Clusters are formed using the simple step procedures summarized below:

Step 1: Initially, let each work center be a cluster and compute C_{ij} for $i,j = 1,\ldots,M$. Set the threshold value to $V = 1$.
Step 2: Merge clusters i and j into a single cluster k if $C_{ij} > V$
Step 3: Remove columns i and j from the cluster matrix and replace them with new row and column y such that, for every existing cluster z, $C_{yz} = \max(C_{iz}, C_{jz})$.
Step 4: If the stopping limit is reached, stop. Otherwise, set $V = 1 - \Delta V$, where ΔV is a user-defined increment. Go to Step 2.

The stopping limit in Step 4 is a user-defined threshold based on factors such as the maximum areas within a cluster or the maximum number of work centers within a cluster. A total of $C < M$ clusters are formed using this procedure where the area of each cluster is defined using the aggregated space requirements of its work centers. Flow volumes between clusters are updated to consist of the total flow between the work centers included within a cluster, and the work centers included in other clusters (i.e., $F_{zy} = \sum_{i \subset z} \sum_{j \subset y} F_{ij}$ for $i,j = 1,\ldots,M$, and $z,y = 1,\ldots,C$; where C is the number of clusters). The extent to which the clustering procedure minimizes flow volumes between work centers is dependent on the material-flow patterns in a problem. The process of selecting an SFC for locating clusters within a block layout is described later.

4.3.2 SFC Selection

In representing the planar space on which a block layout is to be created, the area is subdivided into N unit areas of equal size with the space requirements for each work center defined in terms of an integer number of unit areas. In the cluster-based procedure, the clustering step is followed by the selection of an SFC based on the value of N and the desired perimeter shape for the layout. The use of an SFC to define the assignment of unit areas to the planar space occupied by a layout assures contiguous work centers by requiring that the unit areas of a given work center are located sequentially in the layout (Bozer et al., 1994). The technique also facilitates perturbation of layout alternatives through adjustments in the sequence used to assign clusters or work centers to the layout. This simultaneously assures work-center contiguity and reduces the dimensionality of the problem from $N!\Big/\prod_{j=1}^{M} m_j!$ to $M!$, where m_j denotes the number of unit areas allocated to work center j for $j = 1,\ldots,M$. For example, for a problem with three work centers (or clusters) with 3, 4, and 5 unit areas respectively, the total number of combinations drops from 27,720 to 6 combinations if SFCs are used. The use of SFCs is illustrated with the help of examples in Figure 4.2. Figure 4.2a shows three patterns of SFCs (Hilbert-type, spiral, and sweeping) that could be used to generate alternative layouts. The particular choice of SFC will determine the manner in which clusters or work centers are arranged in the

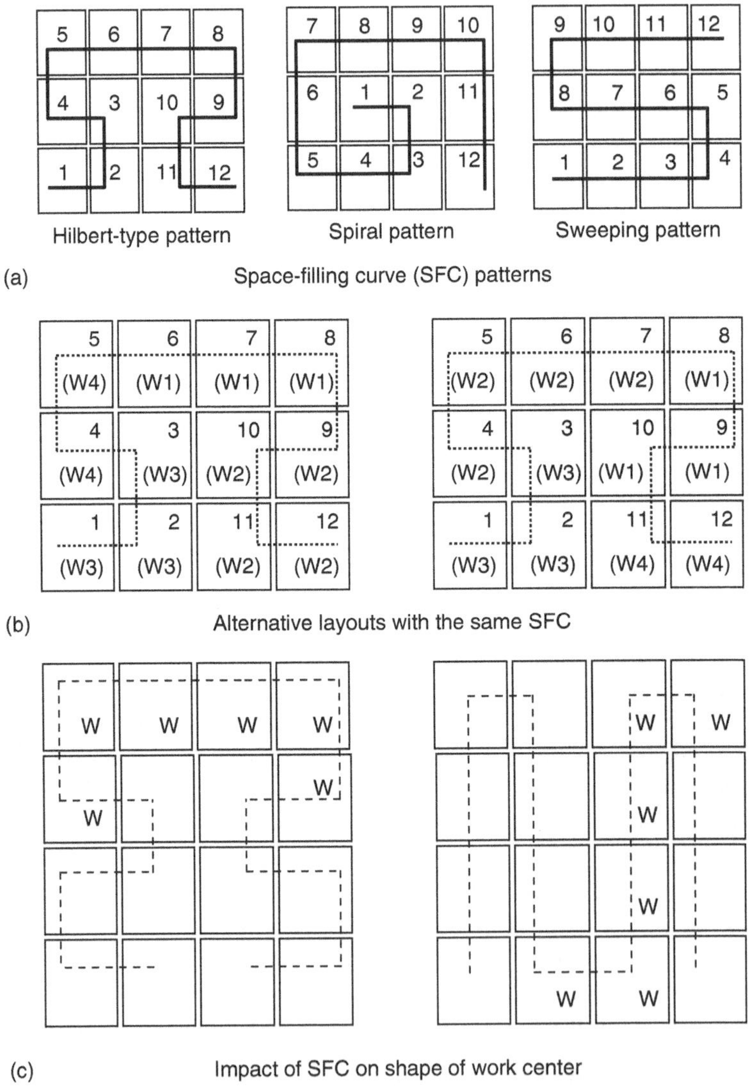

Figure 4.2 Space-filling curves and layouts.

Figure 4.3 Sequencing unit area using sweep-type SFC.

layout. The advantages of these specific types of SFCs are discussed in Meller (1992).

Furthermore, for a given SFC, a candidate layout can be determined as a sequence of clusters or work centers along the curve. For example, Figure 4.2b illustrates two alternative layouts for a problem with four work centers requiring 3, 4, 3, and 2 unit areas, respectively. In both the layouts, the work centers are arranged along a Hilbert SFC. For the layout on the left, the layout sequence is given by (3, 4, 1, 2). The alternative layout on the right is designated by the sequence (3, 2, 1, 4). The impact of the SFCs on work-center shape is illustrated in Figure 4.2c. The figure shows that the shape and location of a work center depend on the sequence with which the SFC visits the grids of the layout floor. Although the choice of the particular SFC restricts the possible shapes a work center can take, the use of SFCs (regardless of the type) ensures the contiguity of the unit areas assigned to particular work centers in a layout solution.

SFC shape also influences individual work-center or cluster shapes and their relative locations. Sweeping-type SFCs are easy to construct although the width of a sweeping curve significantly influences work-center or cluster shape. Following the technique presented in Benjaafar and Sheikhzadeh (2000), this limitation is addressed in the current study by dividing rectangular layouts into bands with a width of 2, 3, or 4 unit areas. An example is illustrated in Figure 4.3 where a planar space comprised of 11 × 27 unit area is partitioned into three unequal rectangles.

4.3.3 Cluster and Work-Center Sequencing

Block-layout alternatives are defined by the sequence of clusters along the SFC and the sequence of work centers within the cluster. Following cluster formation and SFC selection in the cluster-based procedure, the next step is determination of a volume distance minimizing sequence of clusters along the SFC using an SA algorithm. To initialize this search process, a starting sequence is obtained using a greedy procedure of assigning a first cluster randomly followed by the addition of unassigned clusters in an order determined by the magnitude of the volume distance relationship with the last cluster assigned to the sequence. The SA procedure is then applied to optimize the cluster sequence based on the volume distance objective. SA techniques emulate physical annealing through a procedure analogous to heating a substance and decreasing its temperature until it solidifies; the cooling schedule and the time the substance spends at a temperature-state influence the resulting properties of the substance (Chwif et al., 1998). The cooling schedule and the number of iterations to be performed at each temperature, or the epoch length, are user-controlled parameters in SA procedures applied to terminate the search process. SA-based search avoids trapping at local optima by potentially accepting worsening solutions with probability $e^{\delta T}$, where δ is the magnitude of the difference between the objective function value of the incumbent solution and a (worse) candidate solution and T is the state temperature.

Following the SA-based layout technique proposed in Meller (1992), the initial temperature of the SA procedure for the cluster-sequencing problem is set so the probability of accepting a candidate sequence having an objective function value equal to twice that of the initial sequence objective function is equal to 0.001, and the freezing (final) temperature is set to one-thousandth of the initial temperature. In the cluster-sequencing experiments reported in the following section, a linear cooling schedule is applied (i.e., $T_{new} = \alpha T_{current}$), with $\alpha = 0.9$ and an epoch length of 1000 iterations is used. Perturbations of cluster sequences are obtained using pair-wise exchanges of randomly selected clusters.

Once the cluster sequence along the SFC is obtained, the work-center sequence within each cluster forms the basis of the iterative step in the search for an optimal layout. Starting with a random sequence of work centers within each cluster, the material-handling cost for the corresponding layout alternative is computed. Iterations of the search process are then based on exchanging the location of pairs of work centers within a given cluster. That is, layout solutions in the cluster-based procedure are perturbed by randomly selecting a cluster along the SFC. Two work centers within that cluster are then randomly selected. Their locations within the cluster are exchanged and the material-handling cost of the revised layout is computed. Using this perturbation scheme and the material-handling cost objective criterion, the same SA procedure used in the cluster-sequencing step is applied.

The key point in the cluster-based procedure is that only work-center interchanges within the cluster are considered. In contrast, the single-phase procedure bypasses the cluster formation and sequencing steps. Starting with a random sequence of work centers along the SFC, the iterative step in the single-phase procedure is to randomly select any two work centers in the layout. The positions of these work centers along the SFC are then exchanged to obtain a new candidate solution. The SA-search procedure is applied in the same way as in the cluster constrained two-way interchanges used in the cluster-based procedure. The procedure for computing material-handling cost is described in the following section.

4.3.4 Materials-Handling Equipment Costing and Selection

The scope of this study is limited to applications characterized by discrete, palletized unit loads. Even with this restriction, the selection among equipment alternatives for material movements in a facility can be a complex task apt to yield multiple solutions for any specific problem. Many different researchers have investigated material-handling equipment selection based on cost minimization, maximization of utilization, safety, adaptability, flexibility, or various combinations of these and other objectives (Kulwiec, 1985; Matson et al., 1992; Welgama and Gibson, 1996; Peters, 1998). In this study, we propose an approach for selection among industrial trucks spanning the four levels of mechanization illustrated in Figure 4.4, and based on the guidelines presented in Tables 4.1 and 4.2. These guidelines are adapted from widely referenced sources (*Modern Materials Handling Manufacturing Guidebook*, 1985; The Ergonomic Group—Eastman Kodak Company, 1986; Canadian Centre on Occupational Health and Safety, 1997). Although their generalization to all situations is not possible, they provide an easily applied basis for establishing a preliminary, if not reasonable, material-handling design for block-layout alternatives subject to the restrictions of this study.

Given an assignment of equipment types to material moves based on material-flow distances, cost estimates are generated based on the fixed and variable costs associated with alternative equipment categories. Fixed costs are assumed to include initial costs discounted over the expected economic life of the equipment. Labor is the primary variable cost associated with pallet-handling equipment (Ziai and Sule, 1989), and is the only variable cost included in this study. It is assumed to be proportional to operating time and is uniform across the four categories of pallet-handling technologies considered in this study. Additional assumptions used in cost calculations are summarized in the following parameter values:

Figure 4.4 Material-handling equipment spanning four levels of mechanization.

Table 4.1 Fixed and Operating Costs of Material-Handling Equipment

Truck Model	*Full Load Travel Speed (fpm)*	*Capital Cost Per Unit Equipment ($)*	*Cost of Operating on 10,000 VDV ($/hour)*
Hand-pallet truck	150	1,000	4.33
Electric pallet truck	475.5	12,000	3.29
Stand-up rider truck	616	20,000	1.26
Cushion tire lift truck	959.2	35,000	0.91

Table 4.2 Recommended Traveling Distances and Range of Traveling Distance for Material-Handling Equipment

Model	*Traveling Distance (d in ft)*	*Traveling Speed (fpm)*	*Ergonomic Limit on Operator (Cycles Per Shift, L_k)*
Hand-pallet truck	$d \leq 100$	150	200
Electric pallet truck	$100 < d \leq 250$	220	400
Stand-up rider truck	$250 < d \leq 500$	616	400
Cushion tire lift truck	$d > 500$	959.2	400

a_l: Labor availability (0.875, assuming one hour break in the eight hour work shift)
c_l: Labor cost ($13.50/hour)
a_k: Equipment efficiency (0.90)
I: Minimum attractive rate of return (20 percent)
n: Economic life of new equipment (five years)
H_p: Shift length (eight hours)
P_k: Initial cost per unit for equipment type k, $k = 1, \ldots, 4$ (see Table 4.1)
$(A/P,I,n)$: Capital recovery factor, i.e., the interest formula $I(1+I)^n/[(1+I)^n - 1]$
F_k: Unit salvage value for equipment type k (0.2 P_k)
$(A/F,I,n)$: Sinking fund factor, i.e., the interest formula $I/[(1+I)^n - 1]$
V_k: Average movement speed for equipment type k, $k = 1, \ldots, 4$ (see Table 4.2)
c_k: Cost factor for equipment k ($/hour)
t_{ij}: Move time for loads from work centers i and j (hours)
c_{ijk}: Cost of operating equipment type k between work centers i and j ($/hour)

Recognizing that the rectilinear distance between the centroids of work centers i and j changes with each block-layout alternative (denoted d_{ij} for $i,j = 1, \ldots, M$), material-handling costs for candidate layout solutions are computed using:

$$c_k = [P_k(A/P,I,n) - F_k(A/F,I,n)]/2000 \text{ hours/year}$$
$$t_{ij} = 2\, max(f_{ij}, f_{ji}) d_{ij}/V_k$$
$$c_{ijk} = [(t_{ij}/H_p)/a_k]c_k + [(t_{ij}/H_p)/a_l]c_l$$
$$= 2\, max(f_{ij}, f_{ji}) d_{ij}[c_k/a_k + c_l/a_l]/(V_k H_p)$$

Note that the computation of the cost factor, c_k for each equipment type k, is based on standard engineering economy principles (Sullivan et al., 2002). Further, in these definitions, the form of t_{ij} is based on approximating the time to move loads between pairs of work centers using twice the maximum flow volume in either direction. This approximation is intended to capture the effect of recirculation movement of equipment when there is an imbalance in flow volumes between a pair of work centers. It represents one of many possible heuristics to estimate the effects of equipment repositioning without resort to more complex modeling of the flow dynamics in a system. Subsequently, the cost of materials handling for layout an alternative is computed as

$$COST = \sum_{k=1}^{K} \sum_{i=1}^{M-1} \sum_{j=i+1}^{M} c_{ijk}$$

Note that the material-handling cost ($COST$), defined above, depends on the location of the work centers, the flow between them, and the equipment used.

Operator quantities obtained from application of Tables 4.1 and 4.2 guidelines can yield noninteger quantities and are rounded up to the next largest integer when this occurs. It is also assumed that units of all equipment types can be shared between work centers and operators are cross-trained on all equipment types. To further improve the accuracy of cost estimates, operator requirements are increased when necessary to satisfy safety restrictions. Specifically, ergonomic guidelines (The Ergonomic Group—Eastman Kodak Company, 1986) restrict individual operators to a maximum of 200 material-handling cycles per shift when using hand-pallet trucks and 400 cycles per shift when using powered equipment. Therefore, although completion of 200 hand-pallet cycles or 400 powered-equipment cycles may be possible based on the distribution of move lengths in a layout, the number of operators is constrained to be greater than or equal to the ratio of hand-pallet truck transactions over 200, and the number of powered-equipment transactions over 400.

4.4 Performance of the Cluster-Based Procedure

Consistent with representative block-layout problem sizes reported in Meller (1992), performance comparisons of the cluster-based and single-phase procedures are based on sample problems with 40 work centers. In all comparisons, parameters including the total iteration count (i.e., number of material-handling cost calculations), the random number stream used in the SA-search procedure, and the block-layout starting solutions are fixed. A temperature multiplier of $\alpha = 0.90$, a freezing temperature of one-thousandth of the initial temperature, and an epoch length of 1000 result in a limit of 67,000 iterations in the application of the SA-search procedure to each sample problem. A total of 180 sample problems are studied, based on (1) variation in the distribution of material-flow matrix elements (uniform, exponential, and normal), (2) the material-flow matrix density (5 percent, 10 percent, 20 percent, and 50 percent), and (3) the distribution of work-center space requirements. The distribution of work-center space requirements is based on defining a fit parameter, $0 \leq s \leq 1$ (Malmborg, 1999). For a given value of s, the space requirement of work center i in a sample problem with M work centers and N unit areas is given by $N[(i/M)^s - ((i-1)/M)^s]$. Values of s included in the study are $s = 1.0, 0.9, 0.85, 0.8$, and 0.75. The material-flow matrices constructed for the numerical study are such that the 40 work centers can be grouped into five clusters. Three distributions of work centers among the five clusters are considered, namely (8, 8, 8, 8, 8), (12, 4, 12, 4, 8), and (12, 4, 10, 6, 8). In each sample problem, the elements in the flow matrix are randomly distributed ensuring that at least 80 percent of the total material flow is between work centers contained in the same cluster.

4.4.1 Total Cost of Layout Solutions

The results of the numerical study are summarized in Table 4.3. The table lists the cost of the layout solution (denoted by *COST* in Section 4.3.4) determined after 67,000 iterations using the one-phase (highlighted in bold) and two-phase procedures for each of the 180 sample problems. The results indicate that the cost difference between the layout solutions obtained using the single-phase procedure and cluster-based procedures could be as high as 20 percent. However in all the 180 cases, the single-phase procedure provided a layout solution with less cost, indicating that in layout problems with diverse material-handling equipment types, layout solutions minimizing volume distance do not correlate well with those minimizing total costs.

4.4.2 Impact of Parameters on Total Costs

To obtain insights with respect to the impact of parameters such as the distribution of material-flow matrix elements and material-flow matrix density on total cost (*COST*), the results from Table 4.3 are used to plot suitable trade-off curves in Figure 4.5. In particular, results from the case where each cluster has eight work centers are used. From Figure 4.5, it is observed that, as expected, the total costs increase with material-flow matrix density. However, this increase is less significant at higher values of flow density. It appears that this trend is due to the high proportion of the fixed costs of

Table 4.3 Performance Comparison of Single-Phase and Cluster-Based Procedures in Terms of Total Cost (*COST*)

Flow Density (Percent)	*Uniform* $s = 1$	*Uniform* $s = 0.9$	*Uniform* $s = 0.85$	*Uniform* $s = 0.8$	*Uniform* $s = 0.75$	*Normal* $s = 1$	*Normal* $s = 0.9$	*Normal* $s = 0.85$	*Normal* $s = 0.8$	*Normal* $s = 0.75$	*Exponential* $s = 1$	*Exponential* $s = 0.9$	*Exponential* $s = 0.85$	*Exponential* $s = 0.8$	*Exponential* $s = 0.75$
Intercluster work-center distribution (8, 8, 8, 8, 8)															
5	935.1	1021.6	1017.9	1028.5	969.3	934.5	1024.7	1019.6	1033.4	1018.6	935.7	1030.4	1013.2	1049.2	956.3
	874.6	**882.1**	**892.9**	**892.7**	**872.9**	**879.3**	**871.3**	**894.6**	**890.6**	**881.0**	**886.1**	**861.7**	**887.8**	**870.3**	**873.8**
10	3190.8	3215.0	2949.5	3149.9	3128.9	3959.5	3990.7	3639.8	3902.4	3868.1	5367.6	5414.2	4960.0	5303.0	5263.3
	2670.4	**2671.8**	**2719.6**	**2692.7**	**2718.0**	**3283.8**	**3323.7**	**3341.4**	**3399.8**	**3360.3**	**4508.3**	**4512.1**	**4472.5**	**4593.0**	**4604.5**
20	5809.3	5628.1	5664.0	5694.7	5654.1	6934.0	6693.0	6750.7	6774.9	6693.9	9347.2	9039.0	9103.9	9138.2	9062.2
	5158.2	**5123.9**	**5125.0**	**5141.9**	**5137.6**	**6085.1**	**6103.0**	**6088.1**	**6079.5**	**6104.2**	**8206.1**	**8203.8**	**8214.8**	**8212.3**	**8274.8**
50	7597.0	7644.1	7480.7	7567.8	7749.8	8729.3	8777.0	8607.0	8705.6	8912.2	11338.3	11408.6	11153.9	11291.2	11588.0
	7239.5	**7229.3**	**7248.1**	**7247.6**	**7237.5**	**8301.6**	**8340.8**	**8305.3**	**8313.0**	**8356.8**	**10779.7**	**10770.2**	**10762.4**	**10781.4**	**10768.1**
Intercluster work-center distribution (12, 4, 12, 4, 8)															
5	1089.0	984.1	1031.1	1025.9	1025.9	1081.9	1074.9	1029.9	1026.5	1027.2	1056.0	1120.2	1067.5	1042.5	1051.7
	893.5	**888.1**	**893.2**	**897.6**	**902.8**	**879.3**	**906.1**	**901.3**	**890.1**	**896.5**	**887.6**	**890.3**	**864.6**	**865.7**	**887.9**
10	3502.9	3031.6	3265.4	3125.0	3129.3	4328.4	3737.0	4060.6	3846.4	3865.4	5884.6	5075.6	5122.1	5246.3	5274.1
	2844.3	**2764.3**	**2821.4**	**2839.8**	**2803.4**	**3442.4**	**3458.5**	**3472.0**	**3538.0**	**3479.9**	**4724.4**	**4706.1**	**4812.7**	**4720.8**	**4733.9**
20	6137.6	5972.6	5972.3	5910.8	5944.4	7291.0	7104.7	7096.5	7019.8	7054.6	9862.0	9594.8	9580.8	9487.1	9547.8
	5368.1	**5372.3**	**5378.9**	**5397.2**	**5434.2**	**6356.4**	**6364.4**	**6407.1**	**6420.2**	**6470.2**	**8573.5**	**8478.7**	**8550.0**	**8645.9**	**8673.1**
50	8079.6	7486.2	7665.6	7682.3	7748.7	9388.5	8618.9	8832.4	8863.4	8927.7	12252.7	11242.4	11540.8	11592.3	11684.2
	7365.1	**7443.8**	**7359.7**	**7402.7**	**7371.2**	**8472.5**	**8457.1**	**8457.7**	**8564.3**	**8570.5**	**11076.0**	**11001.6**	**11067.1**	**11109.8**	**11086.2**
Intercluster work-center distribution (12, 4, 10, 6, 8)															
5	1039.6	941.6	944.0	960.2	953.1	1038.2	945.3	944.0	973.7	949.4	1031.3	917.7	923.9	967.2	941.6
	889.8	**884.3**	**877.5**	**892.6**	**899.0**	**877.5**	**896.7**	**901.4**	**895.3**	**906.8**	**869.4**	**875.6**	**873.3**	**878.9**	**878.3**
10	3233.2	2981.0	2990.0	3014.0	2977.1	4008.6	3683.0	3708.0	3717.7	3683.6	5444.2	5011.4	5051.2	5055.3	5019.8
	2725.0	**2745.4**	**2791.0**	**2821.6**	**2770.6**	**3371.0**	**3396.3**	**3384.5**	**3379.7**	**3386.4**	**4559.9**	**4582.4**	**4616.7**	**4630.7**	**4693.9**
20	6067.2	5999.8	5534.4	5582.5	5562.5	7222.5	7154.5	6549.0	6612.4	6575.9	9738.0	9665.9	8849.3	8943.6	8883.8
	5215.7	**5208.9**	**5275.1**	**5306.9**	**5294.8**	**6230.8**	**6217.9**	**6225.9**	**6299.5**	**6316.1**	**8418.5**	**8370.7**	**8382.0**	**8448.7**	**8443.4**
50	8107.6	8107.8	7978.1	7698.0	7662.1	9407.7	9402.7	9230.0	8890.1	8842.8	12267.7	12273.4	12046.2	11596.2	11526.0
	7326.4	**7294.4**	**7322.2**	**7365.6**	**7349.3**	**8397.4**	**8427.6**	**8453.1**	**8398.1**	**8481.7**	**10963.1**	**10952.1**	**11021.5**	**11003.6**	**11004.9**

Note: Results of single-phase procedure are in boldface and cluster-based procedure in normal font.

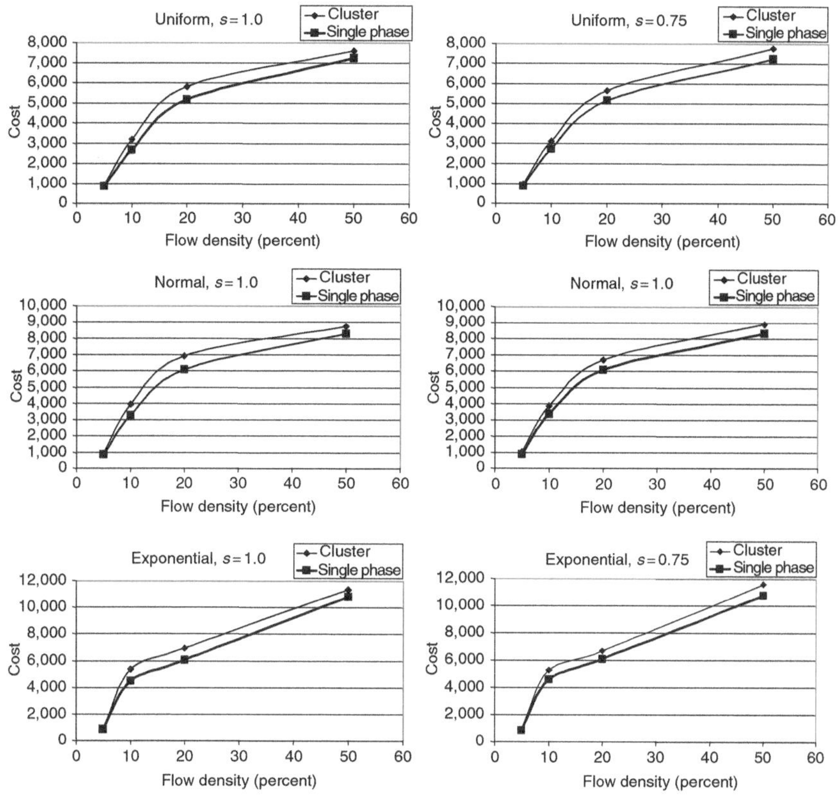

Figure 4.5 Impact of material-flow density on cost.

material-handling equipment in the total cost. In contrast to material-flow matrix density, work-center space requirements (determined by the parameter, *s*) do not have significant impact on the choice of material-handling equipment and hence total costs. The total costs also appear to be more influenced by the overall material-flow matrix density than the particular probability distribution of the material flows between work centers. The total costs show similar trends for the three distributions considered in the experiments (the relatively higher cost for the case of exponentially distributed material flows could be attributed to its higher variance). Similar insights are obtained even when the number of work centers in the clusters was different.

4.4.3 Computational Burden

Table 4.4 compares the solutions obtained after 30,000 and 67,000 iterations of the SA algorithm for both the single-phase procedure (highlighted in

Table 4.4 Percentage Reduction (Δ) in Cost of Layout Solution (*COST*) with Increase in Number of Iterations

Flow Density (Percent)	*Uniform*					*Normal*					*Exponential*				
	s = 1	*s = 0.9*	*s = 0.85*	*s = 0.8*	*s = 0.75*	*s = 1*	*s = 0.9*	*s = 0.85*	*s = 0.8*	*s = 0.75*	*s = 1*	*s = 0.9*	*s = 0.85*	*s = 0.8*	*s = 0.75*
Intercluster work-center distribution (8, 8, 8, 8, 8)															
5 (single phase)	0.0	−0.1	0.0	0.0	−0.2	0.0	0.0	0.0	0.0	0.0	−0.2	−0.1	0.0	0.0	0.0
10 (single phase)	0.0	0.0	0.0	−0.2	0.0	0.0	0.0	0.0	−0.1	0.0	0.0	0.0	−0.1	0.0	0.0
20 (single phase)	0.0	0.0	0.0	0.0	0.0	0.0	0.0	0.0	0.0	0.0	0.0	0.0	0.0	0.0	0.0
50 (single phase)	0.0	0.0	0.0	0.0	−0.1	0.0	0.0	0.0	0.0	0.0	0.0	0.0	−0.1	0.0	0.0
5 (cluster based)	**−11.4**	**−10.5**	**−9.2**	**−10.3**	**−11.7**	**−11.5**	**−12.7**	**−9.9**	**−10.6**	**−10.7**	**−9.1**	**−12.0**	**−10.1**	**−8.2**	**−10.9**
10 (cluster based)	**−10.1**	**−9.4**	**−6.4**	**−8.6**	**−7.8**	**−10.1**	**−8.5**	**−6.3**	**−6.3**	**−7.2**	**−8.8**	**−8.6**	**−7.8**	**−6.9**	**−7.1**
20 (cluster based)	**−4.5**	**−4.6**	**−4.5**	**−4.9**	**−4.8**	**−4.5**	**−4.0**	**−4.0**	**−4.6**	**−4.8**	**−4.9**	**−4.5**	**−4.2**	**−5.3**	**−4.1**
50 (cluster based)	**−3.1**	**−3.3**	**−3.2**	**−3.0**	**−2.4**	**−3.6**	**−3.0**	**−3.4**	**−3.2**	**−3.0**	**−3.2**	**−3.0**	**−3.4**	**−2.8**	**−3.3**
Intercluster work-center distribution (12, 4, 12, 4, 8)															
5 (single phase)	−1.6	−0.5	−0.9	−0.9	−0.3	−1.4	−1.6	−0.6	−0.8	−0.4	−0.9	−1.0	−0.4	−0.8	−0.5
10 (single phase)	−0.3	−0.9	−1.1	−0.8	−1.0	−0.6	−0.2	−0.8	−0.8	−0.7	−0.5	−1.0	−0.4	−1.0	−0.4
20 (single phase)	−0.9	−0.5	−0.4	−0.3	−0.4	−0.9	−0.4	−0.5	−0.3	−0.7	−0.8	−0.4	−0.6	−0.2	−0.5
50 (single phase)	−0.3	−0.4	−0.3	−0.3	−0.2	−0.1	−0.2	−0.1	−0.2	−0.4	−0.2	−0.3	0.0	0.0	−0.1
5 (cluster based)	**−11.2**	**−10.5**	**−8.8**	**−10.1**	**−9.6**	**−13.2**	**−9.3**	**−7.8**	**−9.4**	**−11.1**	**−8.6**	**−9.0**	**−10.9**	**−10.7**	**−9.4**
10 (cluster based)	**−9.2**	**−11.9**	**−10.2**	**−8.7**	**−10.8**	**−11.5**	**−11.1**	**−8.9**	**−9.1**	**−10.2**	**−11.5**	**−8.3**	**−8.4**	**−10.7**	**−10.9**
20 (cluster based)	**−7.2**	**−5.9**	**−6.6**	**−5.8**	**−5.7**	**−6.9**	**−5.8**	**−6.3**	**−6.1**	**−5.6**	**−7.6**	**−7.4**	**−7.4**	**−6.4**	**−6.1**
50 (cluster based)	**−4.4**	**−3.0**	**−4.4**	**−3.7**	**−4.2**	**−4.5**	**−4.3**	**−4.7**	**−3.8**	**−3.5**	**−4.5**	**−4.7**	**−4.1**	**−4.1**	**−4.7**
Intercluster work-center distribution (12, 4, 10, 6, 8)															
5 (single phase)	−1.4	−1.1	−0.3	−1.6	−0.9	−0.6	−0.3	−0.8	0.0	−1.8	−0.9	−0.3	−1.6	−1.0	0.0
10 (single phase)	−0.5	−0.7	−1.1	0.0	−1.0	−0.8	−0.7	−0.5	−0.3	−0.9	−0.8	−0.8	−0.4	−1.0	−0.9
20 (single phase)	−0.3	−0.1	−0.4	−0.4	−0.2	−0.6	−0.4	−0.3	−0.4	−0.7	−0.3	−0.2	−0.4	−0.3	−0.6
50 (single phase)	−0.3	−0.1	−0.1	−0.4	−0.3	−0.2	−0.2	−0.2	−0.2	−0.4	−0.2	−0.1	−0.2	−0.4	−0.2
5 (cluster based)	**−11.1**	**−10.7**	**−9.8**	**−11.3**	**−10.4**	**−13.2**	**−9.2**	**−7.7**	**−8.5**	**−8.8**	**−12.4**	**−10.5**	**−10.9**	**−12.0**	**−9.9**
10 (cluster based)	**−10.9**	**−8.5**	**−7.6**	**−7.3**	**−7.9**	**−10.9**	**−8.6**	**−9.1**	**−9.5**	**−10.4**	**−11.5**	**−9.3**	**−8.9**	**−9.5**	**−8.2**
20 (cluster based)	**−6.5**	**−6.8**	**−5.8**	**−5.0**	**−5.5**	**−6.4**	**−6.0**	**−6.2**	**−4.9**	**−5.1**	**−5.9**	**−6.3**	**−6.3**	**−5.9**	**−5.7**
50 (cluster based)	**−3.5**	**−4.3**	**−3.7**	**−3.1**	**−2.5**	**−4.1**	**−4.1**	**−3.6**	**−4.6**	**−3.2**	**−4.1**	**−4.4**	**−3.6**	**−3.2**	**−3.9**

Note: Δ = [1-cost of solution after 30,000 iterations/cost of solution after 67,000 iterations] × 100.

bold) and the cluster-based procedure. For the single-phase procedure, the additional 37,000 iterations yield solutions that are 2 percent to 13 percent lower in total cost. In contrast, the corresponding cost improvements obtained with the cluster-based procedure are at most 2 percent. To compare the two procedures in terms of the solutions obtained at different iterations, the two objective function values are plotted against the iteration count. Figure 4.6 illustrates such a plot for the case when the material-flow matrix density is 5 percent, and the distribution of elements in the matrix is selected from a sample of uniformly distributed random numbers. The figure indicates that, with the cluster-based procedure, the maximum improvement in the objective function is obtained within the first 5000 iterations. Subsequent iterations yield marginal improvements in the solutions. We believe that this is because only intracluster exchanges are permitted in the second phase. The improvements obtainable from such exchanges are limited by the clustering done in the first phase. Because the

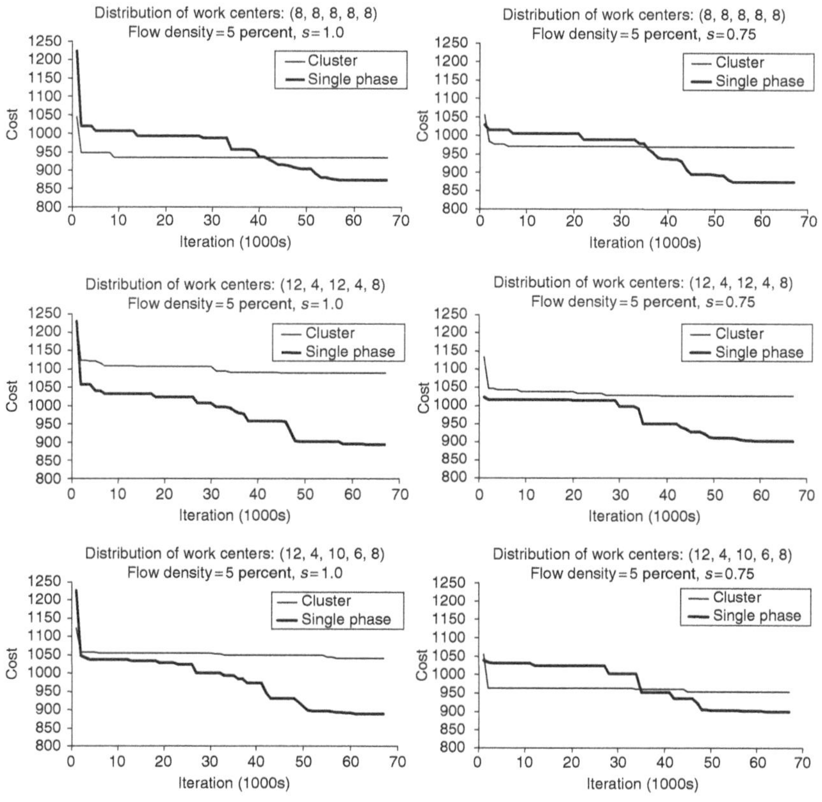

Figure 4.6 Cost of solution at different iterations.

number of work centers in each cluster is limited, the solution algorithm quickly explores possible exchanges and subsequent iterations result in limited improvement and the algorithm is trapped in a local optimum. In contrast, the single-phase procedure does not involve clustering and permits pair-wise exchange of any two of the 40 work centers. Although exploring the numerous possible exchanges is time consuming, the single-phase procedure yields improved solutions. It is interesting to note that in certain cases the solution obtained using the single-phase procedure dominates that of the two-phase procedure at each iteration. The figure also demonstrates that these insights are not sensitive to the distribution of work centers among the five clusters or the space requirements distribution of the individual work centers.

4.4.4 Limitations of the Cluster-Based Procedure

The performance results indicate that the single-phase procedure always provides a better layout solution, indicating that complexity-reduction techniques using surrogate objectives based on volume distance might not always result in layouts that reduce material-handling cost. This is because volume distance calculations typically assume a single type of handling device or unit load while, in reality, materials-handling costs are influenced by multiple types of handling devices with different economic ranges, speeds, fixed costs, and labor requirements. Therefore, the initial step involving clustering of work centers could truncate the solution space significantly causing the subsequent SA-search procedure to be trapped at an inferior solution or local optimum. This limitation does not exist in the single-phase procedure, and therefore is able to search for better solutions. However, the procedure involves significantly higher computational burden. As the results indicate, several iterations of the SA procedure might be necessary to achieve near-optimal solutions. The limitations of the cluster-based procedure suggest that specialized SA procedures might be required to reduce the computation burden associated with block-layout problems involving material-handling costs. This is investigated in the subsequent sections.

4.5 Specialized SA Procedure for Facility Layout and MHS Design

Recall that, unlike greedy procedures, SA algorithms avoid trapping at local optima by potentially accepting inferior solutions during the iterative search process. The probability of acceptance for an inferior solution depends on the magnitude of the difference between the objective function value of an incumbent and candidate solution, and a user-defined parameter described as the temperature at the SA state, T. Representing by S the set of all solutions, and letting f denote the objective function defined on the elements of S, the SA procedure iteratively searches for $x \in S$ that optimizes f over S. If we denote a candidate solution, $y \in N(x)$, where $N(x)$ is the set of feasible neighbors of an incumbent solution x, y is accepted as the next incumbent solution with probability $P_T = \min(e^{(f(x)-f(y))/T}, 1)$. The number of iterations of the SA algorithm required to obtain a solution of acceptable quality is a function of user-defined empirical parameters including (1) the acceptance criteria, (2) the starting or "melting" temperature, (3) the temperature decrement, (4) the final or "freezing" temperature, and (5) the epoch length. In fact, a considerable proportion of the literature on SA methodologies focuses on accelerating SA procedures by developing faster cooling schedules, alternative acceptance strategies, and optimal temperature schedules. Studies reported by Aarts and Laarhoven (1985), Hajek (1988), Huang et al. (1986), Boese and Kahng (1994), and Lam and Delosme (1988) are examples of this work.

In the comparison of single-phase and cluster-based approaches described in Sections 4.3 and 4.4, the parameters of the SA algorithm were set based on the procedure in Meller (1992). In particular, the initial temperature of the SA procedure for the cluster-sequencing problem was set so the probability of accepting a candidate sequence having an objective function value equal to twice that of the initial sequence objective function is equal to 0.001, and the freezing (final) temperature is set to one-thousandth of the initial temperature. In addition, a linear cooling schedule was applied (i.e., $T_{new} = \alpha T_{current}$), with $\alpha = 0.9$ and an epoch length of 1000 iterations. As the results indicated, several iterations of the SA procedure were necessary to achieve near-optimal solutions. The study described below reexamines these input parameters and investigates specialized two-stage SA procedures to solve large-scale, block-layout problems with materials-handling costs.

4.5.1 Two-Stage SA Algorithms

Two-stage SA algorithms consist of a fast heuristic stage and an SA stage. In the first-stage, the early iterations of the standard SA algorithm that are normally executed at high temperatures are replaced by a computationally efficient heuristic procedure. SA is then performed to further improve the quality of the solution obtained from the fast heuristic stage. Using the solution returned by the first-stage procedure as its initial solution, the SA stage can be started from a lower starting temperature. A considerable percentage of the SA algorithm can be omitted if high-quality solutions are returned by the first stage. Two-stage SA has also been investigated extensively in the literature, including studies by Johnson et al. (1989), Rose et al. (1988, 1990), and Varanelli and Cohoon (1993, 1995, 1999).

A major problem in controlling the convergence of both two-stage SA algorithms is the determination of initial conditions including a starting solution and an initial temperature (T_0). The initial temperature should be low enough to differentiate the search process from a random walk but high enough to encourage frequent movement to nearby solutions. Unlike single-stage SA algorithms, a high-quality starting solution obtained using a first-stage heuristic, such as greedy search, can enable a lower starting temperature for the second (or SA) stage of a two-stage algorithm. This is important because the starting temperature can significantly affect the performance of an SA procedure with regards to the speed of convergence and the quality of a final solution obtainable for a fixed number of iterations.

General guidelines for estimating initial solutions are proposed in Johnson et al. (1989), Rose et al. (1990), Dowsland (1995), Rayward-Smith et al. (1996), and Varanelli and Cohoon (1999). A number of general guidelines are also offered in the literature for specifying the starting temperature. Rose et al. (1990) employ Markov equilibrium dynamics for approximating an equilibrium temperature in two-stage SA procedures where the probability distribution for the change in the objective function is approximated through generation of a large number of moves from a first-stage heuristic. The equilibrium temperature is approximated such that the expected change in the objective function equals zero at equilibrium; $E(\Delta f) = 0$, where $\Delta f = f_t + 1 - f_t$.

Varanelli and Cohoon (1999) propose a starting temperature for a two-stage SA procedure at a given temperature-state using the average of candidate solutions observed at the given temperature. Building on numerical studies reported by White (1984), Otten and Ginneken (1988), Hajek (1985), and Aarts et al. (1988), Varanelli and Cohoon (1999) exploit findings that the distribution of objective function values at a given temperature can be approximated using a normal distribution. Denoting the expected value and the standard deviation of the cost at state temperature t_k by E_k and σ_k, respectively, they define $E_k \approx E_\infty - (\sigma_\infty{}^2/t_k)$, and $\sigma_k \approx \sigma_\infty$. Because the minimum-cost solution, i_{BSF}, observed during the kth temperature-state, is at most equal to E_k, it follows that $E_k \geq c(i_{BSF})$. Subsequently, they define $E_k = c(i_{BSF})_k + \gamma_\infty \sigma_k$, where γ_∞ denotes the expected offset of i_{BSF} from E_k and propose the stage two starting temperature given by

$$T_t(i) \approx \sigma_\infty{}^2/(E_\infty - f(i) - \gamma_\infty \sigma_\infty),$$

where

i denotes the current solution

$f(i)$ denotes the average solution found at the current temperature

γ_∞ is set to equal a probability value satisfying

$$P[E_\infty - \gamma_\infty \sigma_\infty < X < E_\infty + \gamma_\infty \sigma] \approx 1 - |L_M|^{-1}$$

with L_M as the approximate size of the neighborhood of the solution at i.

The limiting behavior of SA is also determined by the temperature schedule. For $t = 0, 1, \ldots, \tau$, the sequence of T_t values is called the temperature or "cooling" schedule if T_t approaches 0 as $t \rightarrow \infty$. A number of different temperature schedules are proposed in the literature including fixed schedules, geometric schedules, and logarithmic schedules. Fielding (2000) investigates the impact of temperature schedules and reports that fast cooling algorithms, widely used in SA applications, are generally nonconvergent but tend to outperform fixed-temperature schedules for large problems. To insure high-quality solutions, repeated independent runs are sometimes appropriate although long computational times associated with slow cooling schedules can make this strategy impractical in many applications. In a related study, Cohn and Fielding (1999) find that the number of iterations performed at each temperature, or "epoch length," can have little if any effect on the performance of SA algorithms using geometric cooling schedules. Empirical experimentation performed by the authors on cost-based block-layout problems solved with SFCs has found that a final (freezing) temperature equal to 0.0001 f_0 generally allows adequate time for one-stage and two-stage SA procedures to stabilize at a single final solution (Al-Araidah, 2005).

4.5.2 SA Control Parameter Guidelines for Block-Layout Problems

Next, general guidelines for specifying SA control parameters are proposed for the class of block-layout problems under consideration. In particular, parameters for solving moderate- to large-scale block-layout problems (i.e., 20–50 work centers) are investigated. To investigate rules for SA control parameters for problems with this cost structure, a "best-so-far" (BSF) solution concept is applied to estimate starting temperatures. Specifically, as the temperature is decremented from an initial "melting point" to a final "freezing point," the cost function is evaluated an epoch-length number of times at each intermediate temperature. In each case, the BSF solution is retained as the starting solution at the next temperature. If no improvement in the value of the BSF solution is observed over a given temperature, the neighborhood of the solution space associated with the BSF solution is revisited at the next temperature. The value of the BSF solution at temperature-state t

represents the first-order statistic $X^t_{(1)}$ (or minima) of x for all solutions visited by the algorithm at the temperature T_t for $x = x_1.x_{epoch}$, thus associating different minimums with different solutions. Monitoring the improvement in the value of the BSF solution is equivalent to monitoring data in the left tail of the distribution of cost-function values defined by the solution space. Using the moments of the asymptotic distribution of the first-order statistic (minima), and the distribution of the search space, a starting temperature for the SA stage of the algorithm can be estimated. Building on this idea, the key feature of the guidelines proposed in this study is estimation of a starting temperature as a function of the BSF solution obtained in the first stage of a two-stage procedure using estimates of the moments of the search space and minimum order statistic. This approach has been found to yield tighter limits on the value of the initial temperature compared to the approach proposed by Varanelli and Cohoon (1999) thereby accelerating the convergence process for the class of problems considered in this study.

The relationship between the expected value of the BSF solution at a temperature-state and the temperature at that state is characterized using a regression model of the form:

$$E_t(X_{(1)}) = \beta_0 - \beta_1/T_t - \varepsilon \quad \text{or} \quad T_t = \beta_1/(\beta_0 - E_t(X_{(1)}) - \varepsilon,$$

where

$E_t(X_{(1)})$ is the expected value of the BSF

β_0, β_1, and ε are the regression parameters

To estimate these regression parameters, a single-stage SA algorithm with homogenous temperature schedule ($T_t = 0.9^t T_0$) is utilized where T_0 is the starting temperature that is set equal to the objective function value of the initial solution. Using a single initial solution, ten random number streams are utilized to generate ten output BSF-curves to obtain alternative solutions based on the same temperature schedule. The values of the BSF are then averaged to estimate the expected behavior of the BSF solution for each temperature-state. To estimate the moments of the search space and the first-order statistic, n samples of size m are utilized with the size of the sample (nm) set equal to the epoch length. The estimated mean and standard deviation of the solution population are then given by

$$E_\infty(x) = \_i = 1 \ldots nm \; x_i/nm \text{ and}$$

$$\sigma_\infty(x) = [\Sigma_{i=1\ldots nm} \, (x_i - E_\infty(x))/(nm - 1)]^{-1}.$$

The estimated mean and standard deviation of the first-order statistic are given by

$$E_\infty(X_{(1)}) = [\Sigma_{i=1\ldots n} \, (X^i_{(1)} - E_\infty(X_{(1)}))/(n - 1)]^{-1},$$

where $X_{(1)}$ is the first-order statistic of sample i such that $X_{(1)} = \min(f(x))$ for $x = x_1.x_m$. Empirical studies with test problems in support of this study indicate that the relationship between the BSF solution and state temperature can be accurately characterized using the model:

$$T_t = \sigma_\infty(x)\sigma_\infty(X_{(1)})/[E_\infty(X_{(1)}) - E_t(BSF) - (E_\infty(x) - E_\infty(X_{(1)}))],$$

where results obtained from the first-stage heuristic must satisfy:

$$E_t(BSF) < E_\infty(X_{(1)}) - (E_\infty(x) - E_\infty(X_{(1)})).$$

Applying this result, Figure 4.7 outlines a two-stage solution procedure for controlling the convergence of SA for the class of cost-based block-layout

Initiation
Given work centers, flow volumes, and facility dimensions; select the preferred SFC.

Sampling
Generate n samples of size m to estimate the mean and the standard deviation for the population $E_\infty(x)$ and $\sigma_\infty(x)$ respectively, and those of the first-order statistic $E_\infty(X_{(1)})$ and $\sigma_\infty(X_{(1)})$ respectively.

Greedy Heuristic
Utilizing an initial solution and a random number stream,

Initiate: X_0, *counter* = 1
$X^* := X_0$
    While (counter ≤ Stage 1-iteration budget) do;
        X_{new} = *generate-sequence* (X^*); $\Delta f = f(X^*) - f(X_{new})$; *if* $(\Delta f > 0)$ *then* $X^* := X_{new}$;
    enddo;

SA Algorithm
Use $f(X^*)$, $E_\infty(x)$, $\sigma_\infty(x)$, $E_\infty(X_{(1)})$, and $\sigma_\infty(X_{(1)})$ to estimate the starting temperature T_0.
Initiate: $X := X^*$, $T = T_0$, $t = 0$;
    While $(T_t \leq T_f)$ *do;*
        While (Epoch Counter ≤ N) do;
            X_{new} = *generate-sequence* (X); $\Delta f = f(X) - f(X_{new})$;
            (f is the objective function)
            If $(\Delta f > 0$, *or*, $\exp(\Delta f/T) >$ *random*$(0,1))$ *then* $X := X_{new}$;
            If $(f(X_{new}) < f(X^*))$ *then* $X^* := X_{new}$;
        enddo;
        T_{t+1} = *update* $(T_t) = a^{t+1}T_0$;
    enddo;

X_0	Initial sequence of work centers	α	Quenching coefficient of the temperature schedule
T_0	Melting temperature		
X	Candidate arrangement of work centers	T_t	State temperature
X^*	Optimal arrangement of work centers	T_f	Freezing temperature
t	Temperature-state	N	Epoch length

Figure 4.7 Two-stage, greedy-SA procedure with adaptive temperature (GSA-AT).

problems considered in this study. Using greedy search in stage 1 to obtain initial solutions in combination with this adaptive starting temperature concept in the second stage, the proposed procedure applies a geometric cooling schedule with $T_t = 0.9^t T_0$, a freezing temperature of $T_f = 0.0001 f_0$, and an epoch length of 1000 iterations. In Section 4.6, results are examined for a series of realistically sized test problems to determine whether this procedure improves upon the performance of the general method of Varanelli and Cohoon (1999) by accelerating convergence while maintaining solution quality. In addition, a single-stage SA procedure is applied and used as a further basis of comparison to evaluate the general effectiveness of one-stage versus two-stage procedures.

4.6 Performance Assessment of the Proposed Algorithm

To compare the performance of the proposed two-stage SA algorithm to an analogous two-stage procedure using the method of Varanelli and Cohoon (1999) in stage 2, an initial series of eight block-layout problems are generated. These problems are developed to span three key problem attributes including the number of work centers, the material-flow density, and the distribution of work-center space requirements. The problem set includes problems with 25, 32, 40, 52, and 63 work centers. Four replicates of the 40 work-center problems are included because the techniques reported in this study were conceived to solve block-layout problems of this approximate size. The second attribute differentiating test problems is the density of the material-flow matrix (i.e., proportion of nonzero elements), which varies from 5 percent to 50 percent within the eight test problems. In all cases, the nonzero elements of the material-flow matrix are generated to follow a random distribution. The third problem attributes concerns as how the space is distributed among the work centers within a block layout. As before, the distribution of work-center space requirements is based on defining a fit parameter $0 \leq s \leq 1$ (Malmborg, 1999). For a given value of s, the space requirement of work center i in a sample problem with M work centers and N unit areas is given by $N[(i/M)^s - ((i-1)/M)^s]$. The fit parameter values used in the test problems vary from a minimum of $s = 0.75$ to a maximum of $s = 1$ (where work centers have identical space requirements). Based on these three attributes, test problems can be described by a vector of the form (M, δ, s), where δ $(0 \leq \delta \leq 1)$ denotes the density of the material-flow matrix.

Each of the eight randomly generated test problems is solved using the proposed two-stage greedy-SA adaptive temperature (GSA-AT) procedure, the two-stage greedy-SA procedure of Varanelli and Cohoon (GSAVC), and a single-stage SA procedure. Consistent with preliminary empirical investigations of materials-handling, cost-based layout problems conducted in support of this study, a geometric cooling schedule ($T_t = 0.9^t T_0$), a freezing temperature of $T_f = 0.0001 f_0$, and an epoch length of 1000 iterations are used with all three procedures. For the single-stage procedure, these limits allow for 88 temperature decrements and a total of 88,000 evaluations of the materials-handling cost function.

The results of the experiment are summarized in Table 4.5 for each of the three solution techniques. This table presents the materials-handling cost per period associated with the optimal solution and a count on the number of evaluations of the objective function needed to reach it. The last two columns of the table present the percentage cost improvement in the optimal solution for the GSA-AT technique relative to the GSAVC technique, and the percentage reduction in the number of iterations needed to obtain the optimal solution. The GSA-AT method produces a higher-quality solution relative to GSAVC in three of the eight problems with an average improvement of 1.1 percent. The GSA-AT yields a slightly lower-quality solution in five of the eight problems with an average loss in solution quality of 0.3 percent. Despite roughly equivalent performance with respect to solution quality, the results from the test problems suggest that the GSA-AT procedure is effective in accelerating the convergence process with computational savings ranging between 14.3 percent and 48.9 percent. Across the eight test problems, the average savings is 28.9 percent with a standard deviation of 11.4 percent. This performance can be observed graphically in Figure 4.8 which summarizes the convergence of the three procedures for one of the test problems. As the figure shows, both the GSA-AT and GSAVC procedures converge more rapidly than the single-stage procedure. However, the curves also illustrate how the GSA-AT procedure

Table 4.5 Summary of Results for Eight Test Problems[a]

Problem	*SA*	*Count*	*GSAVC*	*Count*	*GSA-AT*	*Count*	*Cost (Percent)*	*Count (Percent)*
40,0.05,0.75	$868	880	$879	470	$861	240	*2.00*	*48.90*
40,0.10,0.90	$3347	880	$3338	400	$3314	310	*0.70*	*22.50*
40,0.20,0.85	$5337	880	$5308	350	$5317	300	*−0.20*	*14.30*
25,0.16,1.00	$912	880	$919	540	$912	430	*0.70*	*20.40*
32,0.25,1.00	$1838	880	$1843	470	$1848	350	*−0.20*	*25.50*
40,50,0.80	$7202	880	$7211	360	$7220	210	*−0.10*	*41.70*
52,0.23,1.00	$4411	880	$4428	470	$4454	340	*−0.60*	*27.70*
63,0.248,1.00	$7108	880	$7091	360	$7111	250	*−0.30*	*30.60*

[a]Table includes material-handling cost per period ($1000s), count on the number of objective function evaluations (1000s), and percentage cost/computational improvement of the proposed GSA-AT method relative to the GSAVC method.

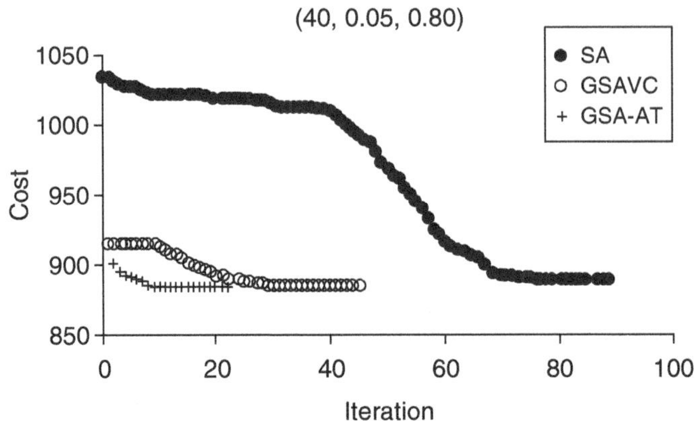

Figure 4.8 Illustration of the convergence process for the three procedures with a sample problem (40, 0.05, 0.80).

effectively avoids the inefficient search pattern that characterizes higher temperatures with the GSAVC procedure. This is the key feature of the GSA-AT method that differentiates it from other procedures for the class of problems considered in this study.

To examine problems with $M = 40$ work centers in greater depth, 20 additional test problems were generated with material-flow matrix density varying at four levels (i.e., $\delta = 5$ percent, 10 percent, 20 percent, and 50 percent), and the fit parameter defining the distribution of work-center space requirements varying at five levels (i.e., $s = 0.75$, 0.80, 0.85, 0.90, and 1.0). The results with respect to the materials-handling cost of the optimal solution and the number of reevaluations of the objective function needed to reach it are provided for each of the three solution methods in Table 4.6. For this problem set, the GSA-AT method produced a higher-quality solution relative to GSAVC in 11 of the 20 problems with an average improvement of 1.0 percent. The GSA-AT yielded a slightly lower-quality solution in 9 of the 20 test problems with an average loss in solution quality of 0.4 percent. Once again, the GSA-AT procedure was successful in accelerating the convergence process for all test problems. The percentage savings in the number of reevaluations of the objective function ranged between 12.8 percent and 71.1 percent with an average savings of 37.7 percent and a standard deviation of 13.7 percent.

4.7 Summary and Conclusions

This chapter summarizes observations from two studies that are both aimed at efficiently generating high-quality solutions for moderate- to large-scale

Table 4.6 Summary of Results for 20 Test Problems with $M = 40$[a]

Problem	*SA*	*Count*	*GSAVC*	*Count*	*GSA-AT*	*Count*	*Cost (Percent)*	*Count (Percent)*
40,0.05,0.75	877.49	880	884.11	469	873.80	308	*1.17*	*34.33*
40,0.05,0.80	877.49	880	878.11	452	871.50	259	*0.75*	*42.70*
40,0.05,0.85	877.96	880	875.19	424	868.27	219	*0.79*	*48.35*
40,0.05,0.90	883.03	880	881.95	435	857.23	309	*2.80*	*28.97*
40,0.05,1.00	881.49	880	875.99	469	864.92	325	*1.26*	*30.70*
40,0.10,0.75	2827.42	880	2836.15	399	2801.09	277	*1.24*	*30.58*
40,0.10,0.80	2800.23	880	2816.96	407	2792.39	253	*0.87*	*37.84*
40,0.10,0.85	2799.62	880	2766.1	382	2810.17	282	*−1.5*	*26.18*
40,0.10,0.90	2801.25	880	2785.6	386	2769.61	244	*0.57*	*36.79*
40,0.10,1.00	2784.67	880	2783.9	396	2773.14	281	*0.39*	*29.04*
40,0.20,0.75	5373.98	880	5359.84	342	5393.54	99	*−0.6*	*71.05*
40,0.20,0.80	5334.07	880	5335.45	311	5354.33	168	*−0.35*	*45.98*
40,0.20,0.85	5336.96	880	5308.36	321	5317.28	280	*−0.17*	*12.77*
40,0.20,0.90	5322.23	880	5315.44	306	5323.12	132	*−0.14*	*56.86*
40,0.20,1.00	5305.93	880	5333.27	328	5320.66	263	*0.24*	*19.82*
40,0.50,0.75	7333.9	880	7354.51	322	7355.71	182	*−0.02*	*43.48*
40,0.50,0.80	7345.34	880	7349.52	310	7339.28	200	*0.14*	*35.48*
40,0.50,0.85	7326.25	880	7324.4	295	7327.14	185	*−0.04*	*37.29*
40,0.50,0.90	7303.52	880	7326.46	291	7335.13	123	*−0.12*	*57.73*
40,0.50,1.00	7313.08	880	7318.46	307	7331.13	220	*−0.17*	*28.34*

[a]Table includes material-handling cost per period ($1000s), count on the number of objective function evaluations (1000s), and percentage cost/computational improvement of the proposed GSA-AT method relative to the GSAVC method.

block-layout problems based on realistic measures of material-handling costs. It considers layout problems where multiple types of materials-handling devices are used within a facility. Although realistic in many cases, cost modeling of this scenario is made significantly more complex by the need to dynamically assign materials-handling devices to individual material movements as layout alternatives are generated through an iterative perturbation process. Both studies simplify the underlying search using SFCs that help to reduce block layout to a sequencing problem. However, further measures need to be taken to reduce the search computations.

The initial efforts focus on the possibility of using cluster-based approaches to reduce the computational effort required to solve material-handling, cost-based layout problems with palletized flow using heterogeneous material-handling equipment. The cluster-based procedure attempts to exploit the potential correlation between material-flow volume distance and material-handling cost to reduce complexity. This approach is compared with a single-phase procedure that uses SA to search for layout solutions that minimize total material-handling cost. Experimental studies indicate that the single-phase procedure always provides a better layout solution than the cluster-based approach, indicating that complexity-reduction techniques using surrogate objectives based on volume distance might not always result in layouts that reduce material-handling cost. The initial clustering step in many cases truncates the solution space significantly, causing the subsequent search procedure to be trapped at an inferior solution or local optimum.

The limitation of the cluster-based approach motivated investigation into the use of specialized SA procedures as an alternative means to improve the computational efficiency of solving moderate- to large-scale block-layout problems. Based on experimentation, a two-stage optimization procedure is proposed that applies greedy search in the first stage to facilitate a lower starting temperature in a second SA stage of the procedure. Test results for realistically sized sample problems indicate that the proposed technique seems to successfully accelerate the convergence to an optimal solution by avoiding inefficient search at higher temperatures.

Given the complexity of the materials-handling cost model described in this study, it is not difficult to envision applications where computational efficiency is an important criterion in solving block-layout problems. The model proposed in this study provides a reasonably promising solution strategy that can be easily applied to generate solutions of comparable quality to alternative methods but at significantly lower computational cost. Investigation into whether the technique yields similar results for other classes of problems is a topic of future research.

References

Aarts, E.H.L. and Laarhoven, P.J.M. van, 1985, "A new polynomial-time cooling schedule," *Proceedings of IEEE ICCAD-85*, Santa Clara, CA, pp. 206–208.

Aarts, E.H.L., Korst, J.H.M., and Laarhoven, P.J.M. van, 1988, "Solving traveling salesman problems by simulated annealing," *Journal of Statistical Physics*, 50, 187–206.

Al-Araidah, O., 2005, *Simultaneous Facility Layout and Materials Handling System Design*, Ph.D. dissertation, Department of Decision Sciences and Engineering Systems, Rensselaer Polytechnic Institute, Troy, NJ.

Al-Araidah, O., Krishnamurthy, A., and Malmborg, C.J., 2006, "A two-stage simulated annealing procedure for block layout problems," *International Journal of Production Research*, 44(20), 4417–4429.

Al-Araidah, O., Krishnamurthy, A., and Malmborg, C.J., 2007, "A comparative study of single and two phase approaches for the layout problem with materials handling costs," *International Journal of Production Research*, 45(4), 951–970.

Askin, R.G. and Standridge, C.R., 1993, *Modeling and Analysis of Manufacturing Systems.* John Wiley, New York.

Balakrishnan, J., Cheng, C.-H., and Wong, K.-F., 2003, "FACTOPT: A user friendly FACility layout OPTimization system," *Computers and Operations Research*, 30, 1625–1641.

Benjaafar, S. and Sheikhzadeh, M., 2000, "Design of flexible plant layouts," *IIE Transactions*, 32, 309–322.

Boese, K.D. and Kahng, A.B., 1994, "Best-so-far vs. Where-you-are: Implications for optimal finite-time annealing," *Systems and Control Letters*, 22, 71–78.

Bozer, Y.A. and Meller, R.D., 1997, "A reexamination of the distance-based facility layout problem," *IIE Transactions*, 29, 549–560.

Bozer, Y.A., Meller, R.D., and Erlebacher, S.J., 1994, "An improvement-type layout algorithm for single and multiple-floor facilities," *Management Science*, 40(7), 918–932.

Canadian Centre for Occupational Health and Safety (CCOHS), 1997, "Pushing & Pulling—Handcarts." Available at www.ccohs.ca/oshanswers/ergonomics/push2.html.

Chu, H.-K., Egbelu, P.J., and Wu, C.-T., 1995, "ADVISOR: A computer-aided material handling equipment selection system," *International Journal of Production Research*, 33(12), 3311–3329.

Chwif, L., Barretto, M.R.P., and Moscato, L.A., 1998, "A solution to the facility layout problem using simulated annealing," *Computers in Industry*, 36(1–2), 125–132.

Cohn, H. and Fielding, M., 1999, "Simulated annealing: Searching for an optimal temperature schedule," *SIAM Journal of Optimization*, 9, 779–802.

Dowsland, K.A., 1995, "Simulated annealing." In: C.R. Reeves (Ed.), *Modern Heuristic Techniques for Combinatorial Problems.* McGraw-Hill, London.

Eastman Kodak Company—Ergonomics Group—Health and Environment Laboratories, 1986, *Ergonomic Design for People at Work*, Vol. 2. Van Nostrand Reinhold, New York.

Fielding, M., 2000, "Simulated annealing with an optimal fixed temperature," *SIAM Journal of Optimization*, 11, 289–307.

Hajek, B., 1985, "A tutorial survey of theory and applications of simulated annealing," *Proceedings of the 24th IEEE Conference on Decision and Control.* Fort Lauderdale, FL, pp. 755–760.

Hajek, B., 1988, "Cooling schedules for optimal annealing," *Mathematics of Operations Research*, 13, 311–329.

Hassan, M.M.D., Hogg, G.L., and Smith, D.R., 1985, "A construction algorithm for the selection and assignment of materials handling equipment," *International Journal of Production Research*, 23(2), 381–392.

Huang, M.D., Romeo, F., and Sangiovanni-Vincentelli, A., 1986, "An efficient general cooling schedule for simulated annealing," *Proceedings of IEEE ICCAD-86.* Santa Clara, CA, pp. 381–384.

Johnson, D.S., Aragon, C.R., McGeoch, L.A., and Schevon, C., 1989, "Optimization by simulated annealing: An experimental evaluation; part 1, graph partitioning," *Operations Research*, 37, 865–892.

Kim, Y.-K. and Kim, K.-H., 1997, "A routing algorithm for a single transfer crane to load export containers onto a containership," *Computers and Industrial Engineering*, 33(3–4), 673–676.

Kochhar, J.S., Foster, B.T., and Heragu, S.S., 1998, "HOPE: A genetic algorithm for the unequal area facility layout problem," *Computers and Operations Research*, 25(7–8), 583–594.

Kulwiec, R.A., 1985, *Materials Handling Handbook*. John Wiley, New York.

Lam, J. and Delosme, J.-M., 1988, "Performance of a new annealing schedule," *Proceedings of the 25th ACM/IEEE DAC*, Anaheim, CA, 306–311.

Liggett, R.S., 2000, "Automated facilities layout: Past, present and future," *Automation in Construction*, 9, 197–215.

Malmborg, C.J., 1999, "Empirical approximation of volume distance distributions for line layout problems," *International Journal of Production Research*, 37(2), 375–392.

Malmborg, C.J., Krishnakumar, B., Simons, G., and Agee, M.H., 1989, "EXIT: A PC-based expert system for industrial truck selection," *International Journal of Production Research*, 27(6), 2339–2354.

Matson, J.O., Mellichamp, J.M., and Swaminathan, S.R., 1992, "EXCITE: Expert consultant for in-plant transportation equipment," *International Journal of Production Research*, 30(8), 1969–1983.

Meller, R.D., 1992, *Layout Algorithms for Single and Multiple Floor Facilities*, Ph.D. dissertation, University of Michigan, Ann Arbor, MI.

Meller, R.D. and Gau, K., 1996, "The facility layout problem: Recent and emerging trends and perspectives," *Journal of Manufacturing Systems*, 15(5), 351–366.

Modern Materials Handling Manufacturing Guidebook, 1985, "Transportation—The key to flexibility and throughput," pp. 55–61.

Muther, R., 1973, *Systematic Layout Planning*. Cahers Books, Boston, MA.

Otten, R.H.J.M. and Ginneken, L.P.P.P. van, 1988, "Stop criteria in simulated annealing," *Proceedings of IEEE ICCD*, Rye Brook, NY, pp. 549–552.

Peters, B.A., 1998, "An introduction to material handling equipment selection." Available at www.mhia.org/et/pdf/mhi/cicmhe/equpguid.pdf.

Rayward-Smith, V.J., Osman, I.H., Reeves, C.R., and Smith, G.D., 1996, *Modern Heuristic Search Methods*. John Wiley, Chichester, UK.

Rose, J.S., Snelgrove, W.M., and Vranesic, Z.G., 1988, "Parallel standard cell placement with quality equivalent to simulated annealing," *IEEE Transactions of CADICS*, 7, 387–396.

Rose, J.S., Klebsch, W., and Wolf, J., 1990, "Temperature measurement and equilibrium dynamics of simulated annealing placements," *IEEE Transactions on CADICS*, 9, 253–259.

Scriabin, M. and Vergin, R.C., 1985, "A cluster-analytic approach to facility layout," *Management Science*, 31(1), 33–49.

Sullivan, W.G., Wicks, E.M., and Luxhoj, J., 2002, *Engineering Economy*. Prentice-Hall, New Jersey.

Tam, K.Y. and Li, S.G., 1991, "A hierarchical approach to the facility layout problem," *International Journal of Production Research*, 29(1), 165–184.

Tomkins, J.A and White, J.A., 1984, *Facilities Planning*. John Wiley, New York.
Varanelli, J.M. and Cohoon, J.P., 1993, "Two-stage simulated annealing," *Proceedings of the 4th ACM/SIGDA Physical Design Workshop*. Lake Arrowhead, CA, pp. 1–10.
Varanelli, J.M. and Cohoon, J.P., 1995, "A two-stage simulated annealing methodology," *Proceedings of the 5th Great Lakes Symposium on VLSI*. Buffalo, NY, pp. 50–53.
Varanelli, J.M. and Cohoon, J.P., 1999, "A fast method for generalized starting temperature determination in homogeneous two-stage simulated annealing systems," *Computers and Operations Research*, 26, 481–503.
Welgama, P.S. and Gibson, P.R., 1996, "An integrated methodology for automating the determination of layout and materials handling system," *International Journal of Production Research*, 34(8), 2247–2264.
White, S.R., 1984, "Concepts of scale in simulated annealing," *Proceedings of IEEE ICCD*. Port Chester, NY, pp. 646–651.
Ziai, M.R. and Sule, D.R., 1989, "Computerized materials handling and facility layout design," *Computers & Industrial Engineering*, 17(1–4), 55–60.
Zollinger, H.A., 1996, "Expanded methodology to concept horizontal transportation problem solutions." In: R.J. Graves, L.F. McGinnis, D.J. Medeiros, R.E. Ward, and M.R. Wilhelm (Eds.), *Progress in Material Handling Research*. Material Handling Institute, Charlotte, NC, pp. 651–663.

Chapter 5

Impact of Cross Aisles in a Rectangular Warehouse: A Computational Study

Gürdal Ertek, Bilge Incel, and Mehmet Can Arslan

CONTENTS

Abstract Order picking is typically the most costly operation in a warehouse, and traveling is typically the most time-consuming task within order picking. In this study, we focus on the layout design for a rectangular warehouse, a warehouse with parallel storage blocks with main aisles separating them. We specifically analyze the impact of adding cross aisles that cut storage blocks perpendicularly, which can reduce travel times during order picking by introducing flexibility in going from one main aisle to the next. We consider two types of cross aisles, those that are equally spaced (Case 1) and those that are unequally spaced (Case 2), which respectively have equal and unequal distances among them. For Case 2, we extend an earlier model and present a heuristic algorithm for finding the best distances among cross aisles. We carry out extensive computational experiments for a variety of warehouse designs. Our findings suggest that warehouse planners can obtain great travel-time savings by establishing equally spaced cross aisles, but little additional savings in unequally spaced cross aisles. We present a look-up table that provides the best number of equally spaced cross aisles when the number of cross aisles (N) and the length of the warehouse (T) are given. Finally, when the values of N and T are not known, we suggest establishing three cross aisles in a warehouse.

5.1 Introduction

Order picking is generally the most significant operation in a warehouse, accounting for approximately 60 percent of all operational costs in a typical warehouse (Frazelle, 2001). Cost of order picking is affected by the decisions regarding the facility layout and the selection of storage and retrieval systems, and by the implemented strategies such as zoning, batching, and routing. Travel cost is typically the largest cost component within order picking, accounting for about 50 percent of the costs associated with order-picking activities (Frazelle and Apple, 1994). Because order picking, specifically traveling, is costly, reducing the travel time spent for order picking can significantly reduce operational costs.

In this chapter, we present the findings of a study that focuses on the strategic layout decisions of how many cross aisles to establish within a rectangular warehouse and how to determine the distances among them. A "rectangular warehouse" can be defined as a warehouse with equi-length

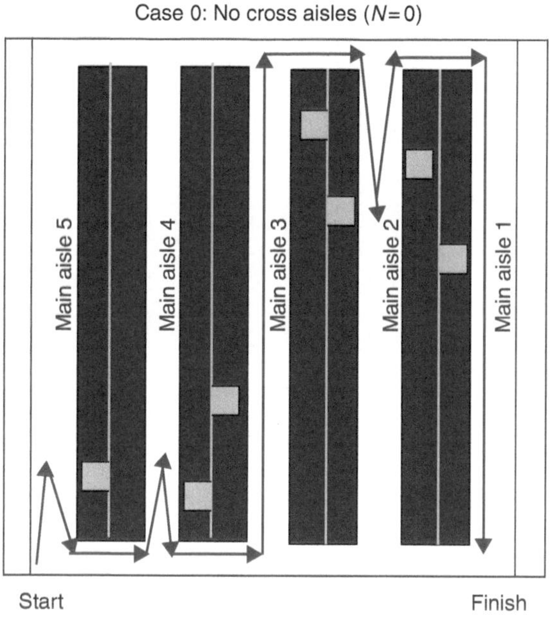

Figure 5.1 Case 0: A rectangular warehouse with four storage blocks, five main aisles, and no cross aisles. (The vectors show a route to pick an order with seven items.)

parallel "storage blocks," separated by aisles in between (see Figure 5.1). A subregion of a large warehouse, where routing decisions are made independently from the remaining regions, can also be considered as a rectangular warehouse, given that it satisfies the structural properties described earlier. A rectangular warehouse may have only "main aisles" which separate the storage blocks vertically (Figure 5.1), or may also contain one or more "cross aisles" perpendicular to the main aisles, which divide the storage blocks horizontally (see Figures 5.2 through 5.4). The main advantage of cross aisles is that they enable savings in travel times, especially during the order-picking operations. Figures 5.1 and 5.2 illustrate an example of a rectangular warehouse where the creation of a cross aisle can reduce the travel distance while picking an order with seven items: The addition of the cross aisle (Figure 5.2) shortens the travel distance by enabling shortcuts from the fifth main aisle to the fourth main aisle and from the fourth main aisle to the third main aisle.

Storage blocks typically consist of steel racks that are installed on the warehouse floor permanently during the construction of a warehouse. Thus, the decisions regarding the quantities and dimensions of storage

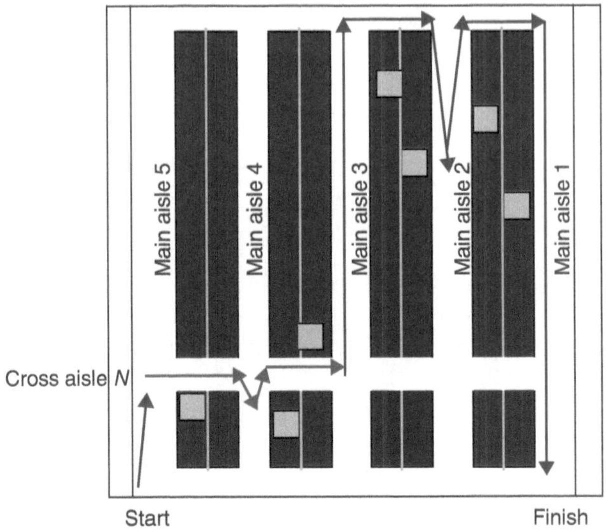

Figure 5.2 The warehouse displayed in Case 0 with an interior cross aisle with shortcuts linking main aisle 5 to main aisle 4 and main aisle 4 to main aisle 3.

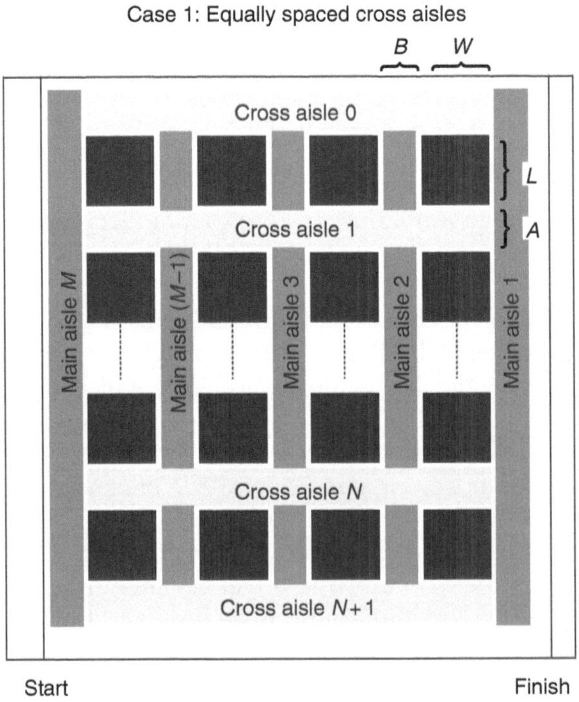

Figure 5.3 A rectangular warehouse with equally spaced cross aisles (Case 1).

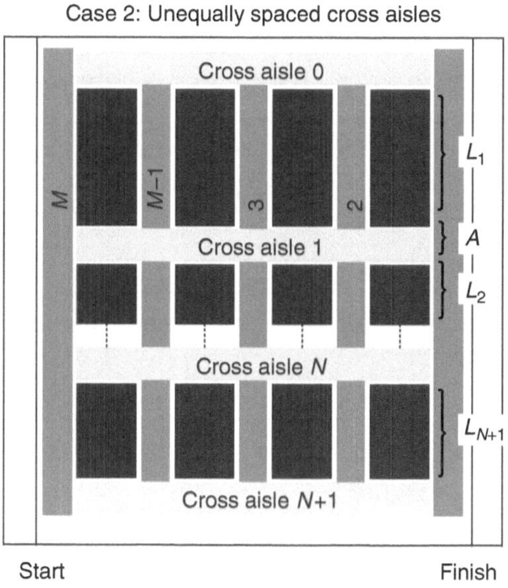

Figure 5.4 A rectangular warehouse with unequally spaced cross aisles (Case 2).

blocks, main aisles, and cross aisles are strategic decisions. These decisions should be made considering many factors, including:

- The physical dimensions of the building
- The characteristics of the materials to be stored, including physical dimensions, weights, shelf lives, pallet sizes, and projected demand patterns
- The characteristics of warehouse equipment such as forklifts and automatic guided vehicles (AGVs)
- The quantity and capabilities of the workforce
- The capabilities of the information system, i.e., the warehouse management system (WMS)

One classic challenge raised by these factors is how to incorporate the interactions between different decision levels in the design and operation of warehouses (Rouwenhorst et al., 2000). For example, the strategic decision of determining the best warehouse layout, the tactical decision of assigning the products to the storage locations in the best way, and the operational decision of determining the best order-picking routes are all interdependent. In our study, we assume that the widths of the storage blocks, the main aisles, and the cross aisles are fixed, and that the items stored in the warehouse all have the same demand frequencies. Even under these simplifying assumptions, the strategic decisions regarding the number of cross aisles and the distances between them (the lengths of storage blocks) have to be made by estimating the average travel distance in order picking under a specific routing algorithm. Hence, in our study, we assume that the routing algorithm and the storage locations of products are predetermined, and focus on the strategic decisions regarding cross aisles.

For a rectangular warehouse, one can identify the following three cases with respect to the number of cross aisles N and the distances in between them:

Case 0: Warehouse with no cross aisles ($N = 0$), as shown in Figure 5.1.
Case 1: Warehouse with N equally spaced cross aisles ($N \geq 1$), as shown in Figure 5.3.
Case 2: Warehouse with N unequally spaced cross aisles ($N \geq 1$), as shown in Figure 5.4.

In our study, we seek answers to the following research questions regarding the rectangular warehouse:

Should the cross aisles be established equally spaced (Case 1) or unequally spaced (Case 2)? In other words, should the storage blocks have an equal length or variable lengths? How much travel-time savings do cross aisles bring? Under which settings do cross aisles bring the most travel-time savings? How many cross aisles should there "ideally" be in a rectangular warehouse? In other words, what is the best number of cross aisles?

To answer these questions, we carry out extensive computational experiments reflecting a variety of warehouse settings with different values for warehouse lengths (T), number of cross aisles (M), and pick densities (D). Based on a thorough analysis of our experimental results, we come up with answers to the aforementioned research questions.

One unique aspect of our research is that we extensively apply the starfield visualization technique from the field of information visualization. In "starfield visualization," various fields of a dataset are mapped to the axes of a colored 2-D or 3-D scatter plot, and to the attributes of the glyphs (data points) such as color, size, and shape. "Information visualization" is the growing field of computer science that combines the fields of data mining, computer graphics, and exploratory data analysis (in statistics) in pursuit of visually understanding data (Spence, 2001; Keim, 2002). The ultimate goal in information visualization is to discover hidden patterns and gain actionable insights through a variety of—possibly interactive—visualizations.

The use of a visualization approach in the analysis of our numerical results will enable us to make important observations and develop managerial insights. To the best of our knowledge, this is the first attempt where data/information visualization techniques are employed to this extent in the warehousing and facility logistics literature.

5.2 Related Literature

Rouwenhorst et al. (2000) present a reference framework and classification of warehouse design and operating problems. Van den Berg and Zijm (1999) provide another review of the warehousing literature that classifies warehouse management problems. Sharp (2000) summarizes functional warehouse operations; database considerations; and tactical, strategic and operational issues in warehouse planning and design.

Within the vast facility logistics literature, there exist studies that solely focus on order-picking routing and order batching for the purpose of reducing travel time. An early study by Ratliff and Rosenthal (1983) solves the routing problem in order picking. Based on the number of aisles, the authors propose an algorithm that solves the problem to optimality. They state that the algorithm computation time grows linearly in the number of aisles, and is thus scalable for solving real-world problems.

Roodbergen and De Koster (2001a) analyze the relationship between warehouse layout and average travel time. They consider a rectangular warehouse in which a single cross aisle divides the warehouse into two equal-length blocks. The authors present a dynamic programming algorithm to determine the shortest order-picking routes, and show that the addition of the cross aisle decreases average order-picking time significantly.

In another study, Roodbergen and De Koster (2001b) compare several algorithms for routing order pickers in a warehouse with more than one cross aisle. They introduce two new heuristics, combined and combined$^+$, and compare them with the S-shape, largest gap, and aisle-by-aisle heuristics in the literature. The authors prove through computational tests that the combined$^+$ heuristic performs best among the five heuristics. A branch-and-bound algorithm is used as a benchmark to compare the performances of the generated heuristics.

De Koster et al. (1999) report a real-world application, where they significantly improve the efficiency of manual order-picking activities at a large retail distribution center in the Netherlands. In the first stage of their study, the authors apply a routing heuristic, which ensures that order pickers pick items from both sides of an aisle. This heuristic alone achieves a 30 percent reduction in travel time, and consequently a saving of 1.2 order pickers. In the latter stage of their study, the authors apply order batching, time-savings method, and a combined routing heuristic (De Koster and Van der Poort, 1998) jointly, and achieve 68 percent reduction in travel distance and a saving of 3 to 4 pickers. This study is the perfect example of how the order-picking strategies and routing algorithms proposed in the literature can be applied in the real world to achieve substantial savings.

Our study is mainly related to the work of Vaughan and Petersen (1999), who consider both layout and routing. Vaughan and Petersen are motivated by the fact that cross aisles can reduce travel distances due to their flexibility in order picking. The authors develop a shortest-path pick sequencing model that is applicable to any number of equally spaced cross aisles (equal-length storage blocks) in the warehouse. Their model assumes that all the items along an aisle are picked before proceeding to the next aisle, and the order picking progresses from the leftmost aisle to the rightmost aisle. This policy is referred to as "aisle-by-aisle policy." The authors compute the optimal routes for a large number of randomly generated picking requests, over a variety of warehouse layouts and order-picking parameters. Their results suggest that when the main storage-aisle length (T) is small, an excessive number of cross aisles can *increase* the average travel distance. This is true especially when the number of storage aisles (M) is small, and when pick density is very small or very large. The authors warn that the savings due to cross aisles diminish, even turn into losses, if the number of cross aisles becomes excessive. This is because the extra distance to traverse the cross aisles increases the travel distances. Additionally, the authors find out that as the main storage-aisle length (T) increases, the optimal number of cross aisles also increases and report that cross aisles are most beneficial for longer warehouses.

5.3 Vaughan and Petersen Model

Vaughan and Petersen (1999) assume certain characteristics with respect to the rectangular warehouses and order-picking policies. Because our study is built on their model, which we will refer to as the V&P model, the following assumptions are also valid for our model:

- There are parallel main aisles, and products are stored on both sides of the main aisles.
- Each order includes a number of items to be picked, which are generally located in various main aisles.
- All the stocks of a particular item are stored in a single location.
- Order pickers can traverse the aisles in both directions and change directions within the main aisles.

- The main aisles are narrow enough to pick from both sides of the aisle without changing position.
- The main aisles are wide enough such that two or more order pickers can operate in the main aisle at the same time.
- There are two natural cross aisles in the warehouse, at the head and rear of the warehouse.
- Cross aisles are not used to store items; they are only used to pass to the next main aisle.
- The items of an order are collected in a single tour.

Block lengths are determined by the locations of the cross aisles that divide main aisles perpendicularly. In our study, the "number of cross aisles" refers to the number of interior cross aisles, which are between the default head and rear cross aisles. We assume that picking routes start and end at the southeast and southwest corners of the warehouse, respectively. Even though some research assumes that order picking ends at the starting point (De Koster and Van der Poort, 1998; Roodbergen and De Koster, 2001), this does not make a great change in travel distance (and thus travel time). Petersen (1997) notes that this change results in at most 1 percent deviation in travel distance.

The dynamic programming algorithm developed by Vaughan and Petersen (1999) finds the optimal route to pick an order under the aisle-by-aisle policy. The complete notation for their so called shortest-path model is as follows:

L: Length of a storage block.
T: Length of the warehouse (equal to the length of main aisles), $T = (N + 1)L$.
M: Number of main aisles.
N: Number of interior cross aisles. (The total number of cross aisles is $N + 2$.)
A: Width of a cross aisle. (This parameter is essential for the calculation of the best aisle-by-aisle route. The model assumes that an order picker walks along the center of the cross aisles.)

This walking pattern is illustrated in Figure 5.5 and the additional distance of $A/2$ to walk to the middle of the cross aisle is reflected in the formulas for $B1_m$ and $B2_m$.

B: Width of a main aisle
C: Width of a storage block

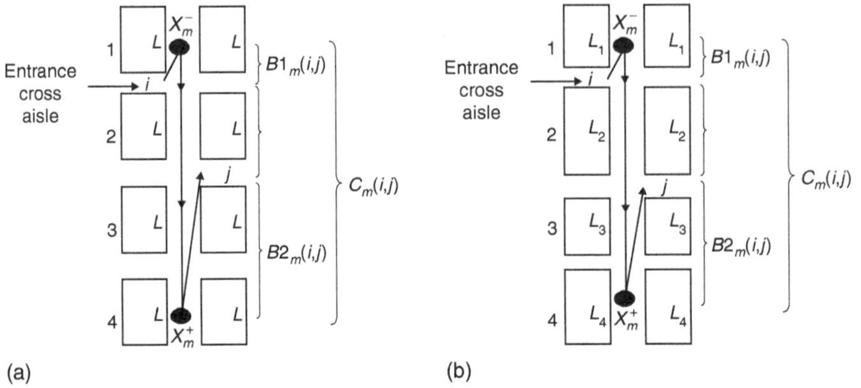

Figure 5.5 (a) A warehouse with storage blocks of equal lengths and (b) a warehouse with storage blocks with unequal lengths. (Arrows represent the total vertical travel distance to pick items in main aisle *m* when the main aisle is entered from the *i*th cross aisle and left from the *j*th cross aisle.)

The notation until now is related to the warehouse layout. The notation below is given for a particular order to be picked:

K_m: The number of items to be picked by the order picker from main aisle $m = 1, 2, \ldots, M$. Thus, the order consists of $\sum_{m=1}^{M} K_m$ items in total.

$X_m(t)$: The location of an item t in main aisle $m = 1, 2, \ldots, M$, and $t = 1, 2, \ldots, K_m$ (undefined if $K_m = 0$) where $0 \leq X_m(t) \leq T$.

(The expressions listed below are demonstrated in Figure 5.5a.)

X_m^+: The location of the item at the south-most location (highest value) in main aisle m (undefined if $K_m = 0$), i.e., $X_m^+ = \max_t \{X_m(t)\}$

X_m^-: The location of the item at the north-most location (smallest value) in main aisle m (undefined if $K_m = 0$), i.e., $X_m^- = \min_t \{X_m(t)\}$

$C_m(i,j)$: The total vertical travel distance required to pick all the items in main aisle m, if main aisle m is entered at cross aisle i and exited to main aisle $m - 1$ at cross aisle j

$B1_m(i,j)$: The length of forward-tracking leg required to pick the items in main aisle m to the north of cross aisle h, $h = \min(i,j)$

$B2_m(i, j)$: The length of back-tracking leg required to pick the items in main aisle m to the south of cross aisle h, $h = \max(i, j)$

$f_m(i)$: The minimum total picking distance required to pick all the items in aisle $m, m-1, m-2, \ldots, 2, 1$ if main aisle m is entered at cross aisle position i

In the V&P model C_m, $B1_m$, and $B2_m$ are calculated as follows:

$$C_m(i, j) = B1_m(i, j) + |i - j|(L + A) + B2_m(i, j) \text{ where}$$

$$B1_m(i, j) = \begin{cases} 0 & \text{for } K_m = 0, \\ 0 & \text{for } X_m^- \geq \min(iL, jL), \\ 2[\min(iL, jL) - X_m^- & \text{for } X_m^- < \min(iL, jL); \\ \quad +A(0.5 + ((\min(iL, jL) - X_m^-)/L))] & \end{cases}$$

and

$$B2_m(i, j) = \begin{cases} 0 & \text{for } K_m = 0, \\ 0 & \text{for } X_m^+ < \max(iL, jL), \\ 2\left[\max(iL, jL) - X_m^+ \right. & \text{for } X_m^+ \geq \max(iL, jL). \\ \quad \left. +A(0.5 + ((X_m^+ - \max(iL, jL))/L))\right] & \end{cases}$$

The dynamic programming equations for each stage are given as follows:

$$f_m(i) = \min_j \{C_m(i, j) + f_{m-1}(j)\}, \text{ where } f_1(i) = C_1(i, N+1).$$

Stages of the dynamic programming are related to the main aisle numbers in the warehouse. The desired shortest-path picking route is determined by evaluating $f_M(N+1)$.

5.4 Modified Model

Now we present our model that allows us to find the best routes according to the aisle-by-aisle heuristic for the case of unequally spaced cross aisles (Case 2). The primary difference between our model and the V&P model is that the storage blocks now have variable lengths L_i (Figure 5.5b) instead of a fixed length of L (Figure 5.5a) where L_i is the length of the ith storage block for $i = 1, \ldots, N+1$. Thus, the length of the warehouse T which is equal to the length of the main aisles can be expressed as $T = \sum_{i=1}^{N+1} L_i$.

Next, we define two new notations that give us the indices of the blocks where the north-most and south-most items within an aisle are located:

Blockof(X_m^+): Index of storage-block L_i in main aisle m where X_m^+ is located for $i = 1, 2, \ldots, N + 1$.

Blockof(X_m^-): Index of storage-block L_i in main aisle m where X_m^- is located for $i = 1, 2, \ldots, N + 1$.

Finally, the C_m, $B1_m$, and $B2_m$ values are calculated based on the modified definitions of block lengths L_i and the warehouse length T which can be expressed as

$$C_m(i,j) = B1_m(i,j) + \sum_{s=\min(i,j)+1}^{\max(i,j)} L_s + |i-j|A + B2_m(i,j)$$

where

$$B1_m(i,j) = 2\left[\min\left(\sum_{s=1}^{i} L_s, \sum_{f=1}^{j} L_f\right) - X_m^- + A(0.5 + \min(i,j) - Blockof(X_m^-))\right], \text{ and}$$

$$B2_m(i,j) = 2\left[X_m^+ - \max\left(\sum_{s=1}^{i} L_s, \sum_{f=1}^{j} L_f\right) + A(0.5 + Blockof(X_m^+) - 1 - \max(i,j))\right]$$

5.5 Algorithms to Identify Best Storage-Block Lengths L_i

Given T, M, N, A, B, C, and D values, the problem of finding the best storage-block lengths L_i is a difficult problem. This is because the length of a tour is found by solving a dynamic programming problem and the locations of the items are uniformly distributed. The objective function to be minimized is the average travel distance (and thus, the average travel time) over all orders, with the optimal travel distance for each order computed through dynamic programming optimization. We, thus, develop and implement two heuristic search algorithms, namely GSA (grid search algorithm) and RGSA (refined grid search algorithm), to find the best L_i values. GSA takes a warehouse (with its T, M, N, A, B, C values), a set of generated orders, and the number of grids as parameters, and identifies an initial solution, which consists of L_i values. RGSA takes the solution of GSA as the initial solution and carries out a search to reduce the average travel distance (i.e., average travel time). These algorithms are given in pseudo-code and are explained in the Appendix.

5.6 Experimental Design

The different values of model parameters that we have used in our computational experiments are depicted in Table 5.1. We investigate 396 scenarios (problem instances) corresponding to 396 combinations of the warehouse length (*T*), the number of aisles (*M*), and the pick density (*D*). In all these scenarios, the *A*, *B*, *C* parameters are respectively set to fixed values of 2.50, 1.25, and 1.25 (m).

One fundamental parameter is the pick density (*D*), which is the average number of items per main aisle. In each scenario, the total number of items to be picked is calculated as the multiplication of the pick density (*D*) with the number of main aisles (*M*). The 11 pick density values listed above are used for calculating the order sizes during the estimation of average route length for each scenario. Thus, the 11 order sizes used in the experiments are 0.1*M*, 0.5*M*, 1.0*M*, 1.5*M*, 2.0*M*, 2.5*M*, 3.0*M*, 3.5*M*, 4.0*M*, 4.5*M*, and 5.0*M*.

The parameter values in our study are selected such that we can extend the experiments of Vaughan and Petersen (1999). Compared to the 126 scenarios (combinations of *T*, *M*, and *D*) in their study, we consider 396 scenarios. In addition, our parameters take values over broader ranges, we calculate average travel distances over a greater number of instances (1000 orders as opposed to 100 instances), and we consider Case 2 besides Case 1.

Table 5.1 Experimental Design

	Factor	*Number of Values*	*Values*
396 scenarios	Length of main aisles (*T*) (m)	6	30, 60, 90, 120, 150, 180
	Number of main aisles (*M*)	6	5, 10, 15, 20, 25, 30
	Pick density (*D*) (items/aisle)	11	0.1, 0.5, 1.0, 1.5, 2.0, 2.5, 3.0, 3.5, 4.0, 4.5, 5.0
	Number of cross aisles (*N*)	1 (for Case 0)	0
		8 (for Case 1)	1, 2, 3, 4, 5, 6, 7, 8
		3 (for Case 2)	1, 2, 3
	(*A*, *B*, *C*) (m)	1	(2.50, 1.25, 1.25)

Note: *A* = width of a cross aisle; *B* = width of a main aisle; *C* = width of a storage block.

For each warehouse (T, M, N, A, B, C) and for each order size ($D \times M$) we apply the following procedure:

Step 1: Generate a set of 1000 orders with $D \times M$ items each: Each item to be picked is assigned to a storage location by first randomly generating a main aisle number, and then, randomly generating the position within that main aisle on the interval (0, T). The locations of the items to be picked in each order are assumed to be uniformly distributed across the warehouse. This assumption can be encountered in related studies (e.g., see Roodbergen and De Koster, 2001).

Step 2: Apply RGSA.

Step 2.a: Apply GSA for the generated set of orders and the given warehouse. For each feasible configuration of storage blocks, the shortest-path dynamic programming algorithm of the V&P model is solved for each of the orders in the set of orders. Average travel distances in the set of orders are obtained and an initial best configuration of storage blocks that provides the minimum average order-picking travel distances is returned.

Step 2.b: Apply the remaining steps of RGSA. Given the initial solution returned by GSA, RGSA works on improving the L_i values with the objective of minimizing average travel distance.

In Step 1 of the aforementioned procedure, the seed used to generate the random numbers is always chosen the same. The result of the experiments is a dataset with 396 rows and the following 17 columns: T, M, N, D, average travel length in Case 0, average travel length in Case 1 for $N=1, 2, \ldots, 8$ (eight distinct columns), average travel length in Case 2 for $N=1, 2, 3$ (three distinct columns), area of the warehouse (for the scenario). We carry out our analysis in Section 5.7 using this dataset. The values in the dataset are computed through a heuristic algorithm (which is not optimal) and through Monte Carlo simulation. Because we are using heuristic algorithms, from now on, the solution which will be referred to as the "best solution" is actually the incumbent solution, which is not necessarily optimal.

5.7 Analysis of Experimental Results

In this section, we analyze, through starfield visualizations, the results of our computational experiments for the 396 scenarios. The first three figures that we discuss in this section (Figures 5.6 through 5.8) are referred to as "colored scatter plots" in exploratory data analysis literature (Hoffman and

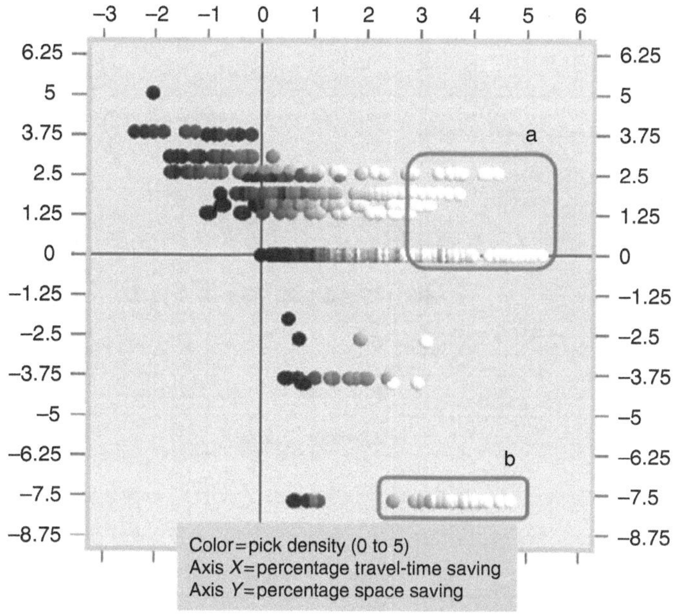

Figure 5.6 Savings with respect to the pick density (*D*).

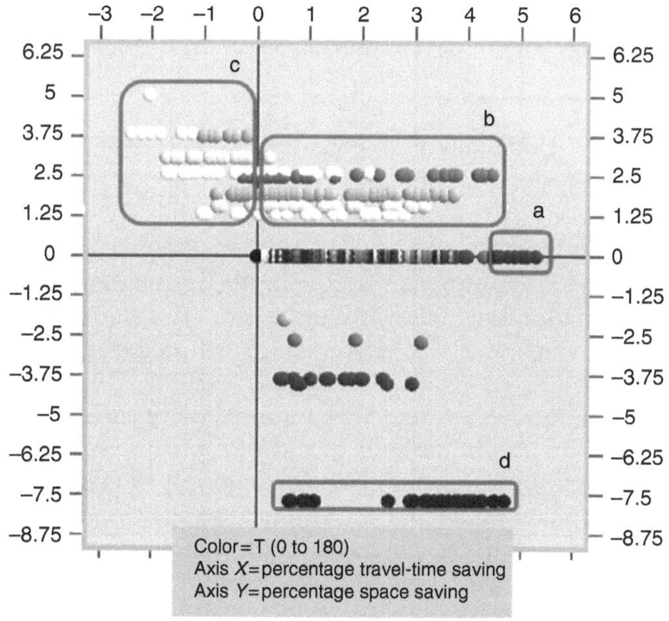

Figure 5.7 Savings with respect to the warehouse length (*T*).

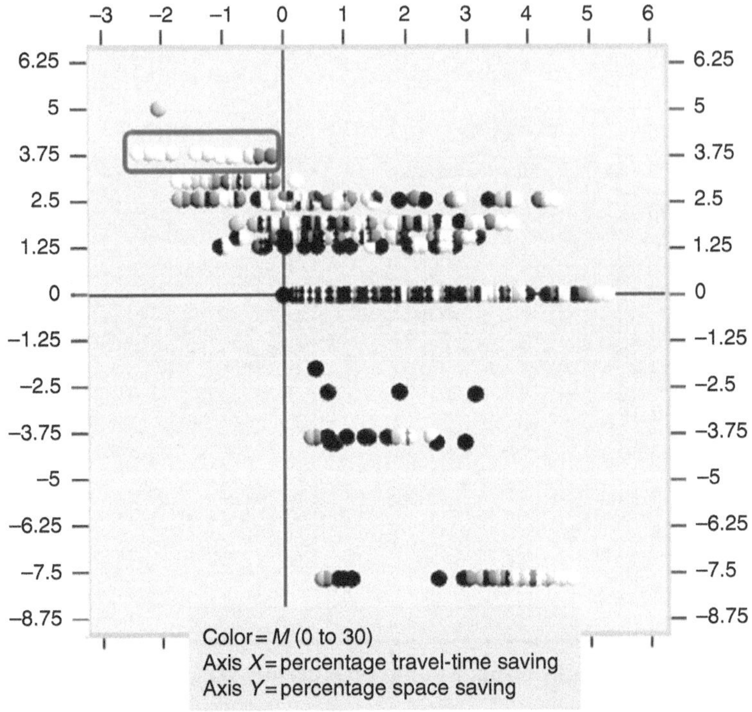

Figure 5.8 Savings with respect to the number of main aisles (*M*).

Grinstein, 2002). According to this naming scheme, Figures 5.9 and 5.10 are referred to as "jittered colored scatter plots," and Figures 5.11 through 5.13 are referred to as "colored 3-D scatter plots." However, rather than using the terminology in exploratory data analysis, we refer to all these plots as "starfield visualizations," following the terminology in the field of information visualization (Shneiderman, 1999). The starfield visualization is an extended version of the scatter plot, with coloring, size, zooming, and filtering.

In each of Figures 5.6 through 5.13, information regarding which parameter is mapped to which attribute of the plot/glyphs is displayed below the plot. For example, in Figure 5.6, *D* (pick density) values are mapped to color of the glyphs; percentage travel-time savings (in Case 2 compared to Case 1) are mapped to the *X*-axis; and percentage space savings (in Case 2 compared to Case 1) are mapped to the *Y*-axis. The range of pick density values is 0 to 5. Lighter colors represent larger values of the mapped parameter, and darker colors represent smaller values of the mapped parameter. All the mappings are linear. Rectangular frames, such as frames (a) and (b) in

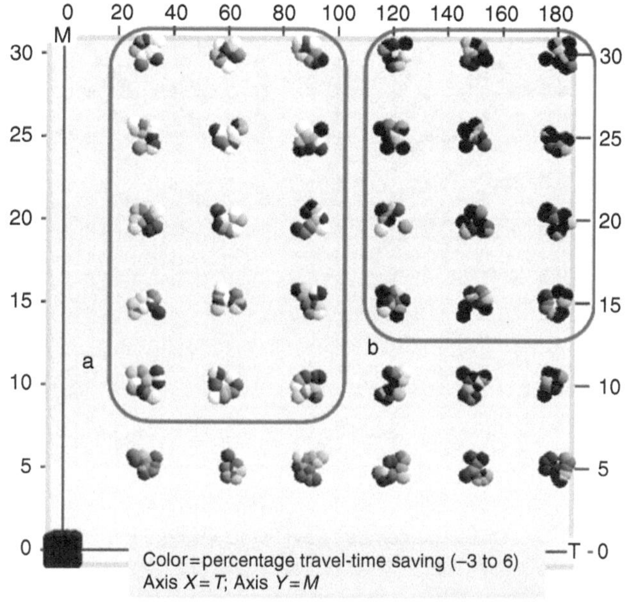

Figure 5.9 Percentage travel-time savings with respect to the warehouse length (T) and number of main aisles (M).

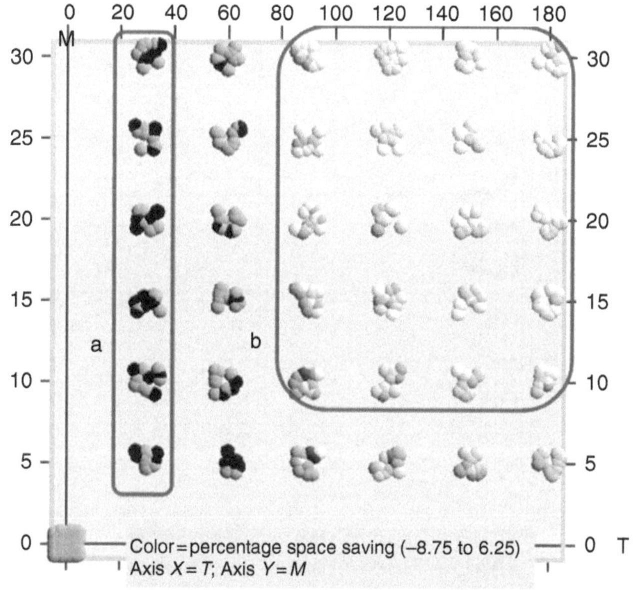

Figure 5.10 Percentage space savings with respect to the warehouse length (T) and number of main aisles (M).

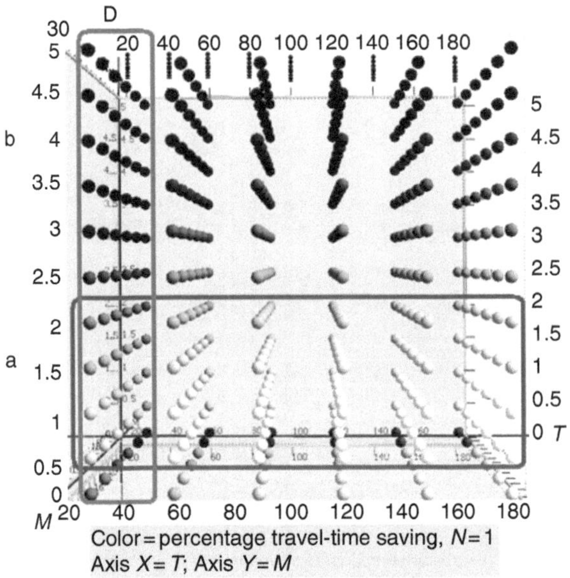

Figure 5.11 Percentage travel-time savings with respect to the warehouse length (*T*), number of main aisles (*M*), and pick density (*D*) for $N = 1$.

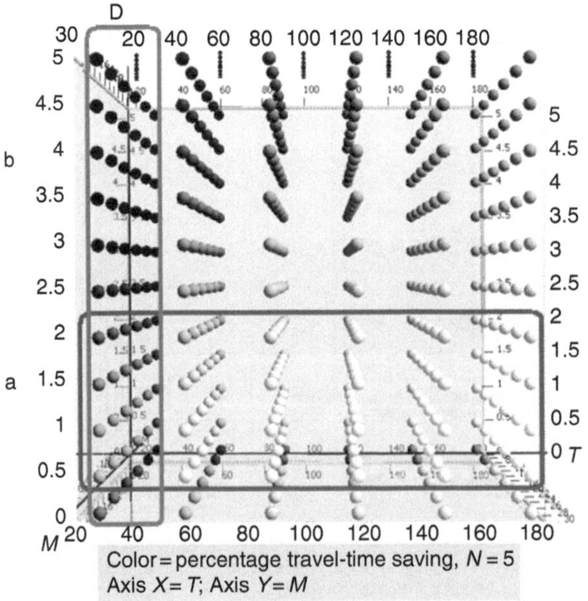

Figure 5.12 Percentage travel-time savings with respect to the warehouse length (*T*), number of main aisles (*M*), and pick density (*D*) for $N = 5$.

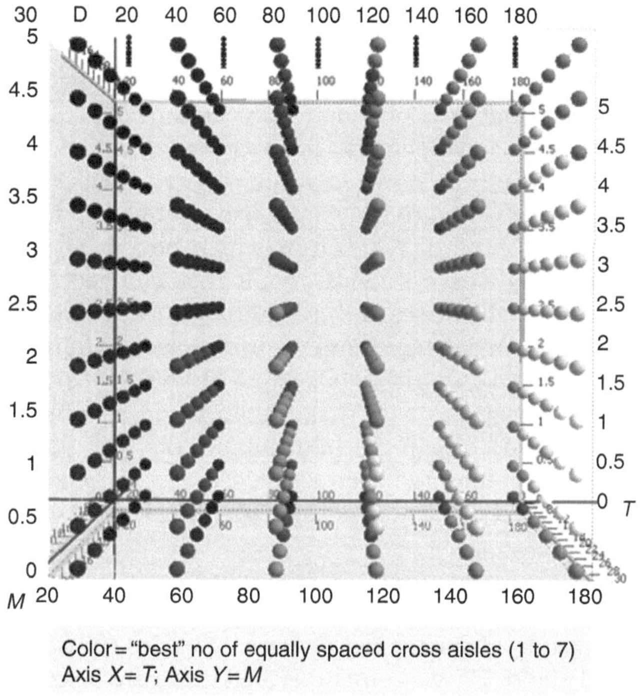

Figure 5.13 The best number of cross aisles (*N*) with respect to the warehouse length (*T*), number of main aisles (*M*), and pick density (*D*).

Figure 5.6, are drawn to highlight specific regions in the plots that exhibit the interesting properties.

5.7.1 Savings in Case 2 Compared to Case 1

Figures 5.6 through 5.10 illustrate the percentage savings gained in layouts with unequally spaced cross aisles (Case 2) compared to layouts with equally spaced cross aisles (Case 1). In Figures 5.6 through 5.8, each scenario is represented by a glyph (data point). The percentage travel-time (distance) savings in Case 2 compared to Case 1 are mapped to the *X*-axis, the percentage space savings are mapped to the *Y*-axis, and various parameters (*D, T,* and *M*) are mapped to colors of the glyphs. In Figures 5.9 and 5.10, each scenario is again represented by a glyph, but this time the *T* values are mapped to the *X*-axis, the *M* values are mapped to the *Y*-axis, and the percentage savings (in travel time and in warehouse space) are mapped to colors of the glyphs.

In Figure 5.6, frame (a) shows that the scenarios with large percentage travel-time savings are all characterized by high pick densities (large D values). Frame (b) shows that under certain scenarios, the percentage travel-time savings are obtained only at the cost of big losses in warehouse space (negative percentage space-saving values on the Y-axis).

In Figure 5.7, frame (a) shows that the scenarios with the largest percentage travel-time savings are for warehouses that have medium T values. Frames (b) and (c) show that scenarios which benefit from Case 2 with respect to percentage space savings are all characterized by large length values (light colors). However, in these scenarios there may be savings as well as losses in percentage travel time, as can be seen in frames (a) and (b), respectively. Frame (d), on the other hand, shows that in instances with shorter warehouses (glyphs with darker colors) the best number of cross aisles for Case 2 is more than the best number of cross aisles for Case 1, and this can result in large percentage losses in warehouse space. These instances are all characterized by high pick densities in frame (b) of Figure 5.6.

In Figure 5.8, the frame shows that the scenarios in which Case 2 results in significant space savings, but also the travel-time losses (negative values on the X-axis) are all characterized by large M values (light tones of gray). From Figures 5.6 and 5.7, we remember that these are also scenarios with low pick densities and largest T values.

The maximum percentage travel-time saving (X value of the rightmost glyph) obtained in the 396 scenarios is 5.28 percent. This result strikingly suggests that unequally spaced cross aisles (Case 2) bring little additional savings in comparison to equally spaced cross aisles (Case 1). In our study, finding the best number and best positions of unequally spaced cross aisles required implementation of a nontrivial algorithm and allocation of significant running times for the computations (approximately ten days in total, most of it for computing the solutions for $N = 3$ in Case 2). Thus, we can conclude that warehouse planners are better off establishing rectangular warehouses with equally spaced cross aisles instead of unequally spaced cross aisles. These results provide answers to the first research question posed in Section 5.1.

Figures 5.9 and 5.10 allow the analysis of savings with respect to T and M, which are mapped to the X- and Y-axes, respectively. Because there are 11 scenarios for each (T, M) pair, jittering is applied to a certain extent to avoid occlusion. So, in these two figures, the glyphs which are clustered together have the same T and M values, but differ in their pick densities (D values).

In Figure 5.9, frame (a) shows that for small values of T, percentage travel-time savings in Case 2 are higher in general (lighter glyphs). For large warehouses, which are highlighted by frame (b), there are less savings, or even losses in Case 2. This may be due to the fact that our algorithm finds the best positions in Case 2 is run only for $N=1$, 2, and 3. We thus believe that development of efficient algorithms for solving Case 2 for larger values of N is critical.

In Figure 5.10, frame (a) shows that shorter warehouses can incur big space losses in Case 2, because for the associated scenarios, the best number of cross aisles required in Case 2 (to minimize travel time) is typically more than the number of cross aisles required in Case 1. Frame (b) shows, as expected, that there are space savings in Case 2 compared to Case 1, because $N \leq 3$ in Case 2 and $N \leq 8$ in Case 1. In general, through our observations in Section 5.7.1, we can conclude that it is sufficient to focus only on Case 1, which we continue to analyze in Sections 5.7.2 and 5.7.3.

5.7.2 Impact of T, M, and D in Case 1

Figures 5.11 and 5.12 show the change in percentage travel-time savings in Case 1 (for $N=1$ and $N=5$) compared to Case 0, under the 396 combinations of T, M, and D (which are mapped to X-, Y-, and Z-axes, respectively). Percentage travel-time saving in each scenario is mapped to color of the related glyph. As shown in frame (a) of both figures, the largest travel-time savings are realized for pick densities (D values) between 0.5 and 2.5. This observation is consistent with earlier findings of Vaughan and Petersen (1999), who state that "the greatest cross aisle benefit occurs at pick densities in the range 0.6 to 1.0 units/aisle," which answers the second research question posed in Section 5.1.

In both Figures 5.11 and 5.12, the impact of M is observed to be negligible, except for $D=0.1$ (warehouses with very small orders). This conclusion is reached by observing that the colors of the glyphs do not change significantly along the Y-axis, except for $D=0.1$. Meanwhile, as shown in frame (b) of both figures, shorter warehouses (with small T values) are most sensitive to changes in D.

Vaughan and Petersen (1999) also state that the "addition of cross aisles generally decreases the picking travel distance on average, with travel distances frequently in the range 70 percent–80 percent, or even less, of that associated with the no cross aisles $N=0$ layout," quantifying the savings that can be obtained. This is another significant finding of our study. The percentage travel-time savings in Figures 5.11 and 5.12 take values between 0 and 35.30. That is, we observe travel-time savings of up to 35.30 percent by adding cross aisles, confirming the earlier findings of the authors.

5.7.3 Best Number of Cross Aisles in Case 1

Figure 5.13 shows the change in the best number of cross aisles in Case 1, which is mapped to color. The parameters *T*, *M*, and *D* are again mapped to *X*-, *Y*-, and *Z*-axes, respectively. Because none of the scenarios has its best *N* value equal to 8, the range of *N* values is between 1 and 7. Also, because the glyphs in the figure do not have the lightest tones of gray, we can observe that the best *N* values are seldom 7 or 6. A count in the experimental results shows that in 389 (all but 7) of the 396 scenarios, the best *N* value is less than or equal to 5. The most frequently encountered *N* value is 4, with 129 scenarios implying that $N = 4$ is the most desirable number of cross aisles.

We can conclude from Figure 5.13 that the main determinant of the best *N* value is the warehouse length *T*, because there are big changes in the color tone as one goes from smaller to larger values of *T*. By judging from the pattern of change in color tone, one can conclude that *M* also has some impact as well. At this point, we can address the third research question stated in Section 5.1 in two parts, both of which are very relevant and important. What is the best number of cross aisles for known values of *T* and *M*? And is there a best number of cross aisles, regardless of *T* and *M*?

The first question is very relevant, because in designing warehouse layouts, warehouse planners typically have a very limited knowledge on future *D* values, while they generally have a good judgment of which values *T* and *M* should take. The answer to the first question is given in Table 5.2,

Table 5.2 The Best Number of Cross Aisles for Different (*T*, *M*) Pairs, Accompanied with the Maximum Travel-Time Losses That One Can Encounter under Any *D*

	$T=30$	$T=60$	$T=90$	$T=120$	$T=150$	$T=180$
$M=5$	1	2	2	3	4	4
	0.52[a]	1.32[a]	0.51[a]	0.30[a]	0.59[a]	0.37[a]
$M=10$	1	2	3	3	4	4
	1.86[a]	0.32[a]	0.23[a]	0.66[a]	0.58[a]	0.38[a]
$M=15$	1	2	3	4	4	5
	2.40[a]	0.86[a]	0.22[a]	0.10[a]	0.35[a]	0.21[a]
$M=20$	2	3	3	4	4	5
	2.55[a]	0.81[a]	0.44[a]	0.10[a]	0.31[a]	0.10[a]
$M=25$	2	3	3	4	5	5
	2.43[a]	0.74[a]	0.42[a]	0.18[a]	0.31[a]	0.12[a]
$M=30$	2	3	3	4	5	5
	2.45[a]	0.74[a]	0.54[a]	0.16[a]	0.27[a]	0.15[a]

[a]Percentage values.

which displays the best number of cross aisles for different (*T*, *M*) pairs, accompanied with the maximum travel-time losses that one can encounter under any *D* values. For example, for (*T*, *M*) = (90, 10), the planner should establish three cross aisles. Among the 11 *D* values tested in our experiments for this (*T*, *M*) combination, establishing a different number of cross aisles resulted in at most 0.23 percent savings compared to three cross aisles.

The second question is very relevant as well, because many warehouse planners would prefer to learn and always remember a single number, rather than having to refer to Table 5.2 in this chapter. Thus, the question is which number of cross aisles to recommend to a warehouse planner regardless of *T*, *M*, or *D* values. The answer to this question is simply three. A warehouse planner can always build warehouses with three equally spaced cross aisles without compromising a significant loss in travel time, in particular, when compared to warehouses with different numbers of cross aisles.

Of course, this result is valid if the *A*, *B*, *C* values are close to the values in Table 5.1, and the warehouse operates under the assumptions of our model, including the usage of the aisle-by-aisle policy for routing. For the scenarios that we analyze in our experiments, the worst travel-time loss for three cross aisles is 6.26 percent (Figure 5.14), and the worst warehouse space (area) loss is 15.38 percent (Figure 5.15). Figures 5.14 and 5.15 show the worst average losses for other values of *N*.

If the warehouse space (area) is the most critical resource, then a warehouse planner can establish two cross aisles, instead of three. In our experiments, having two cross aisles resulted in at most 16.68 percent loss in travel time (Figure 5.14), and 7.69 percent loss in warehouse space (Figure 5.15). Thus, layouts with three (equally spaced) cross aisles are

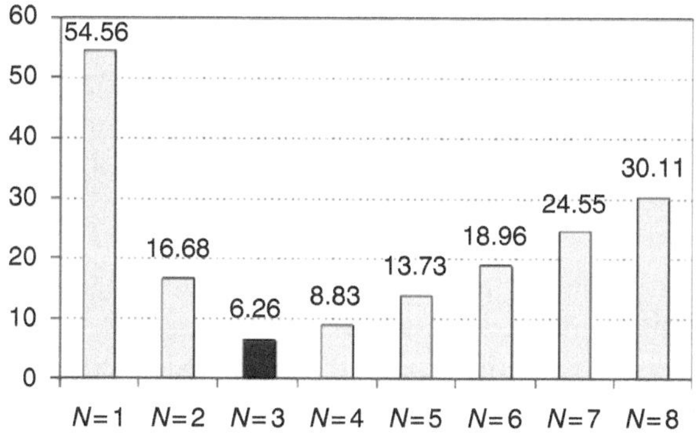

Figure 5.14 The maximum percentage gap—over all scenarios—between the travel time under *N* and the best travel time for that scenario.

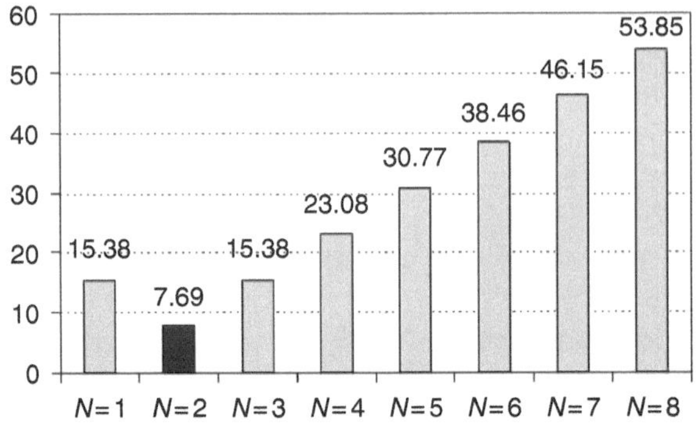

Figure 5.15 The maximum percentage gap—over all scenarios—between the warehouse space under *N* and the warehouse space for the best *N* for that scenario.

robust in terms of travel time, and layouts with two cross aisles are robust in terms of warehouse space.

5.8 Conclusions and Future Work

In this chapter, we presented a detailed discussion of the impact of cross aisles on a rectangular warehouse. We analyzed both equally spaced and unequally spaced cross aisles, which we referred to as Case 1 and Case 2, respectively. For Case 1, we utilized the dynamic programming algorithm presented in Vaughan and Petersen (1999) to determine the optimal order-picking routes under aisle-by-aisle policy. For Case 2, we made modifications to the Vaughan and Petersen (1999) model, including the change of formulas for certain parameters and introduction of new expressions before running the dynamic programming algorithm. We computed the average travel times using Monte Carlo simulation for 396 distinct scenarios, which correspond to 396 different warehouse and demand combinations. Our primary findings are:

- It is more desirable to establish only equally spaced cross aisles than to establish unequally spaced cross aisles.
- Establishing (equally spaced) cross aisles can bring significant travel-time savings and should definitely be considered. We obtained savings up to 35.30 percent in our experiments. Biggest travel-time savings are realized for pick densities between 0.5 and 2.5 ($D \in [0.5, 2.5]$).

- Given the length of main aisles and the number of main aisles (*T* and *M*), warehouse planners can refer to Table 5.2 in this chapter to determine the best number of (equally spaced) cross aisles. If one does not wish to refer to this table, but wishes to learn and remember a single value for the best number of cross aisles, we propose the value of three.

There are several directions for future research relating to our study. We list two of those:

- It is necessary to design faster and better algorithms to identify the best storage-block lengths (L_i) in Case 2, and to validate further our first conclusion.
- It is important to test the robustness of our conclusions under other demand patterns and different routing heuristics.

Our study contributes to the research and practice of warehouse planning/facility logistics by providing actionable insights regarding cross aisles. There is a great potential for research that attempts to solve warehousing problems that require taking interdependent decisions at different time horizons; for example, at both strategic and operational levels. Finally, we suggest the adoption of data analysis techniques from the field of information visualization for discovering knowledge hidden in experimental and empirical data related to warehouse planning.

Acknowledgments

The authors thank Kemal Kiliç at Sabanci University for his extensive help in reviewing the MS thesis of Bilge Incel (Küçük), which in time turned into this chapter. The authors also thank Ş. Ilker Birbil for reviewing the final draft of the paper and suggesting many improvements.

References

De Koster, R. and Van der Poort, E. 1998. "Routing order pickers in a warehouse: A comparison between optimal and heuristic solutions," *IIE Transactions*, 30, 469–480.

De Koster, R., Roodbergen, K.J., and van Voorden, R. 1999. "Reduction of walking time in the distribution center of De Bijenkorf," in M.G. Speranza and P. Stähly (Eds.), *New Trends in Distribution Logistics*. Springer-Verlag, Berlin and Heidelberg, Germany.

Frazelle, E.H. 2001. *World-Class Warehousing and Material Handling*. McGraw-Hill, New York.

Frazelle, E.H. and Apple, J.M. 1994. "Warehouse operations," in J.A. Tompkins and D. Harmelink (Eds.), *The Distribution Management Handbook*. McGraw-Hill, New York.

Hoffman, P.E. and Grinstein, G.G. 2002. "A survey of visualiations for high-dimensional data mining," in U. Fayyad, G. Grinstein, and A. Wierse (Eds.), *Information Visualization in Data Mining and Knowledge Discovery*. Morgan Kaufmann, San Fransisco, CA.

Keim, D.A. 2002. "Information visualization and visual data mining," *IEEE Transactions on Visualization and Computer Graphics*, 8(1), 1–8.

Petersen, C.G. 1997. "An evaluation of order picking routing policies," *International Journal of Operations and Production Management*, 17, 1098–1111.

Ratliff, H.D. and Rosenthal, A.S. 1983. "Order Picking in a rectangular warehouse: A solvable case of the traveling salesman problem," *Operations Research*, 31(3), 507–521.

Roodbergen, K.J. and De Koster, R. 2001a. "Routing order pickers in a warehouse with a middle aisle," *European Journal of Operational Research*, 133, 32–33.

Roodbergen, K.J. and De Koster, R. 2001b. "Routing methods for warehouses with multiple cross aisles," *International Journal of Production Research*, 39, 1865–1883.

Rouwenhorst, B., Reuter, B., Stockrahm, V., Van Houtum, G.J., Mantel, R.J., and Zijm, W.H.M. 2000. "Warehouse design and control: Framework and literature review," *European Journal of Operational Research*, 122, 515–533.

Sharp, P.G. 2000. "Warehouse management," in G. Salvendy (Ed.), *Handbook of Industrial Engineering*. John Wiley, New York.

Shneiderman, B. 1999. "Dynamic queries, starfield displays, and the path to Spotfire." Available at http://www.cs.umd.edu/hcil/spotfire/. Retrieved on November 2006.

Spence, R. 2001. *Information Visualization*. ACM Press, Essex, UK.

Van den Berg, J.P. and Zijm, W.H.M. 1999. "Models for warehouse management: Classification and examples," *International Journal of Production Economics*, 59, 519–528.

Vaughan, T.S. and Petersen, C.G. 1999. "The effect of warehouse cross aisles on order picking efficiency," *International Journal of Production Research*, 37, 881–897.

Appendix

GRID SEARCH ALGORITHM (GSA)

This algorithm returns best$L$, the best block lengths among tested layouts, for a given N. The length of the gridForL array is $(N+1)$ and indicates the number of storage blocks. Moreover, gridsFor$L[i]$ records the number of grids that constitute the length of the ith storage block. If the summation of the elements of gridfor$L$ array is equal to *noOfGrids* value, a feasible storage-block length combination is obtained. When *noOfGrids* = 20 and $N = 2$, for example, then some of the feasible storage-block lengths (L_1, L_2, L_3) would be (1, 12, 7), (11, 4, 5), having the summation of L values equal to *noOfGrids* = 20.

GSA generates feasible configurations of storage blocks systematically and returns the travel distances by solving the modified model that we present for a uniformly distributed set of orders. Average of the travel distances for the order set is taken and the initial best configuration of storage blocks enabling the minimum average order-picking travel distances is labeled as best$L$.

This algorithm generates a greater number of feasible storage-block length alternatives as the number of grids is increased. This results in smaller unit length ($G = T/noOfGrids$). However, the more the number of feasible solution gets, the more will be the computational effort. We observed in our experiments that for the warehouse and order settings described in the next section, *noOfGrids* = 7 is computationally prohibitive (18 days running time including the cases where $N = 4$), and *noOfGrids* is selected as 7.

$\Pi = \{1, \ldots, N+1\}$, $O = \{1, \ldots, \theta\}$

GRID_SEARCH_ALGORITHM (warehouse, orders, noOfGrids)
  G = T/noOfGrids
  for each gridsForL, /* s.t. gridsFor$L[i]$ ≤ noOfGrids, ∀i */
    sumOfGrids = $\sum_{i \in \Pi}$ gridsFor$L[i]$

```
if(sumOfGrids = = noOfGrids&&ARRAY_CONTAINS_NOZERO(gridsForL))
   tempL[i] = gridsForL[i] * G, ∀i ∈ Π
   TempWarehouse.setL(tempL)
 orders[o].setWarehouse(tempWarehouse), ∀o ∈ O
   tempSimulationStatistics = CALCULATE_SIMULATION_STATISTICS(orders)
   tempTravelDistance = tempSimulationStatistics.getAverage()
   if (tempTravelDistance < bestTravelDistance)
      bestL = tempL
      bestTravelDistance = tempTravelDistance
return bestL
```

CALCULATE_SIMULATION_STATISTICS (orders)

```
 travelDistance[o] = getOptimalTravelDistance(orders[o]), ∀o ∈ O
 return statistics for travelDistance data
```

REFINED GRID SEARCH ALGORITHM (RGSA)

This algorithm starts with the result of the GSA as the initial solution, and applies changes in little unit lengths (G) to the initial best configuration of storage blocks (initialBest$L$). In this method, first a range is defined. Half of this range is subtracted from each storage-space length and smaller unit lengths (gridsFor$L[i]$ × G) are added to each storage-space length. The travel distance for the new configuration temp$L$ is calculated for the given order set (orders) and compared with the best result obtained until that time. After trying all feasible configurations of the gridsFor$L$ for the same initial solution and calculating the travel distance for the new storage-block lengths, temp$L$ resulting in the shortest travel distance is assigned as the best configuration of cross aisles, best$L$. Then the range is updated by dividing with the number of grids (*noOfGrids*). Half of this range is subtracted from each storage-space length and smaller unit lengths (gridsFor$L[i]$ × G) are added to obtain new feasible storage-block lengths (temp$L$) and travel distance implied by the updated temp$L$ is calculated for the given order set (orders). The refined grid search is continued until the range declines to a length, which is determined as the smallest range (resolution) to be considered. When the range becomes as small as the resolution, the refined grid search is terminated and the improved configuration of storage-block lengths is assigned as the best configuration of storage-block lengths (best$L$) for the given warehouse and order set.

Any element of gridsFor$L$ can be at most $(N + 1) \times$ *noOfGrids*, because in the RGSA for each storage block, half of the range is subtracted and the length gridsFor$L[i]$ × G is added, for instance:

From the above equations, it is clearly seen that summation of the gridsFor$L$'s elements has to be $(N + 1) \times noOfGrids$. Therefore, an element of gridsFor$L$ is allowed to be $(N + 1) \times noOfGrids$ at most.

The search algorithms result in the best storage-block lengths that give the minimum order-picking travel distance for a problem instance (T, M, N, A, B, C, D) among the tested configurations.

```
REFINED GRID SEARCH ALGORITHM(warehouse, orders, noOfGrids,
resolution)
   initialBestL = GRID SEARCH ALGORITHM(warehouse, orders, noOfGrids)
   range = T/noOfGrids
   iterationNo = 0
   continueFlag = true
   while (continueFlag)
      iterationNo++
      if (iterationNo > 1)      // if not the first iteration
         range = (range/noOfGrids)/2
      G = (range/noOfGrids)/2
      for each gridsForL      // gridsForL[i] ≤ (N + 1) × noOfGrids
         sumOfGrids = Σ_{i∈Π} gridsForL[i]
        if (sumOfGrids = = (N + 1)noOfGrids&&ARRAY_CONTAINS_NOZERO
           (gridsForL))
           tempL[i] = initialBestL[i] − (range/2) + gridsForL[i]*G, ∀i ∈ Π
           tempWarehouse.setL(tempL)
           orders[o].setWarehouse(tempWaarehouse), ∀o ∈ O
           tempSimulationStatistics = CALCULATE_SIMULATION_STATISTICS
              (orders)
           tempTravelDistance = tempSimulationStatistics.getAverage( )
           if (tempTravelDistance < bestTravelDistance)
              bestL = tempL
              bestTravelDistance = tempTravelDistance
   if(range<resolution)
      continueFlag = false
   return bestL
```

Chapter 6

Stochastic Models for Facilities Logistics

Sunderesh S. Heragu, Gang Meng, and Henk Zijm

CONTENTS

Abstract This chapter examines the importance of stochastic models in the design and analysis of facilities-logistics problems. Much of the discussion is geared towards micro-level facilities-logistics problems, that is, layout and logistical problems within the "four walls of a facility." We first present a layout reference model, highlight the impact that stochasticity has on layout problems, and then present models that fully and nearly accurately capture the effects of variability in layout and logistical decisions, and present design tools that will enable a facility analyst to design layouts that perform well with respect to deterministic design criteria as well as stochastic operational-performance criteria.

6.1 Introduction

Structures that are thousands of years old, for example, the Egyptian pyramids, the ruins of the city of Pompeii in modern-day Italy, and Harappa and Mohenjodaro civilizations of the Indus Valley, suggest that the layout problem has been considered by facility designers for thousands of years. The industrial revolution offers some examples of the importance of layout in designing factories. Henry Ford's assembly line is perhaps the first and significant example of factory layout used as one of the means to achieve higher levels of productivity and efficiency. Although we can cite numerous examples to illustrate the importance facility planners and analysts placed on facility logistics, the study of facility-logistics problems via sophisticated mathematical models did not occur until the introduction of the quadratic assignment problem (QAP) in the mid-1950s by Koopmans and Beckmann (1957).

To put a layout problem into context, the entities and activities relevant to the layout problem can be organized into a layered model as shown in Figure 6.1. The reference model has three physical layers: product mix, machine types, and locations on the shop floor. Product mix includes the types of products that need to be produced and their annual volumes (parameters related to demand). Each available machine belongs to a machine type. The number of locations on the shop floor is equal to the total number of machines. The two logical layers of the reference model denote the design activities involved. Process-planning problem maps each product to a sequence of machine types; its output (product routings) includes other production data such as processing time, set-up time, and tooling information. The layout problem is to find a one-to-one mapping of machines on the machine-types layer to locations on the shop floor. The third logical layer (not shown) pertains to the scheduling problem. It assigns the machine types in a product routing to specific machines on specific shop floor locations, and coordinates the timing, sequencing, and prioritizing of all work orders assigned to one machine.

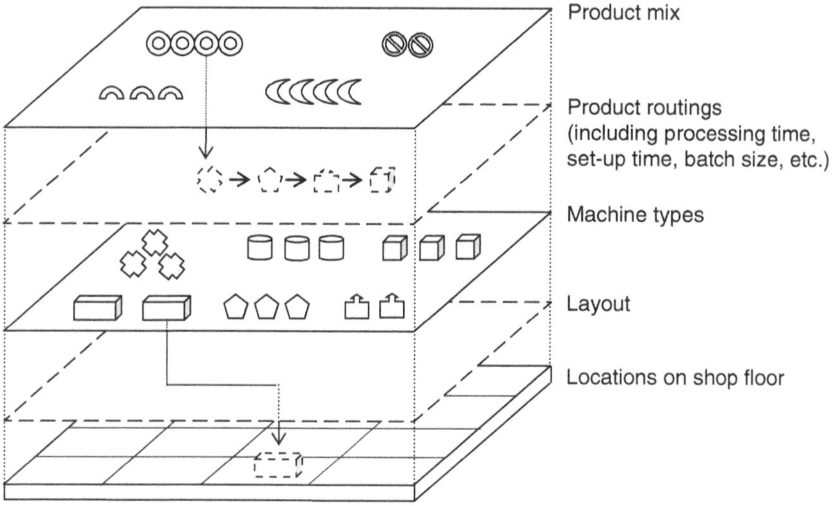

Figure 6.1 Facility-layout problem reference model.

The solution of the layout problem is determined by entities and activities in other layers of the reference model. Define the collection of these entities and activities (i.e., product mix, product routings, machines, and locations on shop floor) as the context of a layout problem. As long as the context is fixed, theoretically speaking, there exists an optimum layout for this context. By definition, the facility-layout problem is a simple assignment of n machines to n locations on the shop floor. What makes the facility-layout problem difficult to solve is the large combinatorial search space, especially when n is large (possibly $n!$ feasible solutions if there are no special restrictions on the locations of specific machines or relative location of a subset of machines) and the construction of a score function which incorporates various business considerations to evaluate the goodness of a layout.

Koopmans and Beckman (1957) first presented the QAP in the context of macro-level facility-location problems, but it has been extensively used to model micro-level factory-layout problems. Since the 1960s there has been a significant volume of research focused on solving the QAP, hence the layout problem, optimally. Assuming the cost to transport a unit load of material through a unit distance is known between each pair of facility locations, the objective of a QAP is to find optimal assignment of all the facilities to locations so that the overall cost of transporting material between each pair of facility locations is minimized. Ever since the QAP was adopted for the micro-level facility-logistical problems, there have been numerous attempts to solve it optimally. But, because the QAP is known to be nondeterministic polynomial time (NP)-complete, optimal algorithms cannot solve problems with more than 15 or 20 facilities and real-world layout problems are much larger. Therefore, numerous attempts have been made to solve the QAP heuristically. Several authors including Kusiak and Heragu (1987), Meller and Gau (1996), and others have surveyed these optimal and heuristic algorithms. A major drawback of adapting the QAP for the layout problem (and hence the techniques developed to solve the QAP-based layout problem) is that the QAP assumes the locations of facilities are known a priori, implicitly assuming that the facilities are of the same area and shape. Because this assumption does not reflect reality—not all facilities will have the same shape and area—Heragu (1992) and others have introduced more realistic modeling representations of the layout problem and appropriate techniques to solve them.

Although the unrealistic equal-area and same-shape assumption was removed in the Heragu (1992) model, another deficiency remains. All the above-mentioned layout models and techniques treat the problem in a deterministic manner. For example, a major assumption made by traditional layout models and algorithms is that the material flows between each pair of facilities are known well into the future (say for a period of five years or more) and that these values do not change. Current realities simply do not allow us to look at these values as being static. The dynamic nature of today's manufacturing and service activities do not allow us to predict material flows that far into the future. Often, we are fortunate if we can predict material flows for the upcoming planning period. Moreover, as pointed out by Heragu and Kochhar (1994), two changes taking place in the manufacturing-design and manufacturing-process technologies enable us to envision layouts that can be changed relatively frequently for relatively low cost.

1. Because we are making products increasingly with lightweight composites that can be engineered to have all the desired tensile- and mechanical-strength-related properties, machines or material-handling systems processing or transporting these composite products need not be heavy or require elaborate foundations.
2. Noncontact manufacturing processes such as electron beam welding and laser cutting also suggest that we do not need heavy machines with elaborate foundations. In fact, Heragu and Kochhar (1994) have proposed lightweight machines with wheels and mounted on tracks that can be clamped in any desired location in a grid of tracks on a factory floor permitting their movement from one location to another as frequently as production changes warrant.

As discussed in Heragu and Kochhar (1994), composites are the primary choice for many discrete manufactured components. Aluminum composites, for instance, can now replace cast iron parts and phenolics are replacing aluminum parts (Arimond and Ayles, 1993; Fujine et al., 1993). Not only are these light, but they can also be engineered to have excellent mechanical properties such as hardness, heat resistance, tensile strength, and vibration absorption. The last property permits machine-tool designers to design functionally equivalent, but lighter tools that do not require an elaborate foundation, making them easily movable. Nonabrasive manufacturing-process technology such as laser cutting, electron beam hardening, and molecular nanotechnology also supports machine-tool designers' quest for making lightweight machining equipment (Asari, 1993). Permanent magnetic chucks that facilitate quick mounting and dismounting of tools have been developed (*American Machinist*, 1993). In fact, these chucks do not magnetize the cutting tool, carry their own energy source, and do not obstruct machining. In addition, these features in and of themselves, support rapid equipment reconfiguration. This trend is likely to continue well into the next two decades. In fact, through a workshop and a Delphi survey, the committee on Visionary Manufacturing Challenges for 2020 has identified adaptable processes and equipment and reconfiguration of manufacturing operations as two key enabling technologies that will help companies overcome two of the six grand challenges or fundamental goals to remain productive and profitable in the year 2020 (National Research Council, 1998). These grand challenges are to "achieve concurrency in all operations" and to "reconfigure manufacturing enterprises rapidly in response to changing needs and opportunities."

Numerous examples where facility layouts are modified on a frequent basis, sometimes every few months, have been cited in Benjaafar et al. (2002). Northern Telecom, in one of its manufacturing facilities, facing constant product-design changes employs conveyor-mounted work cells that can be readily relocated just before a scheduled production or assembly change (*Assembly Magazine*, 1996). A primary advantage of reconfiguring a layout when warranted by changes in product mix and volume is that material-handling cost can be minimized because equipment can be reconfigured to suit the new production mix and volume. Of course, this cost must more than offset the cost of moving equipment from its current location to a new one.

A layout problem becomes more difficult to solve when multiple layout contexts must be considered and the problem has to be solved frequently in a real-time mode. A typical present-day manufacturing company faces constantly changing product volumes and mix, which make it necessary to update layout accordingly, to maintain operational efficiency. Simultaneously, the rapid advances in materials engineering and manufacturing technology have made it practical and economical to switch layout when needed.

Thus, given the new reality that layouts are likely to change very frequently, possibly every few months rather than years, and that at best, we only have knowledge of the production activities during the upcoming planning period—we need to develop a layout only for that period. In addition, due to the short-term life of a given layout and availability of production data for this time period, it is possible to consider optimizing operational-performance measures such as minimizing part cycle times and work-in-process (WIP) inventory.

Notice that it is relatively easy to get detailed data on material flows, machine setup, process and transfer batching, and other operations, relative to the production activities in the next period as opposed to manufacturing activities for the next five years. Although the required data is available, it is well known that this data is not static. For example, the time to transport a load from one machine to another is a function of where the material-handling device (assigned to transfer that load) is located at the time the material-transfer request comes in. Even if we were to assume that the device is always at the same known location, the travel time is determined by congestion and other factors and is highly variable.

Thus, it makes sense to develop stochastic models for layout analysis. Furthermore, because many factors—deterministic as well as stochastic—have an impact on determining the suitability of a layout, we need tools that can design and analyze a layout with respect to static-design criteria and stochastic operational-performance criteria. Developing layouts on the basis of static-design criteria alone is inadequate and, in some cases, could lead to undesirable consequences. For example, Benjaafar (2002) has shown that a layout that minimizes material-transfer costs when considering only loaded trips (static-design criteria), may have much higher WIP inventory (stochastic operational-performance measure) than another layout that may not minimize (loaded) material-transfer costs. In fact, it is quite possible that a layout that minimizes loaded material-transfer costs might be infeasible in the sense that it leads to an infinite WIP accumulation at one or more machines.

To summarize, the potential to frequently alter layouts, therefore, in a sense, transforms the modern layout problem from a strategic problem in which only long-term material-handling costs are considered to a tactical problem in which operational-performance measures such as reduction of product-flow times, WIP inventories, and maximizing throughput rate are considered in addition to material-handling and machine-relocation costs when changing from one layout configuration to the next.

Meng et al. (2004) have proposed a three-phase procedure for the design and analysis of the next-generation factory layouts. In the first phase, multiple, alternate layouts that perform well with respect to static-design criteria are generated using available algorithms for facility layout.

In the second phase, each alternate layout is evaluated with respect to several stochastic operational-performance measures. The analyses from phases 1 and 2 are combined to rank order a set of three or so layouts that perform well with respect to the user-weighted design and operational-performance criteria.

6.2 Literature Review

The traditional layout problem is primarily concerned with layout of machines for a (deterministic) single planning period. The score (objective) function used to evaluate the goodness of a layout only consists of material-handling cost. Using the topological feature of the layout, the traditional layout problem can be further classified as single-row and multi-row layout problem (Heragu, 2006), which in turn can be further classified depending upon size and orientation of machines.

The focus of much of the research in the traditional layout domain has been on efficient solution of a combinatorial optimization problem. Mathematical programming techniques such as nonlinear programming and mixed-integer programming have been used to solve the traditional layout problem (Houshyar, 1989; Bozer and Rim, 1996; Hsieh and Sha, 1996; Kouvelis and Chiang, 1996; Ho and Moodie, 1998; Heragu, 2006). But, because of the computational cost and solution quality, researchers have turned away from mathematical programming-based techniques to random-search heuristics such as genetic algorithms (Rajasekharan et al., 1998; Kochhar and Heragu, 1999) and simulated annealing (Jajodia et al., 1992; Kouvelis and Chiang, 1992; Harhalakis et al., 1996). Although not rooted in mathematical programming, these random-search techniques have proven to be powerful techniques to solve many combinatorial problems including the layout problem rather efficiently. Artificial intelligence techniques such as expert systems have also been used to solve some types of layout problems efficiently (Heragu and Kusiak, 1990; Ngoi Kok Ann and Chua, 1994; Chung, 1999). A detailed review of the layout problem can be found in Kusiak and Heragu (1987).

Assuming (deterministic) production data for multiple future planning periods is available, the dynamic-layout problem attempts to find a sequence of layouts corresponding to the multiple planning periods. Because multiple planning periods are considered, it is necessary to consider the cost of switching from one layout in one planning period to another in the next. The objective function is then to minimize the material-handling cost over all periods and the overall cost of relocating machines in consecutive layouts.

Lilly and Driscoll (1985) explore the two planning-period layout problems. Rosenblatt (1986) discusses a restricted version of the general dynamic layout problem which he calls the multiphase layout selection problem. In each phase, a traditional layout problem is solved to get several candidate layouts. A dynamic programming technique is then used in all phases to select the sequence of layout that minimizes the overall material-handling and machine-relocation costs. If we consider all possible layouts in each phase, we have a general dynamic-layout problem. Kouvelis and Kiran (1991) also use the dynamic programming technique to solve the multiple-period layout problems and propose a state space reduction approach to reduce the search space. Montreuil and Venkatadri (1991) present a proactive methodology for designing dynamic layouts for the expansion phase of a manufacturing system. The layout for the maturity phase of the manufacturing system is first determined, based on designer's envisioning of possible scenarios in a future (maturity) phase. Starting from this final layout, the methodology interpolates backwards until the initial facility-layout plan is obtained. Under the rigid facility assumption, that is, the boundary of a cell in phase p is within the boundary of the cell in phase $p + 1$, they determine the optimal position of each cell in phase p within the boundary of the cell in phase $p + 1$. Montreuil and Laforge (1992) extend Montreuil and Venkatadri's (1991) model in several ways. They relax the rigid cell assumption so as to make the layout more flexible. They also relax the restriction that the model is good only for the expansion (decline) phase of the manufacturing system. Most importantly, by interpolative design, Montreuil and Venkatadri (1991) assume a "linear future" where only one possible scenario can be considered in each intermediate phase. Montreuil and Laforge (1992) explicitly consider multiple scenarios for each intermediate phase in their model and develop a tree-structured future. As an input to the model, the designer needs to model explicitly the phases of evolution of the manufacturing system and the probabilistic future in each phase. The result is a scenario tree whose nodes represent a possible future characterized by the product mix, facility availability, and other factors that might affect the layout design. A linear programming model is built to process the scenario tree to get the dynamic layout. Compared with the backward, interpolative approach proposed by Montreuil and Venkatadri (1991), this model is more general and more intuitive to understand, which makes it suitable to be used in an interactive layout-design environment where a designer can peek into different future periods and do "what-if" analysis.

Robust-layout problem addresses the stochastic single- or multiple-period layout contexts where demand for one planning period is uncertain (thus multiple-demand scenarios exist for each period). It is motivated by the fact that layout design is usually done in the early stage, based on the forecast of future product demands, and this forecast usually turns out to be highly inaccurate. This makes the optimal design of layout problem meaningless. Another situation where a robust layout is desired is in the multiple-period layout domain where the relocation cost is prohibitive and we must therefore use the same layout across all planning periods. To solve a robust-layout problem, we choose only one layout, which may not be optimal for a particular demand scenario or planning period, but optimal or near-optimal considering all possible scenarios and planning periods. Kouvelis et al. (1992) defined robust layout as the layout whose objective value is within p percent of that of the optimal solution considering the actual demands. To generate the robust single-period layout problem, they adapted Gilmore's (1962) and Lawler's (1963) branch and bound procedure for the QAP. During the process of branch and bound, they keep a list of feasible layout whose objective function value is within p percent of the optimal or best-known solution, and fathom only if the lower bound of the branch is p percent higher than the objective function value of optimal or best-known solutions. As they mention, similar modification can be used in any branch and bound algorithm (see Finke et al., 1987 for a survey of such procedures). To generate robust layout for multiple periods, Kouvelis et al. (1992) apply the modified Gilmore–Lawler procedure on all scenarios in all periods to generate a list of solutions. They then look for solutions that are common across all scenarios. Yang and Peters (1998) solved the robust multiple-period layout problems differently. They employed the notion of planning time window which is a number of planning periods in the future. When the time window equals zero, then it is a dynamic-layout problem with a new layout determined for each new period. When the time window size equals the planning horizon, then it is a pure robust layout, wherein one layout is used for all the planning periods. Within each time window, an average flow matrix (corresponding to the periods in the time window) is calculated and used to generate a robust layout for this time window. The concept of combined adjacency graph (a concept similar to design skeleton used in Montreuil and Laforge, 1992) is used to simplify the constraints of the linear programming formulation of the robust-layout problem. Their model considers both material-handling cost and relocation cost, and assumes the layout rearrangement cost is fixed. The machines in their model can have varying sizes but with fixed orientation and load/unload points. A computationally efficient heuristic is used to obtain the final solution. This heuristic outperforms the QAP-based enumeration method for a set of test problems.

6.3 Need for Stochastic Analysis of Layout Problems

We elaborated in Section 6.1 two important reasons why it is important to consider deterministic design objective factors such as minimization of

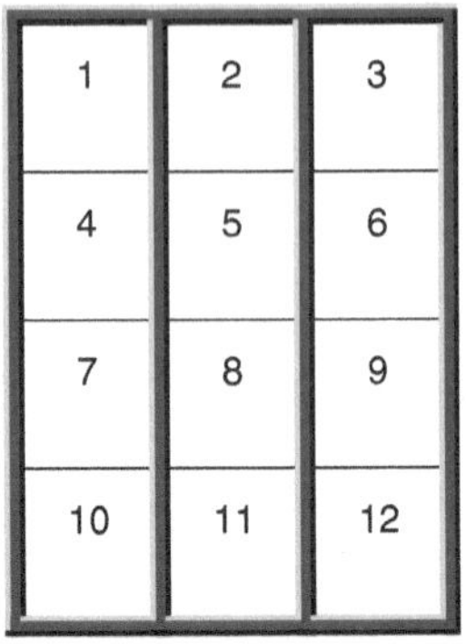

Figure 6.2 One possible layout for a facility manufacturing a single product.

material-handling cost as well as optimization of stochastic operational-performance measures such as WIP inventory, product cycle time, and machine utilization, in evaluating the goodness of a layout. One reason is dictated by the realities of present-day manufacturing—the ability to plan only for the next planning period due to the dynamic nature of manufacturing activities, and the ability to change a layout more frequently due to advances in the manufacturing-process and manufacturing-materials technologies. Another reason we mentioned in Section 6.1, pertains to the fact that focusing on deterministic design factors alone could lead to layouts that are infeasible. We will provide a modified example from Benjaafar (2002) to highlight this point. Consider the layout shown in Figures 6.2 and 6.3. Assume the only product manufactured in this facility visits machines in this sequence—(1, 2, 3, 4, 3, 4, 5, 6, 7, 8, 9, 10, 11, 10, 11, and 12). If we assume that material handling is accomplished by a pool of identical

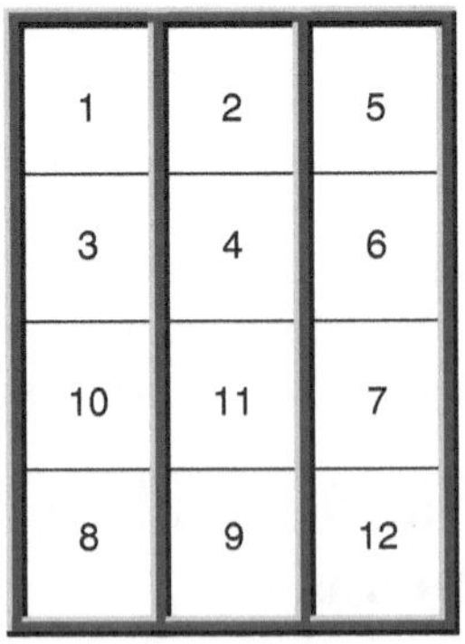

Figure 6.3 Alternate layout for the facility illustrated in Figure 6.2.

devices, realistic values for set-up times, processing and transfer times, and processing and transfer batch sizes—we can use a model such as the one in Meng et al. (2004) to determine whether the average WIP inventory in the layout in Figure 6.2 is much higher than that in the layout of Figure 6.3. Although the former layout minimizes the static-design criteria of minimizing loaded material-handling costs, it performs so poorly relative to stochastic operational-performance measures that it is completely unsuitable due to significant operational inefficiencies.

The above-mentioned example illustrates the fact that placing machines or sets of machines that may have very little loaded trips between them is sometimes advantageous because it minimizes empty material-handling device travel, thereby increasing device utilization and decreasing product wait times as well as WIP inventory. Traditional methods that focus on minimizing static-design criteria (e.g., minimize loaded material-handling costs) completely ignore the effects of idle travel and thus will prefer layout 1 instead of layout 2. In some cases, the WIP inventories of some layouts may be so significant that they could render a particular layout infeasible (Benjaafar et al., 2002).

6.4 Factors Considered in Stochastic Analysis of Layout Problems

Assuming the following facts are known, we can use the parametric decomposition method to estimate operational-performance measures. We only provide brief details here, but additional information can be found in Meng et al. (2008).

- Number and types of machines
- Number and types of discrete material-handling devices
- Number, types, volume, and routing of products to be manufactured in the specified production planning period
- First two moments of external arrival rate for each product
- First two moments of service time for each processing operation
- Set-up time for each operation and transfer for each product

The parametric decomposition method of Meng et al. (2008) incorporates the factors listed in the following subsections. Mathematical details pertaining to those are provided in Meng et al. (2008).

6.4.1 Empty Travel Time of the Material-Handling Device

The natural service time for a processing operation is the average time required for that operation and is usually readily available or can be obtained from the machine responsible for the processing operation. However, the natural service time for a transfer depends not only on the actual (loaded) travel from the originating station to the destination station, but also on the empty travel time from the station at which the material-handling carrier is currently located to the flow-originating station. Although the empty travel time may be small and negligible for intracell transfers, it can have a dramatic impact on intercell transfers. We use the approach developed by Chow (1987) in estimating empty travel time. This approach assumes a first come, first served service discipline, and has been used by others (see e.g., Egbelu, 1987; Fu and Kaku, 1997; and Benjafaar, 2002) to include the loaded and empty travel time for intercell material transfers. To make it more realistic, we assume the loaded travel is stochastic and characterized by two moments—mean and squared coefficient of variation (SCV). Although it can be relaxed, we assume unloaded travel is negligible for intracell transfers.

6.4.2 Setup

Set-up time impacts the two moments (mean and standard deviation) of the effective service time of processing and transfer operations. The natural process or transfer time is independent of the corresponding set-up time. Batch size can vary from one operation to the next. Similarly, the transfer batch size for the same product can vary from one transfer to the next. When x units of a product are processed as a batch, our model assumes that the batch-processing time is x times the processing time of each unit in the batch, but the SCV of the batch-processing time is the same as that of the individual unit in the batch.

6.4.3 Batching

Batching affects the departure process significantly, and modeling this is rather involved. Whitt (1983) identifies three network operations namely superposition, splitting, and departure—and develops expressions that capture these three effects rather well. His model also considers batching but assumes that every product entering a server is batched the same way. Segal and Whitt (1989) consider additional features relative to manufacturing systems, but the batching is once again assumed to be machine specific. Our model is more general and assumes that the batch size is operation specific. In other words, different products visiting the same machine or the same product visiting a machine multiple times could have different batch sizes. Under this more general assumption, and using the notion of relative batch size, Meng and Heragu (2004) consider batching as a fourth network operation (in addition to the three identified by Whitt, 1983) and develop two expressions for the first and second moments of the effective arrival of an aggregate product into a node. The two expressions are shown to be very effective in estimating the mean system-performance measures, by extensive comparison with simulation. Because the derivations of the expressions are extensive, we only present the two expressions below and refer the reader to Meng and Heragu (2004). Although based on Whitt (1983), the Meng and Heragu (2004) formula is significantly different because it considers batching as a fourth network operation and introduces additional factors such as relative batch size and outgoing routing probability.

6.4.4 Machine Failures

The effect of machine failure and non-preemptive machine set up on the first two moments of the service time can be included following the approach in Hopp and Spearman (2000).

6.5 Brief Description of MPA—A Tool for Stochastic Analysis of Layouts

The manufacturing system performance analyzer (MPA), described in more detail in Meng (2002) and Meng and Heragu (2004), is used to evaluate the performance of a layout. It is an extension of Whitt's (1983) queuing network analyzer (QNA). MPA incorporates many realistic manufacturing considerations such as those outlined in the previous subsections—operational and transfer batch sizes, set-up time, empty travel time of the material-handling device and machine failure. Although other analytical models are available (Segal and Whitt, 1989; Benjafaar, 2002), MPA is the most comprehensive and accurate one (Meng and Heragu, 2004). Simulation is an alternative tool for performance evaluation, but it is well known that it is expensive to build and run. MPA not only determines the commonly sought performance measures analytically (and therefore quickly), but also provides the user the option of automatically generating input data for the ProModel® simulation software.

MPA is based on the parametric decomposition (PD) method successfully employed by Whitt (1983) to analytically evaluate key performance measures of a queuing network, that is, a network of queuing systems. As Whitt (1983) puts it, unlike many queuing models which provide exact results for approximate models, PD methods provide approximate results for more exact models. These methods can work with inter-arrival and service times following any general distribution. Given the first two moments (mean and standard deviation) of inter-arrival times of each customer type into the network and its routings as input, the PD method calculates the first two moments of inter-arrival time of an aggregate customer into each node. To do so, Whitt (1983) identified three network operations—departure, split, and superposition (identified in Table 6.1), and approximately computed the effects of these on the first two moments of the aggregate arrival into each node. Each node is a queuing system and represents a server (machine, workstation, or material-handling device). Meng and Heragu (2004) identified an additional network operation, batch or burst, also listed in Table 6.1 and captured its effects on the aggregate arrival. When parts coming from a machine must wait at the next machine to be processed in larger batches, batching occurs. On the other hand, if products coming from one machine are processed in smaller batches at the next, "bursting" occurs. The notation used in deriving expressions for the first two moments of the aggregate arrival into a node and the derivation as well as a detailed explanation can be found in the Whitt (1983) and Meng and Heragu (2004).

The PD method determines the first two moments of the aggregate inter-arrival as well as service times at each node, treats each node as an independent GI/G/m queue with generally distributed inter-arrival and service times, and analytically determines performance measures at each node. It then disaggregates these performance measures to obtain product-specific measures or aggregates the results over the entire network to yield network-specific results. Such a method is approximate but provides results that are very close to those obtained via simulation. Moreover, because the inter-arrival and service-time distributions can be any general distribution, not just exponential, it is more realistic.

Table 6.1 Effects of Basic Network Operations on the Arrival Rates

Operations	*First Moment*	*Second Moment*
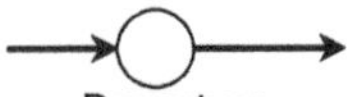 Departure	$\lambda_d = \lambda_a$	$c_d^2 = 1 + (1-\rho^2)(c_a^2 - 1) + \frac{\rho^2}{\sqrt{m}}(c_s^2 - 1)$
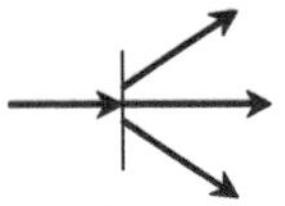 Split	$\lambda_i = p_i\lambda_a$	$c_i^2 = p_i c^2 + 1 - p_i$
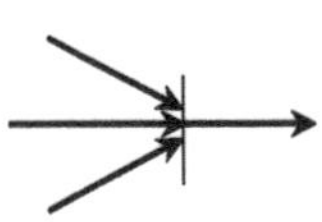 Superposition	$\lambda_a = \sum_i \lambda_i$	$c_a^2 = w\sum_i \left(\frac{\lambda_i}{\sum_k \lambda_k}\right) c_i^2 + 1 - w$
 Batch/burst	$\lambda_{a2} = \gamma_{1,2}\lambda_{d1}$	$c_{a2}^2 = \max\{0, \gamma_{1,2} - 1\} + \gamma_{1,2}c_{d1}^2$

6.6 Case Study

We use part of the data set in Huang and Irani (2000) as the current manufacturing system. We then add several products to introduce a reasonable change in product mix and volume, thus providing a reason to change the layout. The following details are assumed: (1) The shape and size of the machines is not of concern; standard one-by-one space holders are used to illustrate their relative positions. (2) The shop floor space has the fixed dimension of six-by-four units. (3) Work is assigned to specific machine(s) within a machine type using "shortest travel time from previous operation machine" rule, and each machine adopts a first come, first served service priority. Assumption (3) simplifies the effects of scheduling policy on the stochastic performance measures of candidate layouts.

6.6.1 Current Manufacturing System

Table 6.2 shows the routings of products in the current manufacturing system together with the new products (shaded). Note that the numbers in the routings represent machine types, not a specific machine. As shown in the current layout (Figure 6.4), the same type of machine might have different machine numbers. This occurs when duplicate machines are dispersed among different cells. For example machine type J is dispersed in cells 2 and 3, labeled as machines 12 and 15, respectively. When several machines of the same type are placed together within a cell, they are labeled using one machine number. For example the two type D machines in cell 1 are labeled together as machine number 6. The assumption here is that machines of same type placed together share the same incoming queue, thus working as a single station with multiple servers.

6.6.2 Generate Candidate Layouts

Candidate layouts can be generated using an existing software package, variation of an existing layout developed for similar production data, human expertise, or even intuition. Because the current layout is cellular, we build a functional layout from scratch and generate several cellular layouts by varying the current one. In all the layouts, we assume that machine G (#17 in Figure 6.4) cannot be moved due to hard constraints.

The first layout is the pure functional layout (Figure 6.5). Machines of the same type are grouped together into a work center. The relative positions of work centers are determined in such a way that the material-handling cost associated with the layout is the minimum among all pure functional layouts.

Table 6.2 Operation Sequences of Products Produced in the Facility

Product Number	*Sequence*	*Arrival Rate (Per Hour)*
1	A→D→H→I	0.2
2	A→D→G→D→H→G	0.3
3	A→B→D→G→H→I	0.1
4	A→D→G→I	0.3
5	A→F→J→G→I	0.2
6	F→J→G→H→I	0.1
7	F→D→H→I	0.2
8	C→E→B→F→D→H→I	0.1
9	C→E→F→D→H→I	0.1
10	D→G→D→H	0.2
11	F	0.3
12	K→G→L	0.1
13	K→L	0.1
14	K→G→J	0.3
15	A→G→K→J→K→L	0.1
16	A→G→K→J→K→L	0.2
17	K→G→L	0.1
18	F→G→J	0.3
19	L	0.2

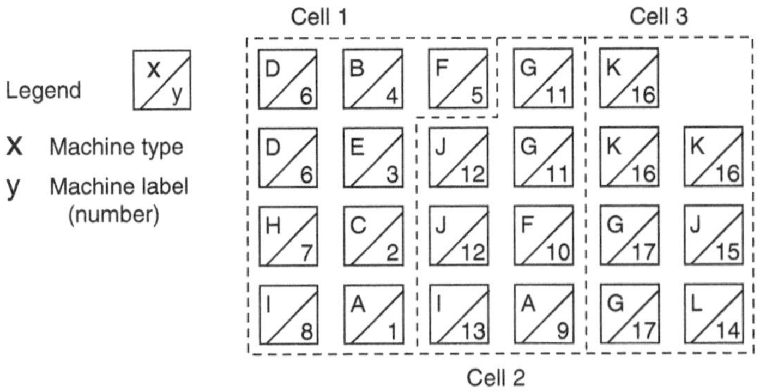

Figure 6.4 L0: Current cellular layout.

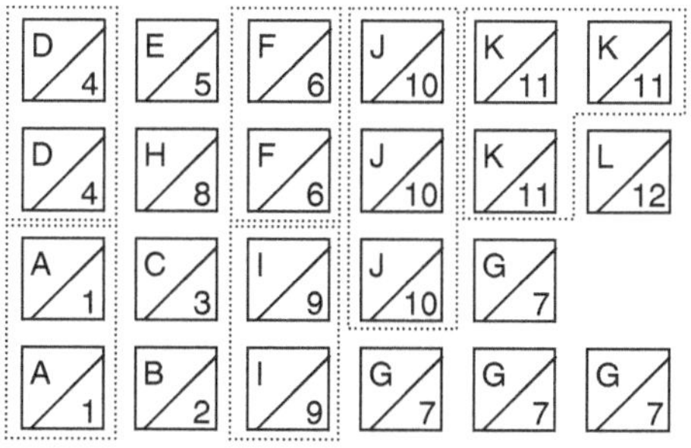

Figure 6.5 L1: Functional Layout.

Layout L2 in Figure 6.6 is a slight variation of the current cellular layout L0 (Figure 6.4). Cell 1 is the same and cells 2 and 3 are rotated 90° clockwise. The idea is to put the two copies of machine A physically adjacent to their own machine types in the adjacent cells. Besides reorientation, layout L3 in Figure 6.7 further reshapes cells to achieve the maximum physical adjacency between machines of the same type. Layout L4 in Figure 6.8 consists of an extra cell where in the machine types shared by different cells are placed. The distance matrix of a layout is determined by the Manhattan distance metric between the mass center of the two work centers.

6.6.3 Choosing between Existing and Candidate Layouts

With the above-mentioned processing and layout data, we can estimate the performance measures relative to each layout. These measures include both deterministic measures such as material-handling cost and stochastic measures such as average waiting time in queue and average queue length.

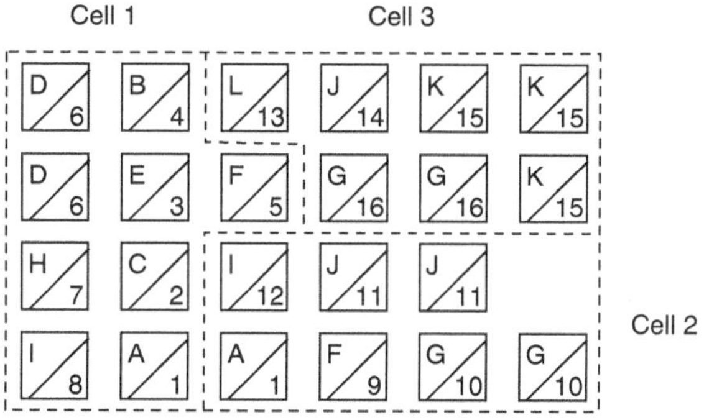

Figure 6.6 L2: Cellular layout with reorientation of cells 2 and 3.

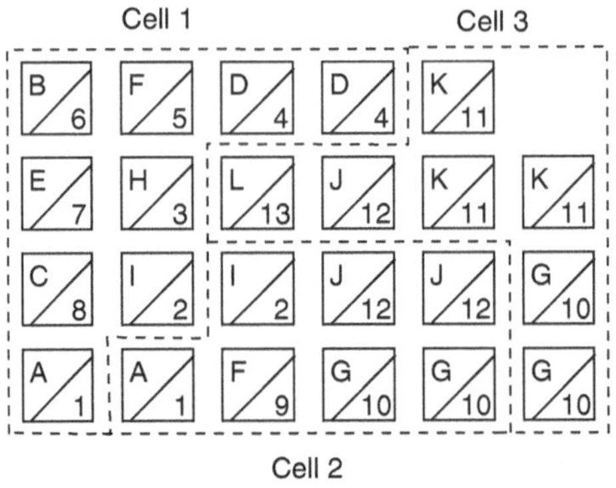

Figure 6.7 L3: Cellular layout with reorientation and reshaping of cells.

To facilitate the decision-making process, some of the measures need to be aggregated into more informative measures. Product lead time is the sum of all processing times spent on the machine, waiting time in queue before each machine, and transfer time between machines. Average WIP inventory level of the shop floor is the sum of the average queue length, over all the machines on the shop floor. Table 6.3 shows the WIP inventory level of the shop floor for each of the five layouts. Table 6.4 shows the material-handling cost and lead time for each product for each of the five layouts. Product-specific material-handling costs are added to get the overall material-handling cost of the system, using the formula $\sum_i T\lambda_i mhd_i$, where λ_i is the arrival rate of product type i and T is the length of the planning horizon, mhd_i is the material-handling cost incurred while handling product i. To relate product lead time with cost, we calculate the delay (estimated completion time minus product due date), then sum over

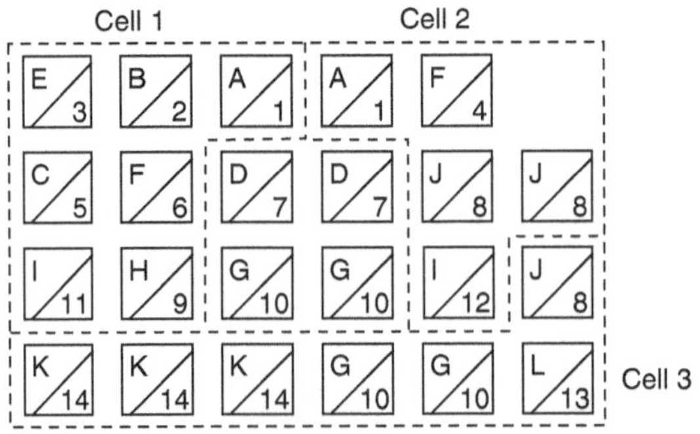

Figure 6.8 L4: Cellular layout with a remainder cell.

Table 6.3 The Overall WIP Level of Different Layouts

	WIP Inventory				
M	*L0*	*L1*	*L2*	*L3*	*L4*
1	3.2	4.6	4.6	4.6	2.3
2	1.5	1.5	1.5	2.6	1.5
3	1.0	1.5	1.0	2.8	1.0
4	1.5	2.5	1.5	2.5	44.8
5	5.1	1.0	11.3	1.6	1.5
6	2.5	4.2	2.5	1.5	1.3
7	2.8	3.4	2.8	1.0	2.5
8	1.7	2.8	0.8	1.5	5.0
9	4.7	2.6	1.5	14.6	2.8
10	2.2	5.0	2.3	3.4	3.4
11	1.3	2.7	4.4	2.7	1.7
12	5.4	1.6	5.4	5.0	9.9
13	1.5	—	35.6	1.6	1.6
14	1.6	—	0.7	—	2.7
15	2.1	—	2.7	—	—
16	2.7	—	2.9	—	—
17	4.4	—	—	—	—
Sum	45.1	33.4	81.4	45.4	82.1

all product types to get the overall lateness of the products. The formula used is $\sum T\lambda_i overdue_i$. The results are shown in the last row of Table 6.4.

Choosing among existing and candidate layouts is a multiple-objective decision problem. Different companies might be concerned with different sets of cost terms. Although most companies use deterministic terms such as material handling and relocation costs as well as stochastic terms such as WIP inventory cost and lead time, some companies might want to include unused space, machine utilization, or cell/machine center shape into consideration.

We have now aggregated the product- and machine-specific cost or performance measures (Tables 6.3 and 6.4) into four systemwide, cost measures: material-handling cost, WIP inventory cost, product lateness penalty cost, and relocation cost (Tables 6.5 and 6.6). In this case study, we use only these cost terms to select the final layout. The next step is to combine the four cost measures into a single one. A layout can then be selected based on this aggregate cost measure.

To help us in choosing among the available layouts, we assume a set of cost measures with a corresponding unit cost vector. For example, let us

Table 6.4 Material-Handling Cost and Product Lead Times Corresponding to Different Layouts

		Material-Handling Cost (Distance)					Due	*Lead Time (Hours)*				
P	λ	*L0*	*L1*	*L2*	*L3*	*L4*	Date	*L0*	*L1*	*L2*	*L3*	*L4*
1	0.2	6.0	6.0	6.5	9.0	4.5	7.6	8.3	7.9	7.3	7.9	6.5
2	0.3	15.5	20.0	25.0	20.0	10.0	9.0	9.3	9.1	9.8	9.1	7.8
3	0.1	13.0	18.0	18.5	16.0	9.5	17.0	17.6	17.3	17.1	17.3	15.9
4	0.3	12.0	10.0	15.5	12.0	6.5	6.4	6.9	6.6	6.4	6.6	5.3
5	0.2	8.0	10.0	8.5	9.0	8.0	42.7	28.7	16.2	26.3	31.2	110.9
6	0.1	9.0	11.0	10.0	10.5	8.0	31.5	15.4	11.9	38.6	27.0	64.6
7	0.2	5.0	6.0	6.0	7.5	6.0	30.1	17.9	9.5	36.3	24.5	62.1
8	0.1	8.0	14.5	9.0	8.5	8.0	35.5	38.8	30.3	48.4	30.2	29.7
9	0.1	8.0	9.5	7.0	8.5	8.0	25.0	28.3	19.8	37.9	19.7	19.2
10	0.2	7.5	13.5	9.5	10.5	6.5	6.2	5.8	6.0	7.1	6.0	6.0
11	0.3	0.0	0.0	0.0	0.0	0.0	13.2	2.5	2.9	2.1	2.9	55.6
12	0.1	4.0	5.0	4.0	6.0	5.0	13.1	9.4	6.2	37.5	6.2	6.2
13	0.1	3.0	1.0	3.0	3.0	4.0	10.4	4.4	4.4	34.6	4.4	4.4
14	0.3	4.0	5.0	4.0	4.0	5.0	14.7	11.4	8.0	38.3	8.0	8.0
15	0.1	11.0	11.5	13.5	13.5	19.5	24.2	25.0	16.0	49.1	16.0	14.7
16	0.2	11.0	11.5	12.5	13.5	19.5	23.4	25.0	16.0	45.4	16.0	14.7
17	0.1	4.0	5.0	4.0	6.0	5.0	13.7	10.0	6.8	38.1	6.8	6.8
18	0.3	3.5	6.5	3.5	4.0	6.0	12.8	12.2	8.7	11.3	23.7	8.2
19	0.2	0.0	0.0	0.0	0.0	0.0	8.0	2.0	2.0	32.2	2.0	2.0
	MHD_cost	132.6	164.0	160.0	161.5	139.0	Overdue	11.2	0.9	208.7	11.8	175.8

Table 6.5 Overall Cost with Unit Cost Vector of (5, 1, 10, 0.1)

		Layout				
Criteria	*Unit Cost*	*L0*	*L1*	*L2*	*L3*	*L4*
WIP	5	45.1	33.4	81.4	45.4	82.1
Material handling	1	132.6	164.0	160.0	161.5	139.0
Overdue	10	11.2	0.9	208.7	11.8	175.8
Relocation	0.1	0.0	32.0	20.0	36.0	56.0
Overall cost		470.0	342.6	2656.0	509.6	2312.9

assume that every distance unit traveled by the material-handling device costs \$5, one unit of shop floor inventory space costs \$1 per hour, every hour of lateness of product delivery incurs penalty of \$10, and that the unit distance cost of relocating a machine is \$0.1. Thus, for this set of cost measures the (WIP, material handling, overdue penalty, relocation) unit cost vector is (5, 1, 10, 0.1). The overall cost of the manufacturing system is the weighted sum of all cost measures. Table 6.5 shows the overall cost while using each of the five layouts with the unit cost vector of (5, 1, 10, 0.1) and Table 6.6 shows those with unit cost vector of (1, 10, 1, 0.1). The cost measures vector is generic, in the sense that any discrete manufacturing system can have the same set of cost measures. But the unit costs are typically company specific, reflecting a manufacturing system's resources, production control policy, and even management strategy. A low WIP inventory unit cost maybe due to the fact that the company has relatively more shop floor space (i.e., shop floor space premium is not high) and a high overdue unit cost reflects the emphasis of the company's eagerness to be responsive to market demand (i.e., penalize production delays). Note that the unit cost of relocation in the two examples is relatively small. There are two reasons. First, for reasons mentioned in Section 6.1, we are assuming that relocation cost in a reconfigurable manufacturing

Table 6.6 Overall Cost with Unit Cost Vector of (1, 10, 1, 0.1)

		Layout				
Criteria	*Unit Cost*	*L0*	*L1*	*L2*	*L3*	*L4*
WIP	1	45.1	33.4	81.4	45.4	82.1
Material handling	10	132.6	164.0	160.0	161.5	139.0
Overdue	1	11.2	0.9	208.7	11.8	175.8
Relocation	0.1	0.0	32.0	20.0	36.0	56.0
Overall cost		1382.5	1677.4	1892.1	1675.7	1653.5

system is relatively small and is a one-time cost, whereas the other costs (WIP, material handling, and overdue penalty) accumulate over time and depend upon production volume.

The unit cost vector can also carry information about the user's solution approach to the layout problem. A unit cost vector of (1, 10, 1, 0.1) emphasizes the importance of material-handling cost over WIP inventory and due-date related penalty costs with a ratio of 10:1, which is close to the scenario of traditional layout problem where only deterministic measures are considered. The unit cost vector of (5, 1, 10, 0.1) emphasizes the importance of WIP and cycle-time-related penalty costs over the relocation or material-handling costs. Obviously, the unit cost vector determines the candidate layout that is finally chosen. As shown in Tables 6.5 and 6.6, different unit costs lead to different choices of layout. When the unit cost vector is (5, 1, 10, 0.1), the pure functional layout L1 has the minimum overall cost. But when the unit cost vector is (1, 10, 1, 0.1), cellular layout L0 has the minimum cost, which suggests keeping the current layout.

6.6.4 Refinement of Selected Layout

The last step is to refine the selected layout before actually applying it to the manufacturing system. One intuitive way is to combine the good features of other competing candidate layouts into the selected one, without jeopardizing benefits of the current layout. With respect to layout L1, one refinement might pertain to the position of machine D. Because there are significant amount of appearances of D→G and G→D in product routings, switching the position of two type G machines with those of machine type B and C will not change the routing of the product, but it reduces material-handling cost of those products that transfer between machine types D and G (see Figure 6.9). Other refinements are possible and the designer must explore appropriate ones before settling on a layout for the next period.

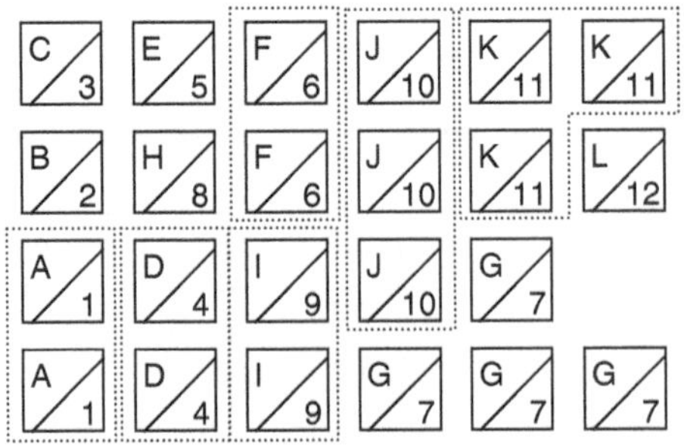

Figure 6.9 Final layout: Switch the position of D with B and C.

6.7 Conclusions

The reconfigurable layout problem only assumes that the production data for the next planning period is available, which is more realistic. While choosing between candidate layouts, reconfigurable layout considers not only deterministic material-handling and relocation costs, but also the stochastic performance measures such as WIP inventory level and product lead time, making it a more comprehensive decision model. As a performance-evaluation tool, MPA fits well in reconfigurable layout framework. It takes arrival, routing, processing, and facility data as input and generates stochastic costs (WIP level, lead time) of the manufacturing system.

Acknowledgment

We gratefully acknowledge support of the research presented in this paper by a National Science Foundation Grant DMI 9908437. Parts of this chapter have been reproduced from Meng et al. (2004).

References

American Machinist (online edition). 1993. U.S. Builder Enters Small Lathe Market, Vol. 137, pp. 10. Retrieved April 22, 2003. <www.americanmachinist.com>

Arimond, J. and Ayles, W.R. 1993. Phenolics creep up on engine applications. *Advanced Materials and Processes*, 143(6), 20.

Asari, M. 1993. Electron beam hardening system. *Advanced Materials and Processes*, 143(6), 30–31.

Assembly Magazine (online edition). 1996. Norstart Custom Telephones rely on flexible conveyor systems. May issue, Retrieved June 2, 2002. <www.assemblymag.com>

Benjafaar, S. 2002. Modeling and analysis of congestion in the design of facility layouts. *Management Science*, 48(5), 679–704.

Benjafaar, S., Heragu, S.S., and Irani, S.A. 2002. Next generation factory layouts: Research challenges and recent progress. *Interfaces*, 32(6), 58–76.

Bozer, Y.A. and Rim, S.C. 1996. A branch and bound method for solving the bi-directional circular layout problem. *Applied Mathematical Modeling*, 20(5), 342–351.

Chow, W.M. 1987. An analysis of automated storage and retrieval systems in manufacturing assembly lines. *IIE Transactions*, 18(2), 204–214.

Chung, Y.K. 1999. Neuro-based expert system for facility layout construction. *Journal of Intelligent Manufacturing*, 10(5), 359–385.

Egbelu, P.J. 1987. The use of non-simulation approaches in estimating vehicle requirements in an automated guided vehicle based transport system. *Material Flow*, 4, 17–32.

Finke, G., Burkard, R.E., and Rendl, F. 1987. Quadratic assignment problems. *Annals of Discrete Mathematics*, 31, 61–82.

Fu, M.C. and Kaku, B.K. 1997. Minimizing work-in-process and material handling in the facilities layout problem. *IIE Transactions*, 29, 29–36.

Fujine, M., Kaneko, T., and Okijima, J. 1993. Aluminum composites replace cast iron. *Advanced Materials and Processes*, 143(6), 34–36.

Gilmore, P.C. 1962. Optimal and sub-optimal algorithms for the quadratic assignment problem. *Journal of the Society for Industrial and Applied Mathematics*, 10, 305–313.

Harhalakis, G., Lu, T., Minis, I., and Nagi, R. 1996. A practical method for design of hybrid-type production facilities. *International Journal of Production Research*, 36(4), 897–918.

Heragu, S.S. 1992. Recent models and techniques for the layout problem. *European Journal of Operational Research*, 57(2), 136–144.

Heragu, S.S. 2006. *Facilities Design*. Universe Publishing Company, Lincoln, NE.

Heragu, S.S. and Kochhar, J.S. 1994. Material handling issues in adaptive manufacturing systems. *The Materials Handling Engineering Division 75th Anniversary Commemorative Volume*, ASME, New York.

Heragu, S.S. and Kusiak, A. 1990. Machine layout: An optimization and knowledge-based approach. *International Journal of Production Research*, 28(4), 615–635.

Ho, Y.C. and Moodie, C.L. 1998. Machine layout with a linear single-row path in an automated manufacturing system. *Journal of Manufacturing Systems*, 17(1), 1–22.

Hopp, W.J. and Spearman, M.L. 2000. *Factory Physics*, Irwin/McGraw-Hill, Burr Ridge, IL.

Houshyar, A. 1989. New solution to a special layout problem. *SME Technical Paper (Series) MS*, MS89–383.

Hsieh, L.F. and Sha, D.Y. 1996. Design process for automated guided vehicle systems: The concurrent design of machine layout and guided vehicle routes in tandam automated guided vehicle systems. *Integrated Manufacturing Systems*, 7(6), 30–38.

Huang, H. and Irani, S.A. 2000. *Hybrid Cellular Layouts for Machining and Fabrication Jobshops*, Technical Report, Department of Industrial, Systems, and Welding Engineering, Ohio State University, Columbus, OH. <http://www-iwse.eng.ohio-state.edu/ISEFaculty/irani/Research/Next Generation.htm>

Jajodia, S., Minis, I., Harhalakis, G., and Proth, J.M. 1992. CLASS-Computerized layout solutions using simulated annealing. *International Journal of Production Research*, 30(1), 95–108.

Kochhar, J.S. and Heragu, S.S. 1999. Facility layout design in a changing environment. *International Journal of Production Research*, 37(11), 2429–2446.

Koopmans, T.C. and Beckman, M. 1957. Assignment problems and the location of economic activities. *Econometrica*, 25, 53–76.

Kouvelis, P. and Chiang, W. 1992. Simulated annealing approach for single row layout problems in flexible manufacturing systems. *International Journal of Production Research*, 30(4), 717–732.

Kouvelis, P. and Chiang, W. 1996. Optimal and heuristic procedures for row layout problems in automated manufacturing systems. *Journal of the Operational Research Society*, 47(6), 803–816.

Kouvelis, P. and Kiran, A.S. 1991. Single and multiple period layout models for automated manufacturing systems. *European Journal of Operational Research*, 52(3), 300–314.

Kouvelis, P., Kurawarwala, A.A., and Gutierrez, G. 1992. Algorithms for robust single and multiple period layout planning for manufacturing systems. *European Journal of Operational Research*, 63, 287–303.

Kusiak, A. and Heragu, S.S. 1987. The facility layout problem. *European Journal of Operational Research*, 27, 229–251.

Lawler, E.L. 1963. The quadratic assignment problem. *Management Science*, 9, 586–599.

Lilly, M.T. and Driscoll, J. 1985. Simulating facility changes in manufacturing plants. *Proceedings of 1st International Conference Simulation in Engineering*, Stratford Upon Avon, UK.

Meller, R.D. and Gau, K.Y. 1996. The facility layout problem: Research and emerging trends and perspectives. *Journal of Manufacturing Systems*, 5(5), 351–366.

Meng, G. 2002. *Open Queuing Network Performance Analyzer For Reconfigurable Manufacturing Systems*. Ph.D. thesis, Department of Decision Science and Engineering System, Rensselaer Polytechnic Institute, Troy, NJ.

Meng, G. and Heragu, S.S. 2004. Batch size modeling in a multi-item, discrete manufacturing system via an open queuing network. *IIE Transactions*, 36(8), 1–11.

Meng, G., Heragu, S.S., and Zijm, W.H.M. 2004. Reconfigurable layout problem. *International Journal of Production Research*, 42, 4709–4729.

Meng, G., Heragu, S.S., and Zijm, W.H.M. 2008. *Two Level Manufacturing System Performance Analyzer*. To appear in *International Journal of Production Research*.

Montreuil, B. and Laforge, A. 1992. Dynamic layout design given a scenario tree of probable futures. *European Journal of Operational Research*, 63(2), 271–286.

Montreuil, B. and Venkatadri, U. 1991. Strategic interpolative design of dynamic manufacturing systems layouts. *Management Science*, 37(6), 682–694.

National Research Council. 1998. *Visionary Manufacturing Challenges for 2020*. National Academy Press, Washington, DC.

Ngoi Kok Ann, B. and Chua, C.K. 1994. Knowledge-based system for strip layout problem. *Computers in Industry*, 25(1), 31–44.

Rajasekharan, M., Peters, B.A., and Yang, T. 1998. Genetic algorithms for facility layout design in flexible manufacturing systems. *International Journal of Production Research*, 36(1), 95–110.

Rosenblatt, M.J. 1986. The dynamics of plant layout. *Management Science*, 32(1), 76–86.

Segal, M. and Whitt, W. 1989. A queueing network analyzer for manufacturing. *Teletraffic Science for New Cost-effective Systems, Networks and Systems*, ITC-12, 1146–1152.

Whitt, W. 1983. The queueing network analyzer. *The Bell System Technical Journal*, 62(9), 2779–2815.

Yang, T. and Peters, B.A. 1998. Flexible machine layout design for dynamic and uncertain production environments. *European Journal of Operational Research*, 108, 49–64.

Chapter 7

Staging Protocols for Unit-Load Crossdocking

John J. Bartholdi, III, Kevin R. Gue, and Keebom Kang

CONTENTS

Abstract To reduce inventory and transportation costs, many distributors use a logistics technique called "crossdocking," in which products in a warehouse move directly from inbound to outbound vehicles, without storage in between. Although products are rarely stored outright, they are often staged on the dock temporarily to facilitate value-added services, to allow efficient loading of outbound trailers, or simply to accommodate unavoidable imbalances in flow. We identify, classify, and compare different protocols for staging pallets in a crossdock, and introduce a new kind of queue, called a "staging queue," with which we model staging in the most common protocols.

7.1 Introduction

Crossdocking is a logistics technique that effectively eliminates the inventory-holding function of a warehouse although still allowing it to serve its consolidation and order-fulfillment functions. The idea is to transfer shipments directly from incoming to outgoing trailers without storage in between. Shipments typically spend less than 24 hours in a crossdock, sometimes less than an hour.

Crossdocks are essentially transshipment facilities to which trucks arrive with goods that must be sorted, consolidated with other products, and loaded onto outbound trucks. From a management perspective, crossdocking is a complex enterprise, involving extensive coordination between the distributor and its suppliers and customers (Schaffer, 1997). The crossdock must know which products are arriving in which trucks, at which times, for which customers, and, if there is a high degree of consolidation—as in the case of a personal computer (PC) distributor matching specific monitors with central processing units (CPU), for example—the crossdock must schedule trucks so as to avoid excessive congestion due to short-term storage.

If a crossdock can have some short-term storage, what differentiates it from a warehouse with a very high turnover rate? The answer is that in a crossdock the destination for an item is known before, or determined upon receipt; in a warehouse, product is stored until a customer is identified. Because the customer is known upon arrival to a crossdock, there is no need to store products as inventory.

Crossdocking is attractive in the distribution industry for two reasons. For some distributors, crossdocking is a way to reduce inventory-holding costs for stock-keeping units with stable, high demand. For others, crossdocking is a way to reduce inbound transportation costs. For example, individual outlets might receive shipments directly from vendors using less-than-truckload (LTL) or small package carriers, leading to excessive inbound transportation costs. Crossdocking is a way to consolidate those shipments to achieve truckload quantities.

7.1.1 Operations Inside a Crossdock

Crossdocks in the distribution and transportation industries take many forms. In the LTL motor-carrier industry, crossdocks are typically long, narrow facilities with truck doors around almost the entire perimeter. Freight is moved by forklifts and by workers pushing carts filled with freight. With very few exceptions, there is no automated material handling.

Crossdocks in retail distribution are typically rectangular and wider than those in the LTL industry to accommodate product staging and value-added services. These facilities may use conveyors and sortation systems, but often most of the material handling is manual. Manual handling is common in retail distribution because it is easy to adjust the capacity of operations in response to seasonal fluctuations and market dynamics. Many retail-distribution centers have small, low-volume crossdocking operations, typically in a corner of the warehouse, with trailer doors on each side, and material flow is relatively isolated from the rest of the warehouse operations. Such crossdocks are not the subject of this chapter: We are interested in high-flow is facilities that are devoted entirely to crossdocking and that require temporary staging of shipments on the dock.

One way to divide crossdocking operations is by the handling units. In "case-pick crossdocking," the distributor receives pallets of product and ships cases, or even individual items ("eaches" in industry parlance). For example, workers at one large retail crossdock we visited receive pallets of product and place them directly into flow rack modules, from which other workers pick cases and send them to the shipping area via a conveyor system. The cases are loaded directly from shipping chutes into outbound trailers, and the product resides in the warehouse for only a few hours.

"Unit-load crossdocking" is strictly pallet in, pallet out, and so may also be called "pallet crossdocking." Warehouse stores such as Costco and Sam's Club use this type of crossdocking because the retail outlets receive, and usually display, pallet quantities. At a typical unit-load crossdock, vendors call in advance to make appointments for deliveries, and the crossdock assigns a time window for the delivery. If the drivers are late, they must make another appointment, typically the following day; if they are early, they wait outside the facility until their appointed time. As soon as the trailer is unloaded, a driver pulls away the empty trailer and another full trailer pulls up. This way, doors at the crossdock are almost always occupied with trailers being unloaded.

Material flow in a unit-load crossdock is relatively simple. Each shipment begins in its arriving trailer and ends in its destination trailer. In the ideal case, workers take the pallets directly from inbound to outbound trailers, which reduces handling cost and keeps the dock clear for improved material flow. In practice, direct transfer rarely happens and pallets are staged because there is a need:

- To perform value-added processes (labeling, pricing, etc.)
- To wait for other items of an order to arrive

- To facilitate building tightly packed loads in the outbound trailer
- To load in reverse order of delivery if there will be multiple stops

Staging shipments on the dock creates problems. First, any staged shipment is handled multiple times, which adds to labor costs and increases risk of damage or loss. Second, it delays shipments in the facility, risking late delivery and consequent penalties. Delayed shipment also risks worker overtime when, as is common in many crossdocks supporting retail, all shipments must be processed everyday. Third, staged freight can create congestion and delay because there is less room for forklift drivers to maneuver. Fourth, staging requires space, which means a larger facility and associated costs.

7.1.2 Research Questions

Patterns of material flow in a crossdock are the result of several design choices, such as where incoming and outgoing trailers are parked, the arrangement of temporary staging areas, how much information is known about shipments upon arrival, and the types of material-handling equipment used (forklifts, conveyors, or gravity flow racks).

The effect of trailer placement on labor costs in a crossdock has been considered by several authors. Tsui and Chang (1990, 1992) consider layouts having all incoming trucks on one side and all outgoing trucks on the other. Crossdocks in the retail-distribution industry are often arranged in this way, so that shipments flow directly from one side of the dock to the other, which facilitates orderly staging of shipments for value-added services. Bartholdi and Gue (2000) found that trailers generating the most worker activity—incoming trailers and the highest-flow destination trailers—should be located in the center of the dock on both sides. In such a layout, freight moves in both directions across the dock and in both directions along the length of the dock, but is generally concentrated in the center region. Gue (1999) considered the effects of assigning incoming trailers to doors dynamically, based on their contents (commonly called the "spotting problem"). He found that when there is freight for relatively few (3–6) destinations per incoming trailer, there is significant benefit to adjusting the layout to account for intelligent spotting. Deshpande et al. (2007) simulated the operation of a crossdock in which incoming trailers are assigned to doors based on the destinations of shipments inside. Bartholdi and Gue (2004) investigated the best shape for crossdocks in the LTL industry. They found that small docks should be rectangular, larger docks should be T-shaped, and very large docks X-shaped.

Previous research has focused on the doors at which freight arrives and at which it departs the crossdock, but has not looked at the details of movement—staging and sorting—as freight crosses the dock. This chapter studies the details of this movement, which we call the "staging protocol." We have found different protocols in industry, which we list and categorize here. We analyze each as an engineering response to the particular information available about a shipment at the time of its arrival at the crossdock, and the particular material-handling systems on the crossdock. We identify a number of design criteria and evaluate each protocol in light of them. Our goal is to understand the advantages and disadvantages conferred by each protocol, and to identify operational environments where each might be appropriate.

In Section 7.2, we describe design criteria for a unit-load crossdocking operation, including both constraints and goals. In Section 7.3, we categorize staging protocols we have encountered in practice and describe how the subject firms executed those protocols and why. In Sections 7.4 and 7.5 we present throughput models for these protocols, including simulation and analytical models for two types of staging systems. We summarize our findings and give suggestions for design in Section 7.6.

7.2 Design Criteria

How should a crossdock organize its material flow to reduce labor cost, support value-added services, and facilitate tightly packed outbound loads? The answers to these questions depend on what information is known about each shipment upon its arrival.

Information about arriving shipments affects material flows in a simple but important way. If freight is allocated to destinations and labeled before arriving at the crossdock, it is possible, at least in principle, for workers to take freight directly from inbound to outbound trailers without intermediate staging. If the freight is not already labeled on arrival, it must be staged on the dock, where other workers provide destination labels. These two types of crossdocking are commonly called "pre- and post-distribution," respectively.

Post-distribution crossdocking incurs double handling of freight but it enables the distributor to postpone allocation until shortly after the freight arrives at the crossdock, by which time the inventory situation of individual stores may have changed. As explained later, some staging protocols are inappropriate for pre- or post-distribution crossdocking. Whichever type of crossdocking (pre- or post-distribution), managers judge the suitability of a staging protocol based on several criteria, the most important of which is generally throughput, or, equivalently, labor cost.

In addition, the staging protocol must support whatever value-added processing is required of the crossdock. It may be that products must be processed by origin (so that products in an incoming trailer receive the same service) or by destination or by some other attribute, such as product type, color, or style. Staging protocols differ in which types of value-added processing they most naturally support.

Finally, staging protocols affect the efficiencies with which trailers are loaded. The percentage volume filled of a departing trailer is called the "load factor"; and it is important to achieve a high load factor, especially for the longer routes, because this reduces the number of trips in the long run. Some staging protocols make it easier to achieve high load factors.

7.3 Staging Protocols

In the "single-stage, sort-at-shipping" protocol, workers pull pallets out of an arriving truck and put them in a lane outside the receiving door (see Figure 7.1). Workers pull pallets out of the other end and deliver them directly to the appropriate outbound trailers. This method is appropriate for post-distribution crossdocking, when pallets must be labeled by destination upon arrival.

The advantage of a sort-at-shipping protocol is that the destination of a pallet need not be known when the worker unloads the freight from the trailer. This relieves the vendor of the burden of labeling pallets before

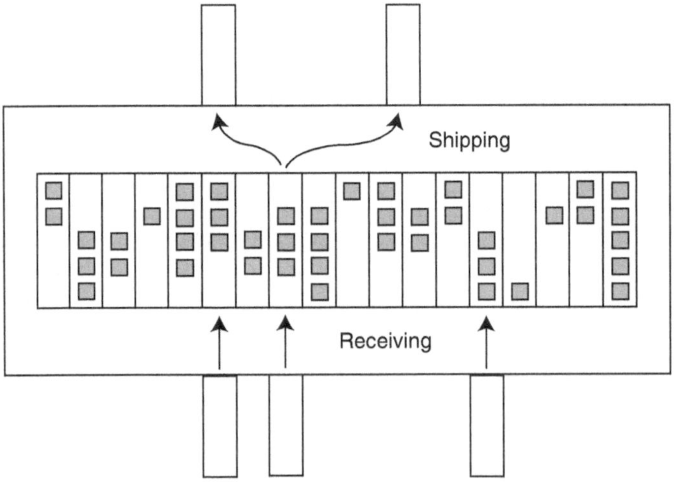

Figure 7.1 A crossdock operating a single-stage, sort-at-shipping protocol. Workers put pallets in lanes corresponding to the receiving doors. Workers in the shipping area sort pallets into appropriate shipping doors.

shipping them and allows postponement of the allocation of freight to destinations. The crossdock may print labels anytime after the contents of an inbound trailer have been sent electronically; and workers apply them after pallets are in the staging area.

We have seen this protocol used by a large retailer of home improvement products. Orders from several vendors arrived "flow loaded" by product type and not by customer (e.g., model A in the nose of the trailer, model B in the middle, model C in the tail). Products were unloaded and staged at the receiving door, where workers sorted products onto pallets by customer, and other workers delivered the pallets to shipping doors, effectively sorting at shipping.

In a "single-stage, sort-at-receiving" operation, workers take pallets directly from receiving door to the lane associated with the proper shipping door (Figure 7.2). Note that this works only if bar codes or other labels have been attached by the vendor. The advantage of staging by shipping door is that workers in shipping have a better view of what freight is available for loading, and so can achieve a tighter pack of freight while loading, thus reducing transportation costs in the long run.

We have seen this protocol at Maritime-Ontario Freight Lines, an LTL carrier in Canada, where the enormous distances make line-haul the largest component of operating cost. Accordingly, it is important to reduce the number of trailers by building tight loads. Pallets at Maritime-Ontario are staged in lanes corresponding to destination so that workers in the shipping area can "cherry-pick" pallets to tightly pack the outgoing trailers.

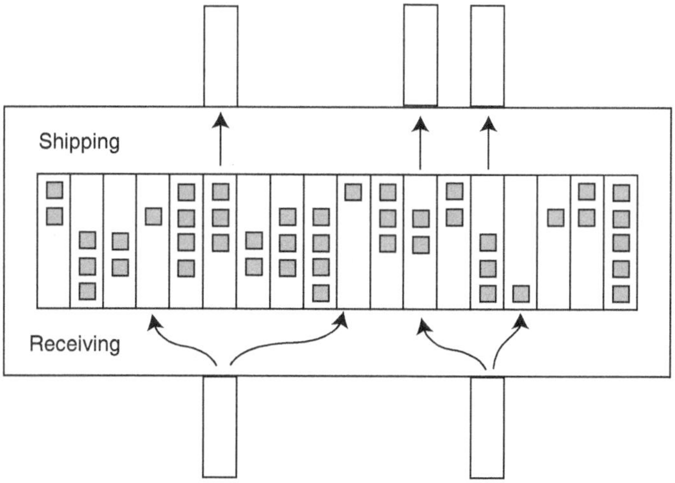

Figure 7.2 The sort-at-receiving protocol, which requires vendors to label incoming freight with its final destination.

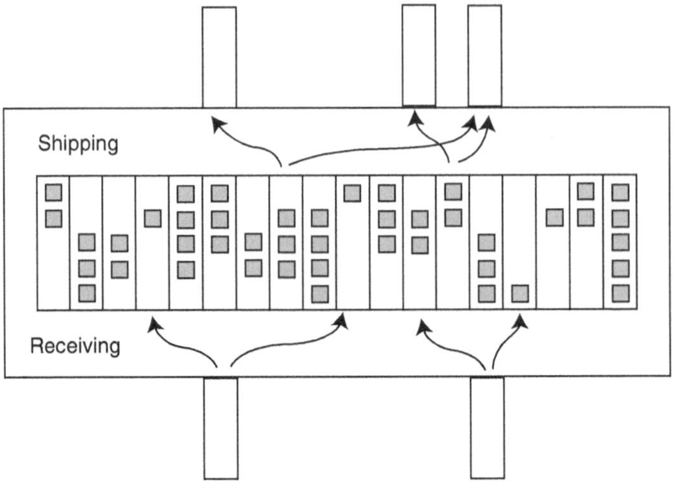

Figure 7.3 The single-stage, double-sort protocol.

Figure 7.3 shows the "single-stage, double-sort" protocol, which combines the previous protocols to allow processing according to a criterion independent of the origin or destination trailer.

We have not seen this protocol in practice, but it might be appropriate for pre-distribution operations where load factor is not a major concern and there is a need for processing according to product type.

Figure 7.4 is a two-stage protocol. "Multistage" protocols make it easier to pack trailers tightly, because workers in shipping can pick from among several pallets in shipping queues, although still allowing value-added processing by other criteria. The disadvantage is that pallets are handled multiple times. Furthermore, the crossdock must be wider to accommodate the additional queues, which increases both fixed costs and labor cost due to travel. (In Section 7.4 we consider the throughput implications of having multiple staging queues.)

This protocol was used at the Costco distribution center near Tracy, California. At the time of our last visit, the facility had negotiated pre-distribution labeling agreements only with its largest vendors. Nevertheless, they had full pre-distribution operations as a goal in hopes of reducing labor costs and cycle time for shipments on the dock. At this facility, it was especially important to reduce cycle times because outbound trailers had specific times at which they had to leave to make to their destinations on time.

We have seen other protocols as well. "Free staging" is used almost universally in the LTL industry. Workers in an LTL crossdock place pallets in the center of the dock opposite or nearly opposite the appropriate shipping

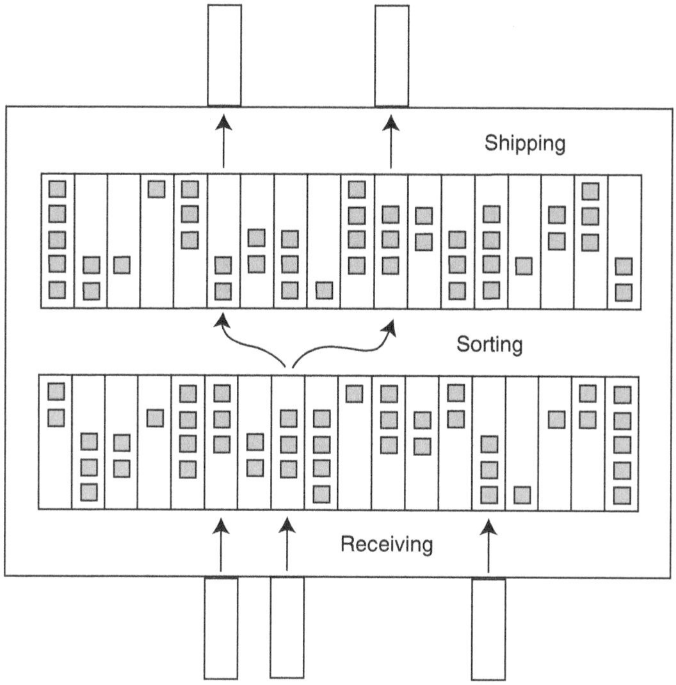

Figure 7.4 A two-stage crossdock. Workers put pallets in lanes corresponding to the receiving doors; a second team of workers sorts pallets into shipping lanes, from which a final team loads them onto outbound trailers.

door. Entry to and from the staging area is from the side toward the shipping door. LTL carriers use this method rather than a single-stage protocol because they usually handle pallets and loose pieces, and they always have to handle oddly shaped freight which is not amenable to a strictly defined staging lane. This method also has the advantage that loading directly into the shipping door, when possible, is very easy because there is a clear path to the door from the receiving area.

In a "stage-by-door" operation, workers put pallets in gaps between shipping doors. This protocol is possible only when the distance between door centers is wide enough to allow pallets between doors, and this is one of the trade-offs of using this method. This method combines the advantage of free staging—that direct loading is easy—with the advantage of sort-at-receiving—that loaders can pick loads for higher load averages. We have seen this method used at a crossdock operated by a third-party logisitics provider for Sam's Club.

Table 7.1 provides a brief visual comparison of four protocols appropriate for retail distribution. Sort-at-shipping requires little in terms of information

TABLE 7.1 Comparison of Staging Protocols

Criterion	*SaS*	*SaR*	*DS*	*MS*
Information requirements	●	○	○	●
Loading efficiency	○	●	○	●
Handling cost	●	●	◐	○
Value-added services	◐	◐	◐	●

● Indicates an advantage; ◐ a slight benefit; ○ a disadvantage.
Note: SaS (sort-at-shipping); SaR (sort-at-receiving); DS (double-sort); and MS (multiple-stage) protocols.

transfer with vendors because pallets are staged outside receiving doors, where they can be labeled after unloading. This is an advantage (●). However, this protocol makes efficient loading difficult (○), because workers must load shipments upon delivery to the shipping door—there is no opportunity to carefully select loads for a tight pack. SaS has acceptable handling cost (●) because pallets are touched only twice (receiving door-to-staging; staging-to-shipping door). Value-added services are possible (◐), but only by origin; that is, distinguishing the service by destination or another criterion is difficult.

Sort-at-receiving requires the crossdock to receive pallets with labels attached, so it has a burdensome requirement for IT coordination with vendors (○). Because pallets are staged according to their destinations, loaders are able to carefully select pallets for tight packing (●). Handling cost is similar to sort-at-shipping (●), and value-added services are possible only by destination (◐).

Double-sort crossdocking seems to combine the disadvantages of the previous two protocols, and even has a slightly higher handling cost (◐) due to the additional travel associated with double-sort. Double-sort has the advantage of allowing a distributor to perform value-added services on a basis other than origin or destination (◐).

Multiple-stage requires little information transfer with vendors (●) and provides the opportunity for high load factors on outbound trucks (●), but this comes at the cost of additional handling (○). Value-added services in a multiple-stage operation are possible both with respect to origin and destination (●).

7.4 Models

As our figures suggest, most unit-load crossdocks have staging areas where pallets are placed on the floor for temporary storage. In the protocols we

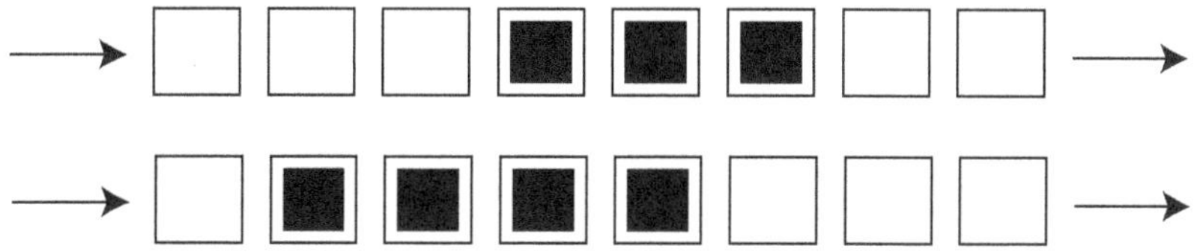

Figure 7.5 How a staging queue works. Top: Pallets occupy positions 3–5. Bottom: After two arrivals and one service, positions 4–7 are occupied.

describe, pallets are sorted into lanes, and these lanes act as finite buffers in which the pallets queue. If the last position in a lane is occupied, the lane is blocked, and this has an effect on throughput.

In an ordinary finite queue, a customer may join as long as the number of customers in the queue is less than the number of positions in the buffer. This is because customers move forward after each service, leaving room for new customers in the rear. We call this a "move-to-front queue." Queues in a unit-load crossdock with floor staging operate differently because pallets in the queue do not move forward after each service. We call this type of queue a staging queue.

Figure 7.5 shows how a staging queue works. Workers deposit pallets in the forward-most empty position in the queue. As the server pulls a pallet from the queue, the remaining pallets do not move and so no room is made for additional pallets to join the queue. (The return aisles of rental car lots are another example of staging queues, in which returning customers park in the forward-most empty space in one of the lanes. As each car is served, an attendant drives away the forward-most car in a lane, but cars in the rear do not advance because they are unoccupied.)

7.4.1 Single-Stage Model

We consider first a staging queue with a single server, which corresponds to the sort-at-shipping, sort-at-receiving, and double-sort protocols. We assume that workers deposit pallets in the forward-most empty position, and the server removes the forward-most pallet from the other side, as in Figure 7.5.

The following observations can be noted in our model:

Lemma 1 *Pallets in a staging queue must be contiguous.*

This is because pallets enter the queue from the rear and occupy the forward-most position, and only the forward-most pallet from the front may be served. Moreover, the block of pallets forms a backward propagating "wave" that either "breaks early" (meaning that it never reaches the last position and blocks the queue) or "beaches" (it eventually blocks the queue until cleared).

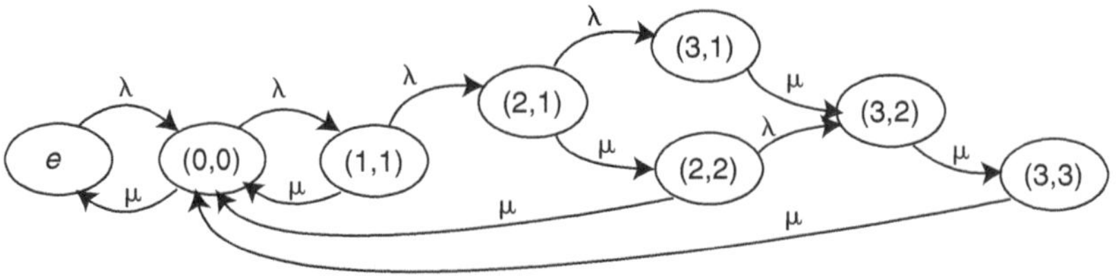

Figure 7.6 The state space diagram for a three-pallet staging queue.

To make our analytical model tractable, we assume that arrivals balk if they find the queue full. In Section 7.5 we use simulation to consider the case of blocking instead of balking.

We model the staging queue as a continuous time Markov chain. Figure 7.6 illustrates the state space for a three-position staging queue.

Formally, we say the system is in state (i, j) when the rearward-most occupied position is i and the forward-most occupied position is j (therefore $i \geq j$). If no positions are occupied and the server is busy, the system is in state (0,0); otherwise, the server is idle and the system is empty and in state e.

Let π_{ij} be the steady state probability that the system is in state (i, j). For the three-pallet case we see the following:

The transition probabilities are

$$\pi_e = \frac{\mu}{\lambda} \pi_{00}, \tag{7.1}$$

$$\pi_{11} = \frac{\lambda}{\lambda + \mu} \pi_{00}, \tag{7.2}$$

$$\pi_{21} = \frac{\lambda^2}{(\lambda + \mu)^2} \pi_{00}, \tag{7.3}$$

$$\pi_{22} = \frac{\lambda^2 \mu}{(\lambda + \mu)^3} \pi_{00}, \tag{7.4}$$

$$\pi_{31} = \frac{\lambda^3}{\mu(\lambda + \mu)^2} \pi_{00}, \tag{7.5}$$

$$\pi_{32} = \frac{2\lambda^3 \mu + \lambda^4}{\mu(\lambda + \mu)^3} \pi_{00}, \tag{7.6}$$

$$\pi_{33} = \frac{2\lambda^3 \mu + \lambda^4}{\mu(\lambda + \mu)^3} \pi_{00}, \tag{7.7}$$

where

$$\pi_{00} = \left(1 + \frac{\mu}{\lambda} + \frac{\lambda}{\lambda+\mu} + \frac{\lambda^2}{(\lambda+\mu)^2} + \frac{\lambda^2\mu}{(\lambda+\mu)^3} + \frac{\lambda^3}{\mu(\lambda+\mu)^2} + \frac{4\lambda^3\mu + 2\lambda^4}{\mu(\lambda+\mu)^3}\right)^{-1}.$$

Arrivals are served anytime the last position in the queue is not occupied, so the effective system throughput is $\lambda_{\text{eff}} = (1 - \pi_{31} - \pi_{32} - \pi_{33})\lambda$.

Notice that if we add a position to the example staging queue (giving it four positions), Equations 7.1 through 7.4 are the same; we need only to derive new equations for what were previously blocking states (π_{3j}) and the new blocking states (π_{4j}), and then recompute π_{00} to obtain the probabilities. Following are the general results.

Theorem 1 *In a staging queue with $n = 1$ position, steady state probabilities are*

$$\pi_e = \frac{\mu^2}{\mu^2 + \lambda\mu + \lambda^2},$$
$$\pi_{00} = \frac{\lambda\mu}{\mu^2 + \lambda\mu + \lambda^2},$$
$$\pi_{11} = \frac{\lambda^2}{\mu^2 + \lambda\mu + \lambda^2}.$$

Proof. Follows directly from the state diagram and some arithmetic.

Theorem 2 *In a staging queue with $n \geq 2$ positions, steady state probabilities π_{ij} are, for nonblocking states, $\pi_e = (\mu/\lambda)\pi_{00}$ and $\pi_{ij} = r_{ij}a_{ij}\pi_{00}$, where*

$$r_{ij} = r_{i-1,j} + r_{i,j-1}, \ (\textit{where } r_{ij} = 0 \textit{ for } i < j,\ r_{i0} = 0, \textit{ and } r_{i1} = 1),$$

$$a_{ij} = \left(\frac{\lambda}{\lambda+\mu}\right)^i \left(\frac{\mu}{\lambda+\mu}\right)^{j-1} \quad (\textit{for } i = 1 \ldots n-1,\ j = 1 \ldots i),$$

$$\pi_{00} = \left(1 + \mu/\lambda + \sum_{i=1}^{n-1}\sum_{j=1}^{i} r_{ij}\, a_{ij} + \lambda/\mu \sum_{k=1}^{n-1}\sum_{j=1}^{k} r_{n-1,j}\, a_{n-1,j} + \lambda/\mu \sum_{j=1}^{n-1} r_{n-1,j}\, a_{n-1,j}\right)^{-1},$$

and for blocking states

$$\pi_{n1} = (\lambda/\mu)\pi_{n-1,1},$$
$$\pi_{ni} = (\lambda/\mu)\pi_{n-1,i} + \pi_{n,i-1} \ (\textit{for } i = 2 \ldots n-1), \textit{ and}$$
$$\pi_{nn} = \pi_{n,n-1}.$$

Proof. For π_e we appeal to the example state diagram in Figure 7.6 directly. The expression for π_{00} comes from $\pi_e + \sum_{ij} \pi_{ij} = 1$ and the recursive expressions for the blocking states. For the remaining nonblocking states, we prove the result by induction. Consider the state (1,1) in the state diagram,

$$\begin{aligned}\pi_{11} &= \frac{\lambda}{\lambda+\mu}\pi_{00} \\ &= (r_{10} + r_{01})\frac{\lambda}{\lambda+\mu}\pi_{00},\end{aligned}$$

which is the result.

Now assume the result is true for $\pi_{i,j-1}$ and $\pi_{i-1,j}$. Notice from the state diagram that for all nonblocking states except state e the relationship $(\lambda + \mu)\pi_{ij} = \mu\pi_{i,j-1} + \lambda\pi_{i-1,j}$ holds, where $\{\pi_{ij} = 0\colon i < j \text{ or } j = 0\}$. Then,

$$\begin{aligned}(\lambda+\mu)\pi_{ij} &= \mu\pi_{i,j-1} + \lambda\pi_{i-1,j} \\ \pi_{ij} &= \frac{\mu}{\lambda+\mu}\pi_{i,j-1} + \frac{\lambda}{\lambda+\mu}\pi_{i-1,j} \\ &= \frac{\mu}{\lambda+\mu}\left(\frac{r_{i,j-1}\lambda^{i}\mu^{j-2}}{(\lambda+\mu)^{i+j-2}}\right)\pi_{00} + \frac{\lambda}{\lambda+\mu}\left(\frac{r_{i-1,j}\lambda^{i-1}\mu^{j-1}}{(\lambda+\mu)^{i+j-2}}\right)\pi_{00} \\ &= (r_{i-1,j} + r_{i,j-1})\frac{\lambda^{i}}{(\lambda+\mu)^{i}}\frac{\mu^{j-1}}{(\lambda+\mu)^{j-1}}\pi_{00}.\end{aligned}$$

For blocking states, we appeal directly to the example state diagram.

Corollary 1 *The effective throughput for an n-position staging queue is* $\lambda_{\text{eff}} = (1 - \sum_j \pi_{nj})\lambda$.

The model has several limitations with respect to crossdocking operations in practice. For example, crossdocks usually have two staging lanes per trailer, giving workers in receiving two queues into which they can drop a pallet. Also, workers arriving to a blocked queue take action to clear the block, such as notifying workers in the shipping area or clearing the block themselves.

We also ignore the effect on arrival and service rates of changing travel distance due to pallets moving in the queue. For our purposes, there are two types of travel in the crossdock: travel to and from the queue, and travel within the queue. Because crossdocks are typically much longer than they are wide, travel within the queue, which is at most the length of 10–15 pallets in practice, is much less than travel to and from the queue, which can be as much as several hundred feet. Thus the effect of changes in travel within the queue has a negligible effect on the arrival and service rates to the queue.

We assume inter-arrival and service times are exponentially distributed, mostly as a matter of analytical tractability; however, we also note that there are many sources of variance for worker rates, and this argues at least for a distribution with high variance. Examples of worker variance include varying travel distances, time to label or inspect pallets, difficulty loading or unloading pallets, and downtime due to breaks or equipment malfunctions.

7.4.2 Move-to-Front Queues

Some crossdocks use flow racks or inclined rollers for pallet storage. In these devices, pallets automatically roll forward after every service, just as milk in a grocery store rolls forward when a carton is removed (see Figure 7.7). How much better is a move-to-front queue than a staging queue?

We can describe the state of a move-to-front queue with a single variable describing which is the rearward-most occupied position. Let a queue containing i pallets be in state i. It is easy to show that for a move-to-front queue with n pallet positions and $\rho = \lambda/\mu$, $\pi_e = \rho^{-1}\pi_0$, $\pi_i = \rho^i\pi_0$ and $\pi_0 = (\rho^{-1} + 1 + \rho + \rho^2 + \cdots + \rho^n)^{-1}$. As before, the system is producing whenever the queue is not full, so effective system throughput is $\lambda_{\text{eff}} = (1 - \pi_n)\lambda$.

Observe that because both queues block with a single pallet:

Lemma 2 *A move-to-front queue with one position is equivalent to a staging queue with one position.*

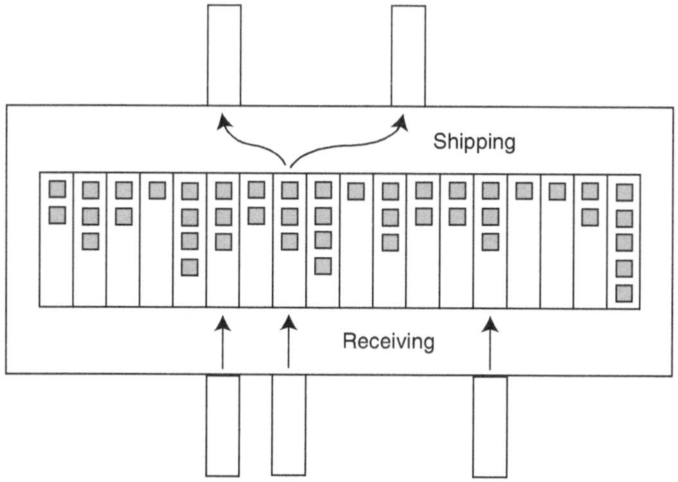

Figure 7.7 When pallets are staged in a move-to-front device such as flow rack, they roll forward to the frontward-most open positions.

Also,

Lemma 3 *For both the move-to-front and staging queues,* $\lambda_{\text{eff}} \to \lambda$ *as* $n \to \infty$.

This is true because as the buffer size goes to infinity, neither system blocks at all, and throughput is as high as possible.

Lemmata 2 and 3 suggest that for very small and very large queues, the staging and move-to-front queues behave similarly. Figure 7.8 shows the percent difference in λ_{eff} between a move-to-front queue and a staging queue for buffer sizes in between, when $\lambda = \mu$.

Remark 1 *A move-to-front queue has greater throughput than a staging queue for all buffer sizes >1 and the maximum percent difference occurs at buffer size 11.*

It is interesting to note that a 48-foot trailer is 12-pallets long, and consequently, staging areas are often about that length as well. The remark suggests that using a move-to-front storage device can increase system capacity by as much as 11 percent, but such devices also have several disadvantages. First, because it occupies floor space, a storage device obstructs material-flow patterns that could otherwise be used to "cut corners" and make workers more efficient. Second, storage devices are fairly inflexible, potentially making changes to the dock difficult. Third, storage devices prevent workers on the shipping side from selecting pallets not at the head of the queue for loading, potentially reducing load average per trailer and increasing transportation costs. Fourth, many storage devices have a high initial cost, whereas floor staging has none.

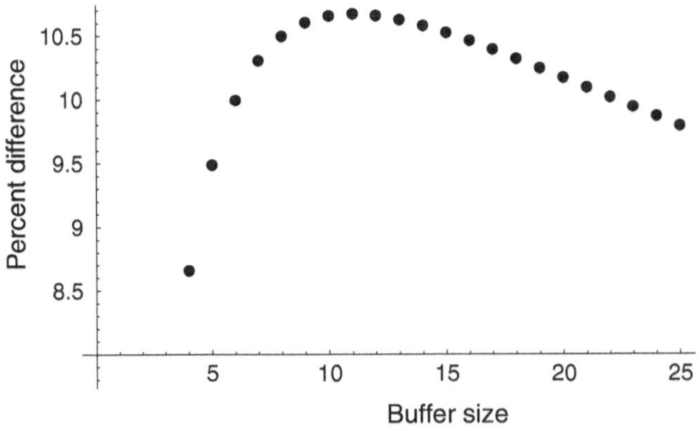

Figure 7.8 The move-to-front queue has higher throughput than a staging queue for all buffer sizes, with the greatest difference at size 11.

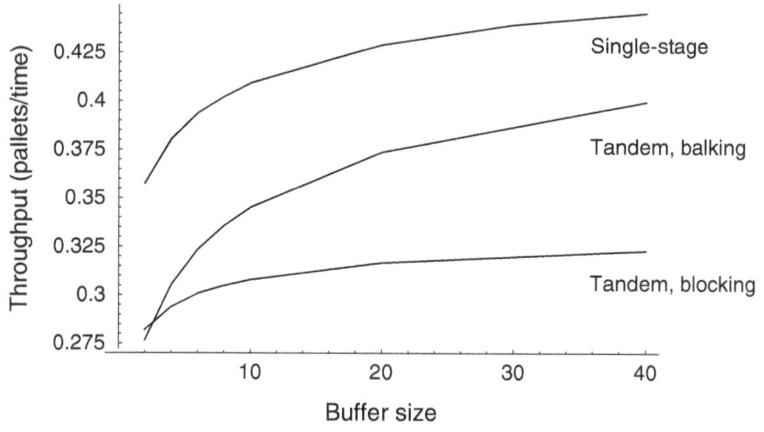

Figure 7.9 A single-stage system has significantly higher throughput than either tandem system with the same number of positions.

7.5 Two-Stage Model

A two-stage system achieves the advantages of both staging-by-receiving and staging-by-shipping, but at what cost? To gain insight, we simulated a tandem staging queue system in which departures from the first queue become arrivals to the second. We ran two scenarios: in the first, arrivals to the shipping queue balk if it is full; in the second, the server for the receiving queue is blocked until the shipping queue is cleared. In each scenario, we set $\lambda = \mu_1 = \mu_2 = 0.5$, where μ_1 and μ_2 are the mean service rates for queues 1 and 2, respectively.

Figure 7.9 compares a single queue with n positions with tandem systems (each queue having $n/2$ positions) for the blocking and balking cases.

Remark 2 *A two-stage staging system has significantly lower throughput than a single-stage system when entities block between stages.*

The implication for crossdock design is that although a two-stage system offers the dual advantages of staging-by-receiving and staging-by-shipping, these advantages come at a cost of lower throughput. In practice this would be realized with higher levels of congestion as throughput increases, or with higher labor costs.

7.6 Conclusions

There are several ways to organize staging within a unit-load crossdock. Each protocol we presented in this chapter has a distinctive set of advantages and disadvantages, which determines its suitability for the information and logistics environment faced by a distributor.

The sort-at-shipping protocol is appropriate for distributors that do not have sophisticated information coordination with their suppliers, or for those with many suppliers, where such coordination is not practical. This protocol has the advantage of single-stage queueing, and so handling costs are relatively low. Sort-at-receiving requires the distributor to receive labeled unit loads, and therefore requires a high degree of information coordination. It too has low handling costs due to single-stage queueing. The dual-sort protocol provides the distributor the ability to perform value-added services according to a criterion other than the origin or destination of the shipment. The multiple-stage protocol combines advantages of sort-at-shipping and sort-at-receiving, but at the cost of additional handling and facility costs.

We found that although staging queues block more often than move-to-front queues, such as those formed by flow rack, the difference in throughput is less than about 11 percent in the worst case. We believe this argues strongly against using move-to-front storage devices for unit-load cross-docking, especially considering the many other disadvantages, such as high initial cost, lack of flexibility, and obstruction of material flows.

Our results also suggest that multiple-stage systems, although having important operational advantages, do suffer significantly lower throughput than an equivalent single-stage system. This makes an important point about the value of information in a crossdocking logistics system. If a firm invests in the information systems and vendor relationships required for pre-distribution operations, it can take advantage of single-stage crossdocking, including higher throughput and lower labor requirements. The operations manager at a two-stage crossdock we visited stated that his firm would prefer to operate a single-stage system, were they able to establish the necessary information links with all of their vendors.

Acknowledgments

We thank the managers and engineers of Wal-Mart, Sam's Club, Costco, The Home Depot, Maritime-Ontario, and Yellow Transport for allowing us to study their crossdocking operations. We also appreciate the support of the Office of Naval Research (Bartholdi, Gue), the National Science Foundation (Bartholdi, Gue), and The Supply Chain & Logistics Institute at Georgia Tech (Bartholdi).

References

Bartholdi, J.J. and Gue, K.R. 2000. Reducing labor costs in an LTL crossdocking terminal. *Operations Research*, 48(6): 823–832.

Bartholdi, J.J. and Gue, K.R. 2004. The best shape for a crossdock. *Transportation Science*, 38(2): 235–244.

Deshpande, P.J., Yalcin, A., Zayas-Castro, J., and Herrera, L.E. 2007. Simulating less-than-truckload terminal operations. *Benchmarking: An International Journal*, 14(1): 92–101.

Gue, K.R. 1999. The effects of trailer scheduling on the layout of freight terminals. *Transportation Science*, 33(4): 419–428.

Schaffer, B. 1997. Implementing a crossdocking operation. *IIE Solutions*, 29(10): 34–36.

Tsui, L.Y. and Chang, C.-H. 1990. A microcomputer based decision support tool for assigning dock doors in freight yards. *Computers in Industrial Engineering*, 19: 309–312.

Tsui, L.Y. and Chang, C.-H. 1992. Optimal solution to a dock door assignment problem. *Computers & Industrial Engineering*, 23: 283–286.

Chapter 8

Performance Analysis of Unit-Load Automated Storage/Retrieval Systems

Yavuz A. Bozer

CONTENTS

Abstract In this chapter, we derive analytical results for unit-load automated storage/retrieval systems (AS/RS) and compare the throughput performance of two storage/retrieval (S/R) machine dispatching policies over three alternative configurations defined by the locations of the input and output points. These two are simple policies that serve the storage and retrieval requests on a first-come-first-served (FCFS) basis either across the two request queues combined or individually within each request queue. We assume the expected S/R machine cycle times are given or can be computed from the rack dimensions and S/R machine travel parameters. We also assume that the storage and retrieval requests arrive independently according to a Poisson process and that the arrival rates are given. The numeric example we present assumes randomized storage; however, the results shown in the chapter can be used with other storage policies, provided the appropriate expected S/R machine cycle times can be computed. Although the conclusions may change from one problem instance to another, examining two rack shapes and two workload levels, our numerical results suggest that while one policy performs slightly better than the other one in terms of the expected S/R machine utilization, in most cases there is no significant difference between them for practical purposes. Also, the additional workload imposed on the S/R machine due to unbalanced systems (i.e., having more storage requests per time unit than retrieval requests or vice versa) seems less than what one would have anticipated. We conclude the chapter with a brief presentation of possible research opportunities for AS/RS.

8.1 Introduction

Automated storage/retrieval systems (AS/RS) have been in use for nearly half a century. Initial AS/RS were designed for pallet in/pallet out applications, which for some reason is known as "unit-load AS/RS" instead of "pallet load AS/RS." As AS/RS technology evolved, however, new types of systems based on alternative unit loads were introduced. Such systems include, among others, the mini-load AS/RS based on captive trays designed to hold small to medium-sized parts, micro-load AS/RS based on tote boxes, and person-onboard AS/RS where the operator travels into the aisle onboard the storage/retrieval (S/R) machine. The above-mentioned systems are typically used either as an S/R system, that is, unit-load in/unit-load out, or as an order-picking system, that is, unit-load in/less-than-unit-load out. For a detailed exposition on equipment types and applications, including photographs, the interested reader may refer to Tompkins et al. (2003). According to Bastian Material Handling (BMH),

> [The] first AS/RS unit load systems were built in the late 1960's, and were heavy, slow, and very complicated to service. Today's technology is engineered with reliability, speed, and cost effectiveness as primary design criteria. The equipment

> regularly operates in 24/7 applications with unparalleled performance when compared with more manual high-density storage methods.

Indeed, AS/RS benefited considerably from major advances in computer technology, integrated circuits, sensors, induction motors, and control systems software/hardware. Today, we find many AS/RS being used not only in manufacturing, distribution, and warehousing but also in more specialized applications such as parking structures, pharmaceuticals, medical supplies, hazardous items storage, and cold-temperature or refrigerated environments.

Daifuku Co. Ltd. installed Japan's first AS/RS in 1966. Subsequently, the company developed the first computer-controlled AS/RS in 1969 and the first preengineered AS/RS package in 1972. Murata Machinery Ltd., on the other hand, developed Japan's first AS/RS for hazardous items in 1971, and Japan's first AS/RS for freezing temperatures (−40°C) in 1973. The company also installed Japan's first "super-high-rise" AS/RS, which stands 50 m (or 166.67 ft) tall.

In the United States, HK Systems (formed in 1995 when Harnischfeger acquired Eaton-Kenway) installed the first AS/RS in 1969. More recently, HK Systems completed the installation of one of the largest AS/RS in the United States for the Stop & Shop distribution center in Freetown, Massachusetts. The system, designed to supply over 300 stores and operate 24 × 7, is based on 77 S/R machines serving over 11,500 slots located in 90 aisles. According to one source, "You could put a glass of wine on top of a pallet and it wouldn't spill.... The (S/R machines) move very smoothly" (HK Systems, 2004). Other smaller-scale AS/RS installed for use in a variety of manufacturing and distribution settings are well documented in various industrial trade magazines.

AS/RS are also used in hospitals and on university campuses, primarily in libraries. One of the first contemporary (mini-load) AS/RS was installed in 1991 at the California State University, Northridge Library, which has over 1.2 million volumes. The items in the AS/RS are stored in 13,260 bins, each measuring 2 ft × 4 ft in a rack structure installed in an 8,000 sq. ft room with a 40 ft high ceiling. It is reported that at the time of the Northridge earthquake on January 17, 1994, almost all of the library's open-shelf collection fell on the floor but none of the books in the AS/RS was damaged, and the bins remained securely in the rack (California State University). An excellent report outlining user experiences with, and design considerations for, AS/RS installed in libraries is presented in Kebabian et al. (2006).

The benefits of AS/RS have been documented in a number of publications. For example, according to BMH, AS/RS benefits include:

- Bringing material to the operator and cutting cycle time by eliminating wait, walk, and search time.
- Reduces work-in-progress inventory. There is better inventory accuracy and better responsiveness, resulting in reduction or elimination of "safety stock" in the overall inventory model. This has the net effect of inventory reduction.
- Dramatically increases operator productivity. The "Part-to-Picker" model of order fulfillment is three to five times more productive than having the picker travel to the part to complete the fulfillment.
- Provides real-time inventory control with instant reports. With near 100 percent accuracy and real-time information about the inventory on hand, achievable commitments can be made to the customer—as opposed to "best efforts promises."
- Improves product quality and productivity. Real-time information, faster response to a need, physical protection, and traceability of material access—all contribute to a better process where time can be spent on improving the quality of the process instead of on expediting material to a point of use.

In ASAP Automation Web site, the following benefits of AS/RS are listed:

- Dramatic improvements in operator efficiency and storage capacity
- Reduction of work-in-progress inventory
- Improvements in quality and just-in-time performance
- Provides make-to-order capability in addition to make-to-inventory production
- Real-time inventory control and instantaneous reporting functionality

Although these benefits are generally well understood and agreed upon in the material-handling community, cost justification of AS/RS is still a complicated matter because the initial capital investment required is substantial, and certain manual systems—coupled with computerized inventory control packages and automatic identification (autoID) devices—compete to provide at least some of the same type of benefits. An excellent comparison of capital and operating costs between an AS/RS and a very-narrow-aisle (VNA) system with operator-driven industrial trucks is presented in Zollinger (1999). The author reminds us that AS/RS have "more subtle advantages" such as:

- More capability than standard inventory control
- The S/R machine does not take breaks
- Reduced training time
- Higher inventory security
- Less product damage

Perhaps quantifying the less obvious or hard to capture advantages, and ensuring that they are included in the cost–benefit analysis, continues to be a challenge for AS/RS or most other automated material-handling systems for that matter. Another challenge is the real or perceived lack of flexibility in the (throughput) capacity of the system. Although the capacity of an AS/RS can be varied by varying the number of operational hours in a day, once a system is installed with a fixed number of aisles and fixed number of S/R machines, varying the capacity by varying the number of truck operators in the system (as one would do with a manual system) is no longer an option. Of course, one may vary the number of active and inactive aisles in an AS/RS but an inactive aisle is idle capacity that comes at a high cost, especially when aisle-captive S/R machines are used.

AS/RS have also been the subject of some criticism in the Lean Manufacturing community. For example, in Baudin (2004), it is asserted that AS/RS "...are far from common in manufacturing warehouses, and, where found, their users express buyer's remorse more frequently than enthusiasm. We have yet to encounter a case where an AS/RS was installed as part of lean manufacturing implementation." The author traces user dissatisfaction back to the acquisition process and budgeting tactics that often result in an AS/RS being designed "*before* the organization has an opportunity to understand and specify its requirements...." Furthermore, he observes the following general problems for AS/RS:

- Lack of visibility: AS/RS have been denounced as black holes where unnecessary inventory disappears from view.
- Lack of flexibility: The operating policies that can be used with an AS/RS are limited to what its control software will support.
- Impact on manual storage and retrieval operations: An AS/RS can spoil its users into not using the normal visible controls in the manual stores they still need to maintain.
- Focus of attention: In plants that have an AS/RS, its effective use becomes the focus of all debates about materials management.

These concerns should serve as "fair warning" to prospective AS/RS users in manufacturing. In fact, some of them are consistent with our own experiences. For example, in one major furniture manufacturing plant we are familiar with, there was a large and dated AS/RS installed in "the middle of" the plant and it had become a "dumping ground" for excess inventory. As part of their Lean Transformation effort, management decided to phase out the AS/RS. On the other hand, however, there are numerous examples of AS/RS being used very successfully to support a pull-based manufacturing system (see, e.g., BMW, 2004). Whether such applications or plants fully meet the definition or standards of Lean Manufacturing is another matter that is open to debate.

8.2 Literature Search

In this section, we present a very brief review of papers published on *unit load* AS/RS. With some exceptions, we include only those papers published approximately since the mid-1990s. Otherwise, the literature on AS/RS is extensive; in fact, in the material-handling research literature, the number of papers published on AS/RS is perhaps second only to the large number of papers published on automated guided vehicle (AGV) systems.

Although a single scheme would not be adequate to classify or categorize all the papers published on AS/RS, most papers address either design issues or operational issues. Papers concerned with design issues often focus on determining the ideal or most appropriate system size (such as number of storage slots, number of aisles, and rack dimensions), while papers concerned with operational issues often focus on optimizing or predicting the performance of a given design. Of course, there are a number of papers that explore the relationship between system design parameters and performance metrics. Furthermore, whether it is a design paper or a performance-focused paper, many papers use either simulation modeling or analytical techniques, with a few papers making use of both methods.

A basic starting point in studying the performance of AS/RS is the expected S/R machine cycle time for single command and dual command trips when the rack dimensions and the S/R machine travel parameters are specified. Following the expected S/R machine cycle times (published in Bozer and White, 1984) for randomized storage and rectangular storage racks with arbitrary dimensions, a number of papers were published to extend the expected cycle times to account for S/R machine operating characteristics (including acceleration and deceleration) and alternative storage policies such as 2-class storage (see, e.g., Hwang and Lee, 1990; Chang et al., 1995; Park, 2006). The impact of the configuration of the rack on the speed profile of the S/R machine was examined in Chang and Wen (1997).

For given system parameters, including the arrival rate of storage and retrieval requests as well as the S/R machine cycle times, analytic *performance evaluation* models are presented in Bozer and Cho (2005), Eldemir et al. (2003), Hur and Nam (2006), and Lee (1997). Such models are typically concerned with the throughput capacity of the system and the state of the S/R queues under certain operational policies specified for the S/R machine. We will return to these papers in Section 8.3 as they form the basis of this chapter. A performance evaluation model, based on a system of state equations, for S/R machines with twin shuttles is presented in Malmborg (2000).

A number of *performance optimization* procedures or algorithms for the S/R machine have also been published. Such papers generally fall into two categories—those that attempt to optimize the operation of the S/R machine while it is busy, and those that attempt to optimize the position of the S/R machine when it becomes idle (also known as the "dwell point" of the S/R machine). One of the early papers that falls into the first category is Han et al. (1987), where the interleave time, that is, the S/R machine travel time between two openings in the rack, is reduced by matching the storage requests with the retrieval requests. A similar study is presented in van den Berg and Gademann (1999) for dedicated storage and a static "block" of storage and retrieval requests. The block consists of a set of storage and retrieval requests specified a priori; as the S/R machine serves each request in the block, new storage or retrieval request arrivals are *not* added to the block. In Yin and Rau (2006), for a class-based AS/RS, the authors use genetic algorithms with simulation to show that the performance of the system can be improved if various sequencing rules are used on a dynamic basis instead of using a single rule throughout the operation of the system.

Papers in the second category are concerned with optimizing the "parking" position of an idle S/R machine, also known as the dwell point of the S/R machine; they include Chang and Egbelu (1997a,b), Egbelu and Wu (1993), Hwang and Lim (1993), Park (1999), Peters et al. (1996), and van den Berg (2002). The dwell point of the S/R machine has an impact only on those storage or retrieval requests that arrive while the S/R machine is idle. As S/R machines and AS/RS, in general, are capital-intensive investments, it would be unusual to have an expected S/R machine utilization less than 80 percent or so. (Of course, the utilization of the S/R machine will fluctuate with the workload but a properly engineered system should not result in significant S/R machine idle time.) As such, despite the number of studies conducted on the subject, the dwell point of the S/R machine is not likely to have a significant impact on the overall performance of the system. Indeed, in Meller and Mungwattana (2005), after comparing several dwell point strategies, the authors conclude that:

> [T]he dwell-point strategy has an insignificant impact on the relative system response time when the system is highly utilized (i.e., the percentage reduction is typically <2% and no more than 5% for highly utilized systems) and has an insignificant impact on the absolute response time (i.e., <10 secs) for typical systems of any utilization level.

Some of the *performance optimization* studies focus on the storage method instead of the S/R machine and compare storage policies such as randomized storage with dedicated (turnover-based) storage, class-based storage, and shared storage. The interested reader may refer to Kulturel et al. (1999), Thonemann and Brandeau (1998), and van den Berg and

Gademann (2000) for the details. In Hsieh and Tsai (2001), the authors investigate a storage policy based on bill-of-material (BOM) information, and argue that developing a storage policy based on the needs of manufacturing operations "can increase not only the performance of the [AS/RS] but also the performance of the manufacturing system."

There are also a number of papers in the literature that are focused primarily on AS/RS design. Some of these design-focused papers are concerned with the S/R machine. For example, a large majority of the unit-load AS/RS papers assume a single-shuttle S/R machine, which implies that the S/R machine can move only one unit load at a time. However, with the introduction of twin-shuttle or multi-shuttle S/R machines, which can move multiple unit loads at a time, a number of papers have focused on system performance and expected cycle times for such S/R machines (see, e.g., Keserla and Peters, 1994; Meller and Mungwattana, 1997; Malmborg, 2001a; Potrc et al., 2004a,b). Not surprisingly, those who compared single-shuttle versus multi-shuttle S/R machines have generally found that the latter type of S/R machine leads to better system performance—provided appropriate heuristics are used to control the sequence of operations performed by the S/R machine. In addition, a travel time model for a new type of S/R machine with one vertical platform and N horizontal platforms (to serve N tiers of the rack) is presented in Hu et al. (2005).

A genetic algorithm-based design procedure to minimize the total AS/RS cost is presented in Lerher and Potrc (2006). "Rule of thumb" heuristics to configure the storage racks are presented in Malmborg (2001b). Models to estimate cycle times (under dedicated storage) and storage space requirements (under randomized and class-based storage) are presented in Eldemir et al. (2004).

A few papers deal with specialized applications such as low-temperature storage, which is an area of interest for most types of perishable-goods storage ranging from laboratory specimens to food (see, e.g., Hwang et al., 1999; Felder, 2003). In Hwang et al. (1999), the authors present a nonlinear, mixed-integer-programming-based design model to minimize the cost of a refrigerated AS/RS.

As such configurations are quite common in industry, most of the studies in the literature assume aisle-captive S/R machines, that is, each S/R machine is dedicated to a specific aisle and, thus, the number of S/R machines in the system is equal to the number of aisles. However, in some cases, the activity level in each aisle may vary considerably (due to seasonality or other reasons) and it may make more sense to use a fewer number of S/R machines that are transferred (via special transfer cars) from one aisle to another as the need arises. Models to study such systems are presented in Hwang and Ko (1988) and Lerher et al. (2006) for a single S/R machine serving multiple aisles.

8.3 Performance Evaluation

In this section, we present a few simple analytic models to evaluate the performance of an S/R machine that operates under one of two basic service policies. The storage and retrieval requests served by the S/R machine are assumed to arrive randomly and independently according to a Poisson process. We consider alternative input/output (I/O) point configurations for the storage rack and their impact on the performance of the S/R machine. We assume randomized storage, although the models we show can also be used with alternative storage policies provided the appropriate values for the expected S/R machine cycle times are provided.

We also assume the S/R machine has a single shuttle which handles one unit load at a time. Because our focus is unit-load AS/RS, our primary concern from a performance evaluation viewpoint is the throughput capacity of the system. The workload is expressed in the number of storage and retrieval requests that arrive per time unit; each request represents a unit load. For brevity, in the remainder of the chapter we will refer to the S/R machine simply as the "S/R." Also, we assume the reader is already familiar with AS/RS basics such as single command and dual command cycles. Otherwise, the reader may refer to Bozer and White (1984), among others.

Suppose the following system parameters and expected S/R cycle times are specified:

L = Rack length (horizontal)
H = Rack height (vertical)
v_h = Horizontal velocity of S/R
v_v = Vertical velocity of S/R
K = (Constant) total load handling time = load pick-up + load deposit time
SC = $E(SC)$ = Expected single command S/R travel time
TB = $E(TB)$ = Expected S/R travel time between two randomly selected points
DC = $E(DC)$ = Expected dual command S/R travel time = $SC + TB$, by definition

As shown in Figure 8.1, we consider three configurations (i.e., Config-A, Config-B, and Config-C) based on the location of the I/O point. Config-A, with the I/O point located at the lower left-hand corner of the rack, is perhaps the most common configuration in practice; it is also one of the configurations studied most often in the literature. For a rack with arbitrary dimensions and user-specified horizontal and vertical S/R travel speeds, both the values of SC and TB can be obtained for Config-A using the results shown in Bozer and White (1984) and subsequent publications, which refined the

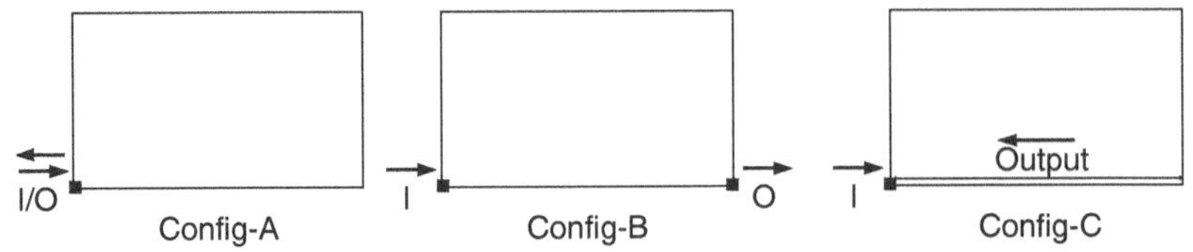

Figure 8.1 Alternative configurations based on the locations of input and output points.

expected S/R travel times due to S/R acceleration and deceleration. Once *SC* and *TB* are derived, it is straightforward to obtain *DC* as $DC = SC + TB$. Note that *TB* itself does not depend on the configuration; that is, once *TB* is determined for a given rack, it applies to any configuration because its value does not depend on the location of the input or output point.

Also note that, due to symmetry, the value obtained for *SC* in Config-A or Config-B is valid as long as the I or the O point is located at any corner of the rack. In fact, using this observation, it is straightforward to study variations of the above configurations by partitioning the rack into smaller sections as long as the I or the O point is located at one corner of each section. For example, in Config-A, if the I/O point is raised vertically by y ft, the rack can be divided into two sections; the lower section would be $y' \times L'$ with the I/O point located at its upper left-hand corner, while the upper section would be $(H - y)' \times L'$ with the I/O point located at its lower left-hand corner. Once *SC* is determined for each section individually, it is straightforward to determine *SC* for the entire rack by taking a weighted combination of the two *SC* values, where the weights are based on the areas of each section.

Config-B is similar to Config-A except, as shown in Figure 8.1, the loads that are retrieved by the S/R are dropped off at the opposite end of the rack. Such a configuration may work better depending on the desired macroflow in the facility. Config-C, on the other hand, is somewhat different because the input point is still located at the lower left-hand corner of the rack but the output point is a conveyor line that runs along the bottom of the rack. The conveyor is installed through the rack openings and runs parallel to the aisle. (The number of openings lost due to the conveyor line is generally very small compared to the total number of openings in the rack.) Where the S/R deposits the load depends on the application but if the conveyor runs only a partial distance into the rack, the S/R often deposits the load at the end point of the conveyor. For Config-C, we assume that the conveyor runs through the entire length of the rack. Unless the queue of loads that have been retrieved by the S/R and deposited on the conveyor is unusually long for some reason, the S/R is likely to be able to deposit the load at any point on the conveyor. To minimize travel time, we assume that when the

S/R picks up a load from the rack, it vertically travels down directly to the conveyor to deposit the load. Of course, Config-C can be modified by running an *input* conveyor through the rack openings on the opposite side of the aisle. The analysis we show here is straightforward to extend to such a configuration with dual conveyors.

Once the appropriate parameters (L, H, v_h, and v_v) are specified, SC and TB can be computed for Config-A and Config-B using past results as we remarked earlier. For Config-C, we can again use past results to compute the expected S/R travel time from the input point to a random point in the rack. However, if the S/R has just deposited a load on the output conveyor, and the next request to be served is another retrieval, the S/R will travel from the deposit point on the conveyor (which is a random point on the conveyor due to the previous retrieval) to a random point in the rack.

Assuming the rack has already been normalized into a $b \times 1$ rack with a scaling factor of T (see Bozer and White, 1984 for details), suppose the random point in the rack and the random point on the output conveyor have coordinate values (x_1, y_1) and (x_2, y_2), respectively. Let t be the travel time from (x_2, y_2) to (x_1, y_1). Following the approach shown in Bozer and White (1984), the cumulative distribution function (cdf) for the travel time is given by the following expression:

$$F(z) = \Pr(t \leq z) = \Pr(\mathrm{Max}[|x_1 - x_2|, |y_1 - y_2|] \leq z). \tag{8.1}$$

Because the two points are sampled independently, this probability can be rewritten as follows:

$$F(z) = \Pr(|x_1 - x_2| \leq z) \cdot \Pr(|y_1 - y_2| \leq z) = G(z) \cdot H(z). \tag{8.2}$$

Both x_1 and x_2 are uniformly distributed between 0 and 1. Hence, as shown in Bozer and White (1984):

$$G(z) = \begin{cases} 0, & z < 0, \\ 2z - z^2, & 0 \leq z \leq 1, \\ 1, & 1 < z < \infty. \end{cases} \tag{8.3}$$

In the y-direction, $y_1 \sim U(0, b)$ while $y_2 = 0$. Therefore, $H(z)$ is simply the cdf of a uniform distribution between 0 and b. Hence, $\Pr(t \leq z)$ is given by the following expression:

$$F(z) = \begin{cases} 0, & z < 0, \\ (2z - z^2)(z/b), & 0 \leq z \leq b, \\ 2z - z^2, & b \leq z \leq 1, \\ 1, & 1 < z < \infty. \end{cases} \tag{8.4}$$

Taking the derivate of $F(z)$ to obtain $f(z)$, and using $E(z) = \int zf(z)\,dz$, we obtain

$$TB' = E(TB') = E(z)T = [(1/3) + (b^2/3) - (b^3/12)]\ T. \qquad (8.5)$$

We will use this expression later when we analyze Config-C.

The performance of the AS/RS depends not only on the configuration of the I/O point(s) but also on the control policy used for dispatching the S/R; that is, the policy to determine which (storage or retrieval) request is served by the S/R and when. Perhaps the most common policy that is also relatively straightforward to model is the first-come-first-served (FCFS) policy. Under the FCFS policy, say, Policy I, each request is served strictly on a first-come-first-served basis regardless of the type of request (storage versus retrieval) and the current position of the S/R. Following the completion of a storage (or retrieval), if both request queues are empty, the S/R becomes idle at the rack (point O). Obviously, there is some loss of efficiency under the FCFS policy because a certain amount of unnecessary empty S/R travel occurs.

An alternative policy, say, Policy II, is described formally in Bozer and Cho (2005). Under this policy, following the completion of a storage, the S/R first checks the retrieval queue. If one or more retrieval requests are found, the S/R serves the oldest retrieval request. Otherwise, it checks for a possible storage request. If none are found, it becomes idle in the rack; if one or more are found, it travels to the input point and serves the oldest storage request. On the other hand, if the S/R has just retrieved a load (which means it is at point O), it first checks for a possible storage request. If one or more are found, the S/R serves the oldest storage request. Otherwise, it checks for a possible retrieval request. If none are found, it becomes idle at point O; if one or more are found, it travels to the rack opening that contains the oldest retrieval request. Note that under Policy II, within each queue, a storage or retrieval request is served on an FCFS basis but when both request types are considered, FCFS is relaxed to potentially reduce S/R travel time.

Although more elaborate policies are possible, and some have been investigated in the literature (see, e.g., Han et al., 1987; Yin and Rau, 2006 as well as the papers we cited earlier for dwell point strategies), such policies almost always require the use of simulation to assess their impact. Also, the improvement obtained by more elaborate policies depends on the I/O configuration, the utilization of the S/R, and the number of storage versus retrieval requests received per hour. To keep the chapter focused on analytic modeling, we will use simple policies and evaluate Policy I on all three I/O configurations, while we evaluate Policy II on Config-A and Config-B. Evaluation of Policy II on Config-C can be performed as future work.

For each one of these scenarios, the storage and retrieval requests are assumed to arrive at each aisle according to an independent Poisson process at a mean rate of λ_s storages/hour and λ_r retrievals hour, yielding a total rate (λ_T) of $\lambda_s + \lambda_r$ requests/hour. In steady state, one would of course expect to have $\lambda_s = \lambda_r$ but there are many instances where the system may perform more retrievals in the morning (or first shift), for example, and more storages in the afternoon (or second shift) or vice versa. Hence, we assume that λ_s is not necessarily equal to λ_r, which also helps keep the results general because $\lambda_s = \lambda_r$ is only a special case from an analytic perspective. For each scenario, we also assume that, given the appropriate parameter values—L, H, v_h, v_v, and acceleration/deceleration data if necessary—the values of SC, TB, TB', and DC have already been determined.

Consider first Config-A under Policy I (FCFS). For this scenario, the stability condition for the S/R can be derived using the approach shown in Chow (1986). Note that there are four possible types of service trips the S/R may perform. Given that it is at the I/O point, the S/R may next perform a storage or a retrieval. Likewise, given that it is at a random point in the rack, the S/R may next perform a storage or a retrieval. Under Policy I, the probability of serving one type of request or another is independent of the current location of the S/R. Therefore, the probability for each type of service trip and the corresponding S/R travel time are given as follows:

	Probability	*S/R Service Time*
S/R at I/O, storage next	$(\lambda_r/\lambda_T)(\lambda_s/\lambda_T)$	$(SC/2) + K$
S/R at I/O, retrieval next	$(\lambda_r/\lambda_T)(\lambda_r/\lambda_T)$	$(SC) + K$
S/R at rack, storage next	$(\lambda_s/\lambda_T)(\lambda_s/\lambda_T)$	$(SC) + K$
S/R at rack, retrieval next	$(\lambda_s/\lambda_T)(\lambda_r/\lambda_T)$	$TB + (SC/2) + K$

where the first term under "probability" is the probability that the S/R is at the given location, and the second term is the probability that a storage or retrieval request is served next.

Given this information, the expected S/R service time per request, $E(S)$, is obtained as

$$E(S) = \frac{\lambda_s\lambda_r}{\lambda_T^2}DC + \left(\frac{\lambda_s^2 + \lambda_r^2}{\lambda_T^2}\right)SC + K. \tag{8.6}$$

In order for the S/R to meet throughput, we must have $\lambda_T \cdot E(S) < 1$; that is,

$$\frac{\lambda_s\lambda_r}{\lambda_T}DC + \left(\frac{\lambda_s^2 + \lambda_r^2}{\lambda_T}\right)SC + \lambda_T K < 1, \tag{8.7}$$

where the third term, $\lambda_T K$, is due to the simple fact that each storage and retrieval request must be picked up once and deposited once by the S/R. (Recall that $\lambda_T = \lambda_s + \lambda_r$.)

Consider next Config-B and Policy I (FCFS). For this scenario, one would expect the S/R to have to "work harder" because, unlike Config-A, the S/R must cover an additional travel distance from point O to point I when it serves a storage request immediately after a retrieval. Suppose the travel time from point O to point I is equal to α, that is, $\alpha = I/v_b$. The stability condition for the S/R can be derived using an approach similar to the one we used for Config-A by updating the service times as follows:

	Probability	*S/R Service Time*
S/R at I/O, storage next	$(\lambda_r/\lambda_T)(\lambda_s/\lambda_T)$	$\alpha + (SC/2) + K$
S/R at I/O, retrieval next	$(\lambda_r/\lambda_T)(\lambda_r/\lambda_T)$	$(SC) + K$
S/R at rack, storage next	$(\lambda_s/\lambda_T)(\lambda_s/\lambda_T)$	$(SC) + K$
S/R at rack, retrieval next	$(\lambda_s/\lambda_T)(\lambda_r/\lambda_T)$	$TB + (SC/2) + K$

Using Policy I in Config-B, the S/R will never become idle at point I because no loads are delivered to point I. Hence, when we say "S/R at I/O," it really means the S/R has just completed a retrieval and physically it is at point O.

Given the new service time for the first case, the expected S/R service time per request, $E(S)$, is obtained as

$$E(S) = \frac{\lambda_s \lambda_r}{\lambda_T^2} DC + \left(\frac{\lambda_s^2 + \lambda_r^2}{\lambda_T^2} \right) SC + \frac{\lambda_s \lambda_r}{\lambda_T^2} \alpha + K, \tag{8.8}$$

which implies that the S/R meets throughput if

$$\frac{\lambda_s \lambda_r}{\lambda_T} DC + \left(\frac{\lambda_s^2 + \lambda_r^2}{\lambda_T} \right) SC + \frac{\lambda_s \lambda_r}{\lambda_T} \alpha + \lambda_T K < 1. \tag{8.9}$$

The additional workload imposed on the S/R due to the separation of the output point from the input point is reflected in the third term that is $(\lambda_s \lambda_r / \lambda_T)\alpha$. In fact, the additional workload imposed on the S/R is maximized when the system is balanced, that is, $\lambda_s = \lambda_r$. As expected, if α is set equal to zero, the left-hand side (LHS) of Equation 8.9 becomes identical to the LHS of Equation 8.7. If $\alpha > 0$ but either one of the λ values is equal to zero, the third term in Equation 8.9 plays no role because the S/R would perform 100 percent single-command cycles and only one of the points (I or O) would be used depending on which λ value is nonzero.

The last configuration we consider for Policy I is Config-C. The analysis is very similar to the previous two configurations except we need the expected S/R travel time between a random point on the output conveyor and a random point in the rack, which we derived earlier as TB'. The S/R service times for Config-C are as follows:

	Probability	*S/R Service Time*
S/R at I/O, storage next	$(\lambda_r/\lambda_T)(\lambda_s/\lambda_T)$	$(\alpha/2) + (SC/2) + K$
S/R at I/O, retrieval next	$(\lambda_r/\lambda_T)(\lambda_r/\lambda_T)$	$TB' + \beta + K$
S/R at rack, storage next	$(\lambda_s/\lambda_T)(\lambda_s/\lambda_T)$	$(SC) + K$
S/R at rack, retrieval next	$(\lambda_s/\lambda_T)(\lambda_r/\lambda_T)$	$TB + \beta + K$

where $\beta = H/(2v_v)$ and $(\alpha/2) = L/(2v_h)$. Also, when we say S/R at I/O, it really means the S/R has just completed a retrieval and physically it is at a random point on the output conveyor.

Given these revised service times, the expected S/R service time per request, $E(S)$, is obtained as

$$E(S) = \frac{\lambda_s\lambda_r}{\lambda_T^2}\left(\frac{\alpha}{2} + \frac{SC}{2}\right) + \frac{\lambda_r^2}{\lambda_T^2}(TB' + \beta) + \frac{\lambda_s^2}{\lambda_T^2}(SC) + \frac{\lambda_s\lambda_r}{\lambda_T^2}(TB + \beta) + K, \quad (8.10)$$

which implies that the S/R meets throughput if

$$\frac{\lambda_s\lambda_r}{\lambda_T}\left(\frac{\alpha}{2} + \frac{SC}{2} + TB + \beta\right) + \frac{\lambda_r^2}{\lambda_T}(TB' + \beta) + \frac{\lambda_s^2}{\lambda_T}(SC) + \lambda_T K < 1. \quad (8.11)$$

Consider next Policy II, which is, generally speaking, more difficult to analyze than Policy I. However, we are interested in the performance of Policy II because it is likely to reduce the workload on the S/R depending on the I/O configuration. The throughput performance of the S/R under Poisson arrival of storage and retrieval requests has been investigated in Bozer and Cho (2005), Hur and Nam (2006), and Lee (1997). In Lee (1997), the queue spaces for storage and retrieval requests are assumed to be finite. Also, single-command and dual-command cycles are assumed to be exponentially distributed, which leads to estimation errors as both cycle times have coefficient of variation (cv) values significantly smaller than 1.0. In Hur and Nam (2006), the storage (or retrieval) queue is assumed to hold only two loads (requests). Although the authors claim that such an assumption makes the model more general, theoretically, storage loads in an AS/RS are not “lost” if a buffer is full, and the retrieval queue is an electronic queue which can hold as many requests as necessary. In fact, with finite queue spaces, the key question concerning whether or not the S/R machine meets throughput goes unanswered. Rather, the performance of the S/R is assessed only indirectly by computing the storage and retrieval requests that are lost due to a full queue.

In Bozer and Cho (2005), the queue space is assumed to be infinite, that is, no storage or retrieval requests are lost. Given the appropriate values for *SC*, *TB*, and *K*, the authors show that, for Config-A under Policy II, as long as $SC > TB$, the S/R meets throughput if the following two inequalities are satisfied:

$$\lambda_s SC + \lambda_r TB + \lambda_T K < 1 \text{ and} \tag{8.12a}$$

$$\lambda_r SC + \lambda_s TB + \lambda_T K < 1. \tag{8.12b}$$

This stability condition is derived by inspecting the status of the storage and retrieval queues at departure instances (when the S/R has just completed serving a storage or retrieval request) but as the authors show in the paper, the result holds for the outside observer as well because storage and retrieval requests arrive in a Poisson fashion. In that regard, the result is exact, but its accuracy, of course, depends on how well *SC* and *TB* are estimated for the given-rack dimensions and S/R parameters.

If $\lambda_s > \lambda_r$, one needs to check only the first inequality because the second inequality would automatically be satisfied if the first one is satisfied. Likewise, if $\lambda_r > \lambda_s$, one needs to check only the second inequality. (In Bozer and Cho (2005), the authors also show the stability condition for a balanced system, that is, the special case where $\lambda_s = \lambda_r$.) The condition that must be satisfied, that is, $SC > TB$ is not a limiting condition. If randomized storage is used, and the I/O point is located at the corner (or even the center) of the rack, it is straightforward to show that $SC > TB$ using the results given in Bozer and White (1984).

We stress that the inequalities given by Equation 8.12 are only for checking the stability of the system; they do not reflect the utilization of the S/R as was the case earlier for the stability conditions we derived for Policy I. As shown in Bozer and Cho (2005), provided the system is stable, the utilization of the S/R can be obtained by computing the roots of a second degree equation whose coefficients depend on the parameter values. The interested reader may refer to Bozer and Cho (2005) for details.

The above-mentioned stability condition applies to Config-A; that is, a single, combined I/O point located at the lower left-hand corner of the rack. Obviously, if the I/O point is moved to an alternative location—for example, it can be raised to a certain height—the stability condition would still apply because one can easily determine the new value of *SC* by partitioning the rack appropriately as explained earlier. Recall that *TB* does not depend on the location of the I/O point.

For Config-A, comparing the stability conditions given for Policy I (Equation 8.7) and Policy II (Equation 8.12) we note that, as expected, the S/R has to work harder under Policy I because one can see that the LHS of the appropriate inequality given by Equation 8.12 is less than or equal to the LHS of Equation 8.7 as long as $SC > TB$. Of course, this result holds true as long as both λ_s and λ_r are greater than zero. If either λ value is equal to zero, the S/R performs 100 percent single-command cycles under either policy and, therefore, both policies lead to the same stability condition/performance. The fact that the S/R performs 100 percent single-command cycles can easily be verified by setting one of the λ values equal to zero on the LHS of Equations 8.7 and 8.12.

Analyzing the performance of Config-B under Policy II is more involved. With this scenario, to serve a storage request right after serving a retrieval request, the S/R has to first travel from point O to point I, which somewhat reduces the efficiency of Policy II. Also, under Policy II, note that the S/R becomes empty (or idle) either at the rack or at point O but never at point I. Using an approach similar to that shown in Bozer and Cho (2005) for Config-B under Policy II, one can see that, as long as $SC/2 > TB$ and $SC/2 > \alpha$, the S/R meets throughput if the following two inequalities are satisfied:

$$\lambda_s SC + \lambda_r (TB + \alpha) + \lambda_T K < 1 \text{ and} \tag{8.13a}$$

$$\lambda_r SC + \lambda_s (TB + \alpha) + \lambda_T K < 1. \tag{8.13b}$$

Using the results shown in Bozer and White (1984), it is straightforward to show that for Config-B the first condition, that is, $SC/2 > TB$, is indeed satisfied. However, the second condition, that is, $SC/2 > \alpha$, is more limiting; it requires α to coincide with the shorter in time side of the rack, and it is not satisfied for all rack shapes. The two conditions, taken together, imply that $SC > TB + \alpha$. Therefore, we need to check only one of the inequalities; that is, if $\lambda_s > \lambda_r$, we need to check only the first inequality; otherwise, we need to check only the second inequality. We remind the reader that the values of SC, TB, and α are not independent because they each depend on the dimensions of the rack and the S/R parameters as shown in Bozer and White (1984).

For Config-B under Policy II, using a different approach, it may be possible to derive an alternative stability condition that does not require to have $SC/2 > \alpha$. Although such an alternative stability condition may exist, we did not attempt to derive it. Also, the stability conditions given by Equation 8.13 may still be valid if $SC/2 \leq \alpha$. That is, it may be sufficient to just use Equation 8.13a if $\lambda_s > \lambda_r$, and use Equation 8.13b if $\lambda_s \leq \lambda_r$. Further research is needed to investigate Config-B under Policy II.

8.4 Numerical Examples

We present a few numerical examples to illustrate the stability conditions we derived in Section 8.3. We consider all three configurations under Policy I, and only Config-A under Policy II. Two storage racks are used to construct the examples. The first rack is 24 ft high and 150 ft long; the second rack is 30 ft high and 120 ft long. Note that the total rack area is fixed at 3600 sq. ft. The S/R travels at a speed of 100 and 400 fpm (ft per min) in the vertical and horizontal directions, respectively. We assume that $K = 0.18$ minutes.

Given the above parameter values, and the results shown here and in Bozer and White (1984), for the first rack we obtain $b = 0.64$ and $T = 0.375$, which yields $SC = 0.4262$ minutes, $TB = 0.1473$ minutes, $TB' = 0.1680$ minutes, and $DC = 0.5735$ minutes. For the second rack we obtain $b = 1.00$ (square-in-time) and $T = 0.30$, which yields $SC = 0.4000$ minutes, $TB = 0.1400$ minutes, $TB' = 0.1750$ minutes, and $DC = 0.5400$ minutes. (More accurate values for SC, TB, TB', and DC can be computed if acceleration and deceleration of the S/R is taken into account.)

The above-mentioned two racks are examined under two workload levels—a "low" workload level of 60 requests/hour and a "high" workload level of 100 requests/hour. Each workload level is divided into three possible cases as follows:

Low level 1:	30 storages/hour and 30 retrievals/hour
Low level 2:	42 storages/hour and 18 retrievals/hour, that is, $\lambda_s > \lambda_r$
Low level 3:	18 storages/hour and 42 retrievals/hour, that is, $\lambda_r > \lambda_s$
High level 1:	50 storages/hour and 50 retrievals/hour
High level 2:	70 storages/hour and 30 retrievals/hour, that is, $\lambda_s > \lambda_r$
High level 3:	30 storages/hour and 70 retrievals/hour, that is, $\lambda_r > \lambda_s$

Note that case 1 (low or high) represents a balanced system, while cases 2 and 3 (low or high) represent unbalanced systems. In case 2 (low or high), there are 2.33 times more storage requests than retrieval requests; in case 3 (low or high), there are 2.33 times more retrieval requests than storage requests. Due to symmetry in the stability conditions, except for Config-C, we expect to see the same results between cases 2 and 3 as the same ratio of 2.33 was maintained.

The results are shown in Table 8.1. The left-most column in Table 8.1, except for the row labeled "II-A," shows the expected S/R utilization for each case. Row II-A represents the left-hand side of Equation 8.12, that is, the appropriate stability condition for Config-A under Policy II. The values in row $\rho_{\text{II-A}}$ were computed using the results shown in Bozer and Cho (2005). Note that, because the arrival rates, λ_r and λ_s, are expressed in

Table 8.1 Expected S/R Utilization and Stability Condition Values

	$b = 0.64, T = 0.375$						$b = 1.00, T = 0.300$					
	Low Workload			*High Workload*			*Low Workload*			*High Workload*		
	Bal.	$\lambda_s > \lambda_r$	$\lambda_r > \lambda_s$	*Bal.*	$\lambda_s > \lambda_r$	$\lambda_r > \lambda_s$	*Bal.*	$\lambda_s > \lambda_r$	$\lambda_r > \lambda_s$	*Bal.*	$\lambda_s > \lambda_r$	$\lambda_r > \lambda_s$
$\rho_{I\text{-}A}$	.5365	.5476	.5476	.8941	.9127	.9127	.5150	.5254	.5254	.8583	.8757	.8757
II-A	.4668	.5225	.5225	.7779	.8709	.8709	.4500	.5020	.5020	.7500	.8367	.8367
$\rho_{II\text{-}A}$	.5296	.5428	.5428	.8504	.8845	.8845	.5092	.5213	.5213	.8221	.8521	.8521
$\rho_{I\text{-}B}$	.6302	.6264	.6264	uns.	uns.	uns.	.5900	.5884	.5884	.9833	.9807	.9807
$\rho_{I\text{-}C}$	.5255	.5550	.4997	.8759	.9250	.8329	.5212	.5396	.5096	.8687	.8994	.8494

Note: Bal. = balanced and uns. = unstable.

arrivals/hour, the expected S/R cycle times given above in "minutes," must be converted to "hours" before checking the stability conditions.

Each case examined in Table 8.1 is feasible except for Config-B under Policy I with a high workload and $b = 0.64$. Obviously, in this particular case, the S/R is unable to handle the additional workload generated by the separation of the I/O point. However, changing the rack shape to $b = 1$ reduces the horizontal distance from the output point to the input point, and thus allows the S/R to handle the workload, albeit at an expected utilization of approximately 98 percent.

In fact, as shown in Table 8.1, for Config-A and Config-B, the S/R has to work harder (i.e., the expected S/R utilization increases) under either policy when the shape of the rack changes from square-in-time ($b = 1$) to flat ($b = 0.64$). Although this result confirms the advantage of square-in-time racks from a cycle-time perspective, the change in the expected S/R utilization is quite small in most instances. Considering the higher cost associated with taller racks, especially in seismically active regions, it is doubtful that square-in-time racks represent the most desirable shape overall. Of course, available land/floor space, and the clear ceiling height (if the system is going to be installed within an existing facility) will also play a role in determining the appropriate rack dimensions.

For Config-C, we note that $b = 1$ again outperforms $b = 0.64$ for balanced systems and storage-heavy systems. However, for retrieval-heavy systems, $b = 0.64$ performs slightly better due to the *output* conveyor that runs along the bottom of the rack. (Recall that we obtained $TB' = 0.1680$ minutes for $b = 0.64$, and $TB' = 0.1750$ minutes for $b = 1.00$.) However, as before, the differences in expected S/R utilization are quite small. Obviously, no statistical significance testing needs to be conducted because the results given in Table 8.1 are analytical results that are also exact results to the extent that the model assumptions and expected S/R cycle times are satisfied.

The results shown in Table 8.1 also confirm that Policy II performs better than Policy I for Config-A. The difference between the two policies is most noticeable for a high workload level and a balanced system. If the system is unbalanced, however, or the workload level is low, the difference between the two policies is mostly negligible. Considering that Policy I is considerably easier to model, explain, and analyze, the results may suggest that FCFS performs reasonably well for all practical intents and purposes. Of course, we make this statement without taking into account other opportunities to improve system performance such as storage–retrieval matching schemes (which, by definition, must relax FCFS) cited in Section 8.2. However, we also note that the improvement gained by such schemes decreases if the system is unbalanced or the workload level is such that the storage and retrieval request queues are fairly short.

The results in Table 8.1 also suggest that the impact on the expected S/R utilization of having significantly more (almost 2.5 times more) storage or retrieval requests is not as large as one may have anticipated. This is generally "good news" for AS/RS users and designers because the relative values of λ_s and λ_r may not be known with sufficient accuracy *a priori*. Furthermore, their relative values may fluctuate throughout the day.

8.5 Conclusions

Although the conclusions may change from one problem instance to another, examining two rack shapes and two workload levels for balanced and unbalanced systems, our numerical results suggest that, while Policy II (which essentially looks for opportunities for dual-command cycles) performs slightly better than Policy I in terms of the expected S/R machine utilization, in most cases there is no significant difference between the two policies for practical purposes. Also, the additional workload imposed on the S/R machine due to unbalanced systems (i.e., having almost 2.5 times more storage requests per time unit than retrieval requests or vice versa) seems less than what one would have anticipated.

We believe future research in AS/RS needs to proceed in two directions. The first direction of research is a continuation of the type of work performed thus far, that is, looking for better ways to model, design, or operate AS/RS while assessing the impact of hardware developments such as multishuttle S/R machines. Despite the good work presented in the literature, such studies need to continue because AS/RS applications and technology are constantly evolving. However, studies of this type, no matter how valuable, consider the AS/RS essentially as a "stand-alone" system. The second direction of research needs to consider the AS/RS as part of a larger system such as a supply network or manufacturing system. When considered in this light, we believe the larger impact of AS/RS can be captured and modeled more effectively. The study presented in Hsieh and Tsai (2001)—where the authors argue that developing a storage policy based on the needs of manufacturing operations "can increase not only the performance of the [AS/RS] but also the performance of the manufacturing system"—is a good example of the second type of studies needed, although we expect that there would be instances where optimizing one or more components of the AS/RS may in fact have an adverse effect on the larger system. Understanding how AS/RS interact with the larger systems they are intended to serve, modeling possible trade-offs to avoid suboptimal designs and operational policies, and looking for "win–win" combinations (where optimizing the AS/RS serves the larger system better) appears to be a promising direction to pursue for future research in AS/RS.

Developing a better understanding of how an AS/RS interacts with the larger system is also likely to have a significant impact on decision making (AS/RS versus manual systems) and cost justification of the large investment required for an AS/RS. For example, though many sources cite the fact that AS/RS improve inventory accuracy, it is difficult to find models that quantify the degree and type of such improvements and their impact on the larger system, whether it is a distribution or manufacturing system. Furthermore, such an understanding would avoid some of the criticism directed towards AS/RS by the Lean Manufacturing community. Obviously, installing an AS/RS is going to make matters worse for a "mass manufacturer" dealing with long delays and excess inventory. However, if a manufacturing company understands the basic premises of Lean and uses the techniques effectively, the accuracy and speed of an AS/RS may become a powerful competitive weapon. If the concern is lack of inventory visibility, it would be a straightforward matter to build a large, computerized, color display to show the total inventory level in the AS/RS at all times for all to see.

Last, but not least, there is increasing concern worldwide about global warming. By keeping humans out of the warehouse, AS/RS can offer substantial energy savings and thus reduce carbon emissions especially in extreme climates or during hot/cold seasons. An interesting article discussing the environmental aspect of automated material-handling systems is presented in *Modern Materials Handling* (2007). According to one quote attributed to the vice president of business development at a major vendor, in comparing a "...300,000 square foot facility with automated high-rise storage to a 1 million square foot conventional warehouse,...when [one factors] in all the lighting, HVAC and vehicles required for the larger facility, the automated facility [is] just 20% of the cost of running a conventional warehouse." Although this individual admits that "...many $50 million automation projects" are not going to be justified on "sustainability alone," he also believes "there is a growing awareness of the impact of energy and emissions," and the customers are "fast-tracking" some plans "...if there's a sustainability aspect to the project."

References

Baudin, M, *Lean Logistics: The Nuts and Bolts of Delivering Materials and Goods.* Productivity Press, New York, 2004.

Bozer, YA and Cho, M, Throughput performance of automated storage–retrieval systems under stochastic demand, *IIE Transactions* 37 (4): 367–378, 2005.

Bozer, YA and White, JA, Travel-time models for automated storage–retrieval systems, *IIE Transactions* 16 (4): 329–338, 1984.

Chang, DT and Wen, UP, The impact on rack configuration on the speed profile of the storage and retrieval machine, *IIE Transactions* 29 (7): 525–531, 1997.

Chang, DT, Wen, UP, and Lin, JT, The impact of acceleration/deceleration on travel-time models for automated storage/retrieval systems, *IIE Transactions* 27 (1): 108–111, 1995.

Chang, SH and Egbelu, PJ, Relative pre-positioning of storage/retrieval machines in automated storage/retrieval systems to minimize maximum system response time, *IIE Transactions* 29 (4): 303–312, 1997a.

Chang, SH and Egbelu, PJ, Relative pre-positioning of storage/retrieval machines in automated storage/retrieval system to minimize expected system response time, *IIE Transactions* 29 (4): 313–322, 1997b.

Chow, WM, Design for line flexibility, *IIE Transactions* 18 (1): 95–103, 1986.

Egbelu, PJ and Wu, CT, A comparison of dwell point rules in an automated storage–retrieval system, *International Journal of Production Research* 31 (11): 2515–2530, 1993.

Eldemir, F, Graves, RJ, and Malmborg, CJ, A comparison of alternative conceptualizing tools for automated storage and retrieval systems, *International Journal of Production Research* 41 (18): 4517–4539, 2003.

Eldemir, F, Graves, RJ, and Malmborg, CJ, New cycle time and space estimation models for automated storage and retrieval system conceptualization, *International Journal of Production Research* 42 (22): 4767–4783, 2004.

Felder, RA, Automated storage and retrieval at −80°C: Managing and documenting specimen security, integrity, and accessibility, *American Laboratory* 35 (1): 22, 2003.

Han, MH, McGinnis, LF, Shieh, JS, and White, JA, On sequencing retrievals in an automated storage–retrieval system, *IIE Transactions* 19 (1): 56–66, 1987.

Hsieh, S and Tsai, KC, A BOM oriented class-based storage assignment in an automated storage–retrieval system, *International Journal of Advanced Manufacturing Technology* 17 (9): 683–691, 2001.

Hu, YH, Huang, SY, Chen, CY, Hsu, WJ, Toh, AC, Loh, CK, and Song, TC, Travel time analysis of a new automated storage and retrieval system, *Computers & Operations Research* 32 (6): 1515–1544, 2005.

Hur, S and Nam, J, Performance analysis of automatic storage/retrieval systems by stochastic modeling, *International Journal of Production Research* 44 (8): 1613–1626, 2006.

Hwang, H and Ko, CS, A study on multi-aisle system served by a single storage-retrieval machine, *International Journal of Production Research* 26 (11): 1727–1737, 1988.

Hwang, H and Lee, SB, Travel-time models considering the operating characteristics of the storage and retrieval machine, *International Journal of Production Research* 28 (10): 1779–1789, 1990.

Hwang, H and Lim, JM, Deriving an optimal dwell point of the storage–retrieval machine in an automated storage–retrieval system, *International Journal of Production Research* 31 (11): 2591–2602, 1993.

Hwang, H, Chang, IH, Moon, SW, and Lim, JM, Economic design of refrigerated automated storage and retrieval systems, *Engineering Optimization* 32 (2): 249–266, 1999.

Kebabian, A, Resetar, D, Thomas, RL, and Skinner, G, Multiple approaches to an automated storage and retrieval system (ASRS): Successes earned and lessons learned, IUG 2006, Denver, Colorado, May 19–22, 2006. (Available at http://library.sonoma.edu/about/ars-sup-2006.pdf)

Keserla, A and Peters, BA, Analysis of dual-shuttle automated storage–retrieval systems, *Journal of Manufacturing Systems* 13 (6): 424–434, 1994.

Kulturel, S, Ozdemirel, NE, Sepil, C, and Bozkurt, Z, Experimental investigation of shared storage assignment policies in automated storage/retrieval systems, *IIE Transactions* 31 (8): 739–749, 1999.

Lee, HF, Performance analysis for automated storage and retrieval systems, *IIE Transactions* 29 (1): 15–28, 1997.

Lerher, T and Potrc, I, The design and optimization of automated storage and retrieval systems, *Strojniski Vestnik—Journal of Mechanical Engineering* 52 (5): 268–291, 2006.

Lerher, T, Sraml, M, Kramberger, J, Potrc, I, Borovinsek, M, and Zmazek, B, Analytical travel time models for multi aisle automated storage and retrieval systems, *International Journal of Advanced Manufacturing Technology* 30 (3–4): 340–356, 2006.

Malmborg, CJ, Interleaving models for the analysis of twin shuttle automated storage and retrieval systems, *International Journal of Production Research* 38 (18): 4599–4610, 2000.

Malmborg, CJ, Estimating cycle type distributions in multi-shuttle automated storage and retrieval systems, *International Journal of Industrial Engineering—Theory Applications and Practice* 8 (2): 150–158, 2001a.

Malmborg, CJ, Rule of thumb heuristics for configuring storage racks in automated storage and retrieval systems design, *International Journal of Production Research* 39 (3): 511–527, 2001b.

Meller, RD and Mungwattana A, Multi-shuttle automated storage/retrieval systems, *IIE Transactions* 29 (10): 925–938, 1997.

Meller, RD and Mungwattana A, AS/RS dwell-point strategy selection at high system utilization: A simulation study to investigate the magnitude of the benefit, *International Journal of Production Research* 43 (24): 5217–5227, 2005.

Park, BC, Optimal dwell point policies for automated storage/retrieval systems with dedicated storage, *IIE Transactions* 31 (10): 1011–1013, 1999.

Park, BC, Performance of automated storage/retrieval systems with non-square-in-time racks and two-class storage, *International Journal of Production Research* 44 (6): 1107–1123, 2006.

Peters, BA, Smith, JS, and Hale, TS, Closed form models for determining the optimal dwell point location in automated storage and retrieval systems, *International Journal of Production Research* 34 (6): 1757–1771, 1996.

Potrc, I, Lerher, T, Kramberger, J, and Sraml, M, Simulation model of multi-shuttle automated storage and retrieval systems, *Journal of Materials Processing Technology* 157 (Special Issue): 236–244, 2004a.

Potrc, L, Lerher, T, Kramberger, J, and Sraml, M, The design of automated storage and retrieval systems using a simulation modeling approach, *Strojniski Vestnik—Journal of Mechanical Engineering* 50 (11): 504–529, 2004b.

Thonemann, UW and Brandeau, ML, Note. Optimal storage assignment policies for automated storage and retrieval systems with stochastic demands, *Management Science* 44 (1): 142–148, 1998.

Tompkins, JA, White, JA, Bozer, YA, and Tanchoco, JMA, *Facilities Planning*, Third edition, Wiley, 2003.

van den Berg, JP and Gademann, AJRM, Optimal routing in an automated storage/retrieval system with dedicated storage, *IIE Transactions* 31 (5): 407–415, 1999.

van den Berg, JP and Gademann, AJRM, Simulation study of an automated storage/retrieval system, *International Journal of Production Research* 38 (6): 1339–1356, 2000.

van den Berg, JP, Analytic expressions for the optimal dwell point in an automated storage/retrieval system, *International Journal of Production Economics* 76 (1): 13–25, 2002.

Yin, YL and Rau, H, Dynamic selection of sequencing rules for a class-based unit-load automated storage and retrieval system, *International Journal of Advanced Manufacturing Technology* 29 (11–12): 1259–1266, 2006.

Zollinger, H, AS/RS application, benefits and justification in comparison to other storage methods: A white paper, Zollinger Associates, Inc., 1999. (Available at http://www.mhia.org/PSC/pdf/asrswhitepaper.pdf)

Bibliography

ASAP Automation; available at http://www.asapauto.com/solutions/automation-technologies/AS-RS/default.asp

Bastian Material Handling (BMH); available at http://www.bastiansolutions.com/

BMW: Custom Cars on Demand, *Modern Materials Handling*, February, 2004.

California State University, Northridge; available at http://library.csun.edu/About_the_Library/asrs.html

Daifuku Co. Ltd.; available at http://www.daifukuamerica.com/daifuku/dac/fada/about.asp

HK Systems Announces Largest AS/RS Project in North American Material Handling History, 6/23/2004; available at http://www.supplychainmarket.com/content/news/article.asp?docid = {E95ACE3A-5AB2-443F-B1E9-D83C7AAD9184}&VNETCOOKIE = NO

Modern Materials Handling, 2007, Going green with automation; available at http://www.logisticsmgmt.com/article/CA6439720.html?text = as%2Frs

Murata Machinery Ltd.; available at http://www.muratec-l-system.com/en/division/history.html

Chapter 9

Carousel Storage Systems

Elkafi Hassini

CONTENTS

Abstract Carousels have been used in virtually all physical storage environments. In this chapter, we report on a wide range of carousel applications. One of the important issues in operating carousels is the management of storage location and order picking. The rest of the chapter is devoted to providing a brief history about the interest in studying storage-location and order-picking problems, and an extensive literature review of the different models that have been studied in this area. We conclude by providing a list of challenging open problems.

9.1 Overview and Motivation

In its 'Holiday 2000 Fulfillment Report' on Internet-based companies, the consulting firm Bizmetric has found that the shortest time between when a customer issues an order and when that order is shipped was ten hours and five minutes (achieved by drugstore.com), whereas the longest time was four days, twelve hours, and nine minutes (by Target.com) (Bizmetric, 2000). In a similar study for the year 2001, the numbers were ten hours and four minutes (by Ashford.com) and three days, fifteen hours, and seven minutes (again by Target.com). If anything, these statistics indicate that order-fulfillment speed is now recognized as an important measure of competitiveness. Though seemingly fast, these fulfillment times need to be improved further for a firm to stay competitive. In fact, a study in 2000 by the consulting group Accenture found that 67 percent of holiday-season shipments ordered online were not received as ordered, and 12 percent did not arrive before Christmas. As most of the online business-to-customer (B2C) orders are small (often constituted of single items referred to as "eaches"), warehouse managers would have to focus more on improving the operations of material-handling systems that handle such small orders. This need for a more efficient management of small orders has been echoed in many practitioners' forums. In the words of Art St. Onge, a material handling consultant: "There is an enormously growing industry that needs to have some really creative thinking applied to it in the development of some new technologies to handle eaches" (Maloney, 2001).

This chapter reviews the major findings in the literature on storage location and space allocation in one of the most commonly used material-handling equipment for small parts: carousel storage and retrieval systems (henceforth referred to as carousel systems). Carousels are used in all links of a supply chain, in both production and service environments, and their usage has been on the rise since their inception in the late 1950s.

With the current information age developments, a thorough study of storage-location and order-picking management in carousel systems has never been more urgent for business managers. With e-commerce becoming a common practice, companies are finding it vital to optimize the operation of their warehouses. In addition to replenishment and stuffing, companies have to also pick customers' order themselves. The process of customer "shopping" is being simplified: whereas in a classical retailer setting the customer "handles" both information and products, now the information comes to the customer online, where decisions on buying are also made, and the retailer assumes the picking, sortation, and delivery of orders. The numbers, indeed, show an explosion in Internet-technology adoption and development, which would in turn mean rethinking the way warehousing and material-handling activities are performed.

To gain a competitive edge in such environments, companies are counting on fast deliveries as exemplified in the mission statement of a third-party warehousing company:

> We define our business as providing customers with warehousing and related services designed to offer 48-hour delivery time to any significant market in the continental United States.... This [operating their own warehouses] will ensure the availability of efficient operating facilities at real estate cost substantially below appraised value.... (Ackerman, 1997)

Canedy (1999) has reported that an Internet-based company is offering only next-day service after it had failed to keep its initial promise of a three-hour delivery time limit. To survive in such hectic conditions an efficient control of warehousing and material-handling activities is vital.

Two important features of B2C e-commerce are: (1) orders are eaches; and (2) the ability of a customer to keep track of an order online. These two features have a decisive impact on the choice of warehousing and material-handling equipment. The equipment has to be suitable for processing eaches and it has to be easily linked to an online order-tracking system. Carousel storage systems are thus a perfect match, and this is one of the reasons their usage is on the rise. In fact, in the United States, carousel sales are second to trucks' sales and experts believe they increased by 20 percent in 1998 (Foster, 1999). A report on a 1997 survey on logistics in Canada (Mcgillivray and Geiger, 1997) found that the largest percentage growth in material-handling equipment is in carousels, which were expected to double in the period 1997–2001.

Managers are realizing that as Internet orders grow, carousels may provide the solution for product storage and order management (e.g., Pletsch, 1998). Quoting the words of a manager of a large material-handling equipment company, Foster (1999) wrote:

> These companies [those starting or switching to e-commerce] used to ship pallet loads, and never really considered carousels for picking individual items. Rocke [manager] says, Now, they see fulfilling individual customer orders right out of their warehouses. Carousels is the only technology that can do what they need.

Another reason for the wide adoption of carousels is that they have been designed to be easily operated by a computer processor that can in turn communicate with other processors in the supply chain. For example, a carousel processor would take orders data from a central warehouse management system (WMS) (or other supply-chain software) and then relay inventory data back to it once an order is processed.

Carousels have been used in virtually all physical storage environments. Recently, they are being successfully introduced in hospitals. As carousel usage in health services increases (e.g., Anonymous, 1991, 1997d; Calgary General Hospital, 1992; Berglund, 1997; Paris, 1988), the need for operating algorithms that would guaranty super-quick deliveries is paramount. For example, some hospitals are opting for centralized warehousing for drugs and equipment and it would be vital for them to provide prompt responses to orders from surgeons in surgical wards or to patients in critical health conditions.

Storage-location and space-allocation problems are part of warehousing management within the area of logistics. These problems arise both in manufacturing and service environments. In manufacturing environments, the need for storage assignment and order picking arises in distribution centers as well as in production premises where small parts are brought closer to the work stations. In addition to Operations Researchers, storage-location problems have also received the attention of Computer Scientists who investigate optimal polices for the storage of computer files. This joint interest will be highlighted in our review of relevant storage-location literature. Before we survey the literature, we define some key terminology and give a brief history about the interest in studying storage-location and order-picking problems, as well as the major developments in the use of carousel systems. We conclude by providing a list of open problems.

9.2 Definitions and Conventions

Some readers might be unfamiliar with a technical term; some others may know several definitions (not necessarily consistent) for a term. To avoid such hindrances, we include definitions of the major important terms that we believe are necessary for comprehending the remainder of the chapter.

"Warehousing" is the act of stocking, mixing, consolidating, and distributing products. Recently, companies are moving into decentralized warehousing, where goods are moved closer to processing facilities. In such settings, carousels are in high use, due to the space saving and convenient product delivery they provide.

"Material(s) handling" refers to the transfer of material, whether through production and service processes or in storage. In our study, we restrict our attention to material handling carried out by the carousel (while rotating) and the picker (while picking and placing).

"Carousel storage and retrieval systems" are storage systems that bring the required part to the picker. As such, they belong to the materials-handling equipment family of part-to-picker systems as opposed to picker-to-part systems (e.g., bin-shelving systems). They are suitable for storing small items, for example, items that are less than case loads in size. There are four main types of carousels.

- Horizontal carousel: It consists of a series of wire baskets or storage containers linked, horizontally and vertically, in a closed loop that is mounted on an oval horizontal track. It can be either top or bottom driven. We will refer to the horizontal series of containers as a layer, or level, and to the vertical ones as a bin. A bin is usually divided into shelves on top of which containers are placed. When a specific item is required, all the carousel levels rotate simultaneously to position the bin that contains the item in front of the picker. The picker has access to one bin at a time. Figure 9.1 illustrates a typical horizontal carousel system.

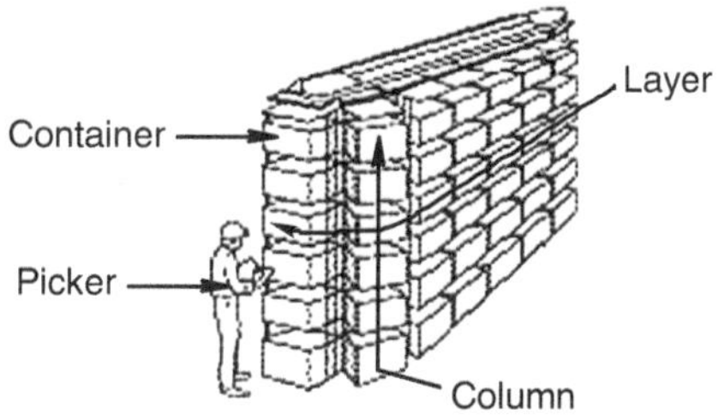

Figure 9.1 A top-driven horizontal carousel.

- Vertical carousel: Parts are stored in bins that are placed on horizontal shelves. These shelves are enclosed in a metal sheet leaving but with a window that allows the picking and placement of parts. The shelves rotate vertically in a closed loop thus allowing the exposure of all bins.
- Horizontal rotary rack: Like a horizontal carousel with the extra option that each level can rotate independently of other levels.
- Twin-bin carousel: like horizontal carousels with the extra feature of allowing the picker to pick from two adjacent bins.

"Order Sortation" is the task of separating one customer's order items from other customers' items. It can be done while or after picking orders.

"Order picking" is the act of picking customer orders from a storage area. Sortation might be needed after (pick-and-sort) or during (sort-while-pick) picking.

A "picker" is a human being or a robot that performs the action of picking and placing parts. The picker is responsible for implementing the storage and order control policies.

By "storage-location" assignment we mean the assignment of stock items to carousel shelves. It is to be distinguished from storage allocation which concerns itself with assigning families of products to storage zones or storage equipment. For example, most "storage allocation" studies focus on allocating products in a "forward-reserve storage" scheme (a storage area is reserved for picking and another for replenishment).

9.3 History

9.3.1 Pioneers

The interest in studying storage-location and order-picking problems started in the 1970s, in both Operations Research and Computer Science, and has been on the rise ever since. Gudehus (1973; cited in van den Berg, 1999) appears to be one of the first researchers to study order-picking problems. In Operations Research, Hausman et al. (1976), Graves et al. (1977), and Schwarz et al. (1978) were among the first to study storage-location problems. In Computer Science, Pratt (1972) was among the first to study the problem of optimally locating files to minimize their access time (distance).

Why did interest in storage-location problems arise only in the 1970s, and why was this interest simultaneously shared by Operations Researchers and Computer Scientists? If we reflect a little on the major relevant events that occurred in the 1970s, the answer becomes evident. In the 1970s, three major events took place: first, the oil crisis prompted companies to look for more efficient production control means—through inventory control, for example. Second, the availability of larger computer mass storage systems enabled businesses to maintain greater data files and to introduce integrated information systems such as materials requirements planning (MRP). Third, was the emergence of global competition through the Japanese just-in-time (JIT) management philosophy that preaches the elimination of all nonvalue-adding production activities. All combined, these events put more pressure on warehouse managers to improve their response times, to store less, and for a shorter time. The increase of information flow and storage availability in computer systems, in turn, prompted Computer Scientists to investigate methods for decreasing access times in computer mass storage systems.

9.3.2 Carousel History

According to the *Oxford English Dictionary* the use of the word "carousel" goes back to as far as 1650. It was derived from the French language and initially it meant "little war." Its use dates back to the age of the Crusades,

> . . . when European knights encountered ring spearing, a popular sport and training exercise among Arabian and Turkish horseman. The practice involved riding full tilt toward a ring suspended from a pole or tree, then running the tip of a lance through the center of the ring. (Valenti, 1997)

When the idea was brought back to Europe it developed into a sport whereby carousel referred to a wood horse that was linked to a center pole, which noblemen used to ride upon and throw lances through a brass ring, thus exercising for real knights' tournaments and saving their real horses for the actual competition (Homes, 2007).

With age, these wooden horses developed into today's merry-go-rounds.

In production systems, the use of carousel dates back to the 1960s. Apparently the idea of using the carousel in production systems evolved from the overhead circular conveyor system used for hanging garments in shops. They were first introduced as conveyor carousels: A conventional conveyor with bins attached on top of it (i.e., just like a single-level horizontal carousel). The name carousel was coined for this system, mainly because of its continuous rotation, just like the merry-go-round that keeps turning round and round.

The first industrial carousels made in the 1960s were like overhead garment conveyors with suspended wire baskets (bins) and shelves. In the

1970s, heavier-duty, bottom-driven versions were introduced that were capable of carrying weights in excess of 1000 lbs per bin. In the 1980s, sophisticated computer-controlled carousels were introduced and integrated into other automated storage and retrieval systems (Kulwiec, 1985). Nowadays, carousel systems are designed in wide varieties and generally have the following parameter settings (Kulwiec, 1985; Ackerman, 1997):

- Bins can be of different sizes but the most common are 21, 24, 30, or 36 in. in width; 14, 18, 20, or 22 in. in depth; and 2 to 10 ft in height, but are generally 6, 7, or 8 ft. The design of bins often allows for flexibility of choice in shelf designs.
- Bin carrying capacity varies from 800 to 1500 lbs and larger capacities can be made if needed.
- The number of bins can vary from 16 to 140.
- Normal rotation speed is in the range of 60 to 80 ft per minute (fpm).
- Carousel length varies from 20 to 80 ft. Length depends on the available space and the desired picking rate.

Operator pick rates reach up to 600 lines per hour and, when automated, the pick rate can reach up to 1500 lines per hour (Anonymous, 1996d; Feare, 1999a).

Single carousel prices range from \$15,000 to \$40,000 (Allen, 1992). Small carousel systems range between \$75,000 to \$250,000 (Luton, 1997; Schulz, 1999). Large and advanced systems cost up to \$5 million (Schulz, 1999). (All values are in U.S. dollars.)

Carousels are often placed side by side only a few inches apart, unless an emergency service aisle is placed in between adjacent carousel units. To take advantage of available ceiling height, carousels are often multi-tiered. In such instances, mezzanine designs are employed to access upper carousels.

9.4 Carousels at Work

We outline the different modes of carousel operation and provide an extensive list of carousel applications.

9.4.1 Carousel Operation

Table 9.1 summarizes some of the typical carousel configurations. Carousels can rotate in either a unidirectional or bidirectional fashion. They can func-

Table 9.1 Possible Carousel Configurations

Factor	*Levels*
Layout	• Single carousel • Pods of $k = 2, 3,$ or 4 carousels
Number of layers	• Single layer • Multiple layers
Bin design	• Single bin • Twin bin
Direction of rotation	• Unidirectional • Bidirectional
Operational mode	• Anticipatory • Nonanticipatory
Picker's motion	• Negligible (one-dimensional) • Nonnegligible (two-dimensional)
Picking station	• Fixed • Mobile

tion in a nonanticipatory mode, where the carousel stays in the same position after finishing a request; or in an anticipatory mode, where the carousel may be positioned, if time permits, in a favorable location for the next request. Carousels are often used in groups, either juxtaposed or multitiered.

Items are picked (or placed) from (or into) the carousel in several ways. An operator can manually pick the necessary items. In such situations, light-tree displays are often used to guide the operator to the appropriate pick location and the quantity needed. These light-trees are often placed between carousels. If a single carousel is higher than the reach of an operator or a mezzanine layout is used (e.g., with multitiered carousels) the operator might use a ladder (Goetze, 1998) or a lift platform (Anonymous, 1997f). The lift platform usually has both horizontal and vertical movement options. Lifts have a standard speed of 120 fpm. Robotic devices for automated insertion and extraction are also available. Usually the robot picks a tote and brings it to the reach of an operator to pick the necessary quantity. The robots can go up to heights of 28 ft (double- or triple-tiered carousel configurations) and can travel at a vertical speed of up to 500 fpm.

9.4.2 Carousel Advantages

There are several advantages for installing horizontal carousels:

- They reduce errors in picking (Anonymous, 1997b; Luton, 1998) and increase inventory accuracy.

- They eliminate walk and search time.
- They reduce counting and restocking times.
- They reduce man hours.
- They allow for efficient use of space, for example, when multitiered.
- They increase productivity with the use of automatic extractors.
- They are easy to integrate within an enterprise's computerized production control system.
- Carousels have several ergonomic benefits for the worker (Trunk, 1993; Anonymous, 1994). They permit the operator to use his or her golden zone—the area between the worker's shoulders and knees that allow for easier picking without excessive bending. In addition, the worker walks less within a carousel system.
- Carousels empower the worker by giving him or her the responsibility of intelligently controlling the automatic order-picking operations.
- They offer security if enclosed.

They can be used in several environments (see references in Section 9.4.3) such as:

- Progressive assembly, where carousels rotate continuously to move parts between work stations
- Point-of-use storage, such as in production kitting
- Order picking, when less-than-case-loads are distributed, such as in mail-order houses, and when batch ordering becomes advantageous
- Maintenance storage
- Burn-in testing for electronic components
- Office applications where the carousel would carry things like files, x-rays, computer tapes, cartridges, and other media
- Bulky items storage and retrieval, where carousels are usually bottom-driven to allow for more rotation power

9.4.3 Carousel Applications

Horizontal carousel storage systems are used in numerous environments. Table 9.2 provides the major businesses that employ carousels, the main function (in addition to storage) they serve, and a sample of references for the reader who wishes to get more details on a specific application.

9.5 Literature Review

In this section, we report on the literature that is relevant to our research topic. Literature that relates to our solution methodology will be outlined

Table 9.2 Carousel Applications

Sector	*Function*	*References*
Airlines	Maintenance	Kardex System, 1991; Air Canada, 1994
Book Publishing	Distribution	McGraw-Hill Ryerson Ltd., 1993; Anonymous, 1995a,b, 1998a; Briggs, 1997
Distribution	Distribution	Canadian Tire Corp. Ltd., 1993; Forger, 1999
Entertainment	Distribution	Witt, 1995; Trunk, 1996b; Allnoch, 1997; Anonymous, 1997c
Food; Manufacturing	Production	Trunk, 1996a
Groceries	Replenishment	Robertson, 1998, 1999a; Schulz, 1999
Health Care; Manufacturing	Distribution	Auguston, 1995b; Anonymous, 1996e, 1997b, 1999; Sciex Ltd., 1996; Veterinary Purchasing, 1996; Diamond Phoenix Systems, 1997; Robertson, 1999b
Health Care; Services	Replenishment	Paris, 1988; Anonymous, 1991, 1997d; Calgary General Hospital, 1992; Berglund, 1997
Mail-Order Houses	Distribution	Anonymous, 1996c; Trunk, 1996b
Manufacturing; Electronics	Distribution	Anonymous, 1976b, 1980, 1989, 1996b, 1997a; Witt, 1989; Torok, 1992; Globe Electric, 1994; Field, 1997; Rix, 1998
	Production	Anonymous, 1982, 1997e; Gupta, 1982; Auguston, 1995b; Timex Canada Inc., 1996; LeBaron and Hoffman, 1998
Manufacturing; Mechanical	Distribution	Subaru Auto Canada Ltd., 1993; Auguston, 1996; Field, 1997; Grand and Toy Ltd., 1999
	Production	Anonymous, 1976a; Klitz, 1983; Kevin, 1985; Kobuki, 1987; Auguston, 1995a; ESAB Welding and Cutting Products, 1996; Wisnia, 1997; Thomas, 1999
	Tooling inventory	Foulds and Wilson, 1993; Gaboune et al., 1994a,b; Anonymous, 1996a; Laporte et al., 1998
Military	Training	Jacobs et al., 2000
Space Industry	Production	Wilson, 1985; Trunk, 1996a; Feare, 1999b

after our discussion of the specific problems. Going from the general to the specific, we adopt the following taxonomy in reporting our literature review:

1. Warehousing: We include literature that discusses warehousing problems in a general setting and warehousing problems in a carousel storage system in particular. Survey studies will appear in this section.
2. Storage location: Studies that looked into storage-location problems in carousels as well as major important results in storage location in general warehouses are reported.
3. Interface of storage location and order picking: Studies that have simultaneously studied the storage-location and order-picking problems are mentioned here.
4. Storage-space allocation: Survey of major works that dealt with storage-space allocation, with emphasis on carousel systems.
5. Related applications: The problems we study have interesting applications in other fields. In this section, we report on the major works from other fields that dealt with problems that are analogous to ours.

Though we did our best to collect all the literature relevant to storage location and space allocation in carousel systems, we only present a selection of works in general WMS.

9.5.1 Warehousing

9.5.1.1 General Systems

Comprehensive Models: Gray et al. (1992) present a comprehensive model for warehouse design and operation. They include decisions such as warehouse layout, equipment and technology selection, item location, zoning, picker routing, pick-list generation, and order batching. They develop a multistage hierarchical decision approach to solve the resulting model.

Survey Studies: Matson and White (1982) reported on the early works done by Operations Researchers in the area of material handling. About ten years later, Cormier and Gunn (1992) reviewed warehousing literature that focused on throughput capacity (comprises picking, batching, and storage policies), storage capacity, and warehouse design. They suggested that further research should be carried out to develop methods for dynamically improving warehouse performance through better storage and picking strategies. Recently, van den Berg (1999) published a survey on the planning and control of warehousing systems. He decomposed the problem of warehousing management into a hierarchy of three major levels: strategic, tactical, and operational. He then reported on the major studies that have been carried in each of these levels. One of his major findings is that many warehouses, even those using warehousing management software, still use simple procedures and that these procedures can be significantly outperformed if a more thorough and structured method is developed for studying warehousing problems. In addition, van den Berg pointed out that few works have presented optimal solutions to the problems they study. He suggested that trying to develop new models for warehousing problems might lead to better, if not optimal, solutions. Rouwenhorst et al. (1998) also provide a recent survey on warehousing design and control problems. They view the three decision levels defined by van den Berg (1999) in a three-dimensional framework, the axes of which are processes, resources, and organization. They highlighted the need for research that would integrate several models and methods to tackle complex warehousing design and control problems. van den Berg and Zijm (1999) outlined several types of warehousing systems, discussed the type of problems encountered in such systems, and provided examples of models that have been developed to solve these problems. They highlighted the need for research that takes into account the interactions between warehousing and inventory decisions.

Warehouse Design: Sung and Han (1992) consider the problem of determining the size of an automated storage/retrieval system (AS/RS) based on demand for stored items and a management-set protection criterion. Park et al. (1999) study the effect of buffer size on system output in an end-of-aisle order-picking system. Sargent and Kay (1995) present a costing model that can be used to help management decide whether or not they should move from a centralized to a decentralized storage system. Rosenblatt et al. (1993) recursively use optimization and simulation techniques to study the problem of minimizing the costs of operating automated warehouses within certain performance criteria.

9.5.1.2 Carousel Systems

Among the first works to introduce automated carousel systems to academia appear to be those of Weiss (1980) and Klitz (1983). Weiss describes carousel storage systems and the emerging computer technology that made it possible to automate them. Klitz describes a simulation model for logistics and manufacturing of diskettes. His aim was to analyze how many carousels were needed for the system.

System Performance: Building on the results for the patrolling repairman problem developed in Mack (1957), Mack and Murphy (1957), and Bunday and Mack (1973)—Koenigsberg and Mamer (1982) study the efficiency of a single server serving a single carousel system. Later, Koenigsberg (1986) and Kim and Koenigsberg (1987) study the efficiency of a single robot serving two or more carousels. Their work is later extended by Bunday and El-Badri (1988), who study the performance of a single robot serving a multiple number of carousels. In their settings, a server moves between carousels to perform picking and replenishment operations. It should be emphasized that the studies in Bunday and El-Badri (1988), Kim and Koenigsberg (1987), and Koenigsberg (1986) are focused on analyzing the utilization of the loader/unloader and the overall efficiency of the carousel system. They do not discuss storage policies. Furthermore, the server (a robot) is mobile. Hamacher et al. (1998) study the expected performance of a carousel system that allows for the parallel operations of several layers. Park and Rhee (2005) study the performance of carousel systems when items are stored according to the organ-pipe arrangement (OPA).

Only recently have researchers considered performance analysis in pods of carousels. Park et al. (2003) look at the case of two carousels. Vlasiou et al. (2004) extend the results given by Park et al. (2003) by considering more general distributions for the pick times. Meller and Klote (2004) looked at the throughput in $n \geq 2$ pods of carousels.

Travel Time Models: Travel time models are useful in comparing alternative storage and picking scenarios. Lee and Hwang (1988) derive the expected travel time per operation cycle in a single carousel served by a single storage/retrieval device. In addition, they estimate the number of requests processed per unit time and present a nonlinear integer model for optimally designing such systems. Rouwenhorst et al. (1997) study the performance of a similar system. However, they consider a stochastic model of the carousel operation and develop expressions for its throughput rate and mean response time. Su (1998) investigates the performance evaluation of carousels of three types: (1) a single carousel, (2) multiple stacked carousels, with a single motor (i.e., only one carousel moves at a time), and (3) multiple stacked carousels each with its own rotary motor. He develops expressions for expected cycle times for single- and double-command cycles in both unidirectional and bidirectional carousel configurations. He uses simulation to validate his analytic results.

Design Issues: Trevino et al. (1994) present a model for economic design of carousel systems given storage-space and throughput requirements. Egbelu and Wu (1998) discuss optimal positioning policies for idle extractors in carousel systems.

9.5.2 Storage Location

9.5.2.1 General Systems

Analysis of Storage Policies: Heskett (1963, 1964), Harmatuck (1976), and Kallina and Lynn (1976) were among the first to consider the problem of assigning items to storage locations to maximize order-picking efficiency. They introduced the cube-per-order (CPO) index storage rule which assigns items with the smallest ratio of volume to demand frequency closer to the picking area. The optimality of the CPO rule is established for cases of single-command (Harmatuck, 1976) and double-command (Malmborg and Bhaskaran, 1990) cycles. Malmborg and Krishnakumar (1989) investigate storage policies for multiple-aisle warehousing systems where multiple storage and retrieval transactions are performed by the storage/retrieval device. They develop specific conditions for the optimality of CPO in such settings. Jarvis and McDowell (1991) use assignment-like algorithms to assign products to storage locations to minimize order-picking time. They note that if the warehouse is not symmetric (e.g., aisles are not symmetrically located around the dock) assigning the most frequently picked items to the nearest aisle (e.g., CPO-like design) will not necessarily minimize the average travel distance. Recently, van den Berg (1996) considers the problem of finding optimal class allocations that minimize the mean single-command cycle time in a general warehouse setting.

Hausman et al. (1976) study the operating performance of three storage assignments: random, class based, and dedicated. They consider only single-command cycles. Graves et al. (1977) extend the study of Hausman et al. (1976) to include dual-command cycles. Schwarz et al. (1978) use simulation to verify the results of Hausman et al. (1976) and Graves et al. (1977). Lee (1992) studied the dedicated storage policy in a man-onboard AS/RS. His aim was to minimize the total travel time required to pick a given number of orders per period. Goetschalckx and Ratliff (1990) consider duration-of-stay (DOS)-based storage assignment and use simulation to compare it to class-based, dedicated, and random storage policies. Kaylan and Medeiros (1988) consider a mini-load AS/RS in a production setting. They compare the performance of dedicated storage and closest-to-next (CTN) location storage using simulation. The CTN rule stores items at the empty location which is nearest to the next workstation. It was found that the CTN rule worked best, but its effectiveness decreases as the work-in-progress level increases. Recently, Kulturel et al. (1999) study the performance of class-based and DOS-based storage assignments. They assume that a continuous review (Q, r) inventory model is used. Muralidharan et al. (1995) introduce a new storage policy for AS/RS that combines both the random and class-based storage policies. Initially, products are stored in a random manner and then, once the crane is idle, the frequently requested items are moved closer to the input/output (I/O) point while the less demanded items are moved away from the I/O point. They show that this new policy performs better than classical storage policies in terms of service and waiting time. They use simulation to study the performance of the new storage strategy.

In some warehousing systems, storage areas are divided into a forward area, where items are picked, and a reserve area, that replenishes the former. van den Berg et al. (1998) study forward-reserve allocation policies that would minimize the expected amount of labor during picking operations in such systems.

When certain types of orders are more frequent than others, it is intuitive to think that those more frequent orders should receive special attention when designing a storage scheme. van Oudheusden and Zhu (1992) investigate storage layouts for man-on-board AS/RS rack systems when some orders are recurrent.

Correlated Storage: Frazelle and Sharp (1987) were among the first researchers to deal with correlated storage assignments. They describe its implementation, benefits, applications, and limitations. In Frazelle and Sharp (1989), they report on how one can measure the degree of correlation between products in practice. Kim (1993) studies the problem of storage location of correlated items in mini-load AS/RS. In addition to material-handling costs, his model also includes inventory costs. Landers et al. (1994) present a conceptual framework and software architecture for a dynamic reconfiguration of an in-the-aisle order-picking system. They allow for variable capacity storage, stock-splitting among zones, and correlated demand within product families. Sadiq et al. (1996) suggest using clustering techniques to identify family items for storage location in a single-aisle facility.

Inventory–Storage Interaction: Wilson (1977) was the first to study the interaction of product-storage location and inventory decisions. He formulated the problem as a linear assignment model. Hodgson and Lowe (1982) deal with the same problem, but they formulate it as a continuous layout model. Hackman and Rosenblatt (1990) consider the problem of deciding which items, and in what quantity, to assign to a limited number of storage locations. Malmborg and Deutsch (1988) later develop an evaluation model for the dual command cycle. They include warehouse layout, expected inventory, and order-picking costs. Finally, Kim (1993) considers both material-handling and inventory costs in his study of the problem of storage location of correlated items in mini-load AS/RS.

9.5.2.2 Carousel Storage Location

Storage Location: Lim et al. (1985) were among the first to study the storage-location problem in carousels. They dealt with the single-carousel situation and presented the OPA storage policy, but did not prove its optimality. In short, OPA arbitrarily assigns the most demanded item to a random bin and successively places around it the next two most frequent items, each on an opposite side, until all items are placed. Later, Fujimoto (1991) studied the OPA in single carousels and used tabu search to heuristically provide arguments for its optimality. Vickson and Fujimoto (1996) considered a single non-anticipatory and bidirectional carousel system. Items' request frequencies are assumed to be known and independent. They show the optimality of a simple greedy scheme for grouping products to carousel bins. Optimality of the OPA for the same carousel system functioning in an anticipatory fashion is later reported by Vickson and Lu (1998). The optimality of the OPA for a single nonanticipatory carousel was independently proven by Seshadri et al. (1994), Bengü (1995), and Vickson and Fujimoto (1996). Seshadri et al. (1994) show the optimality of OPA for both anticipatory and nonanticipatory operational modes. In addition, they show that OPA minimizes the mean queueing delay and time spent in system by requests that arrive to the system as per an arbitrary renewal process. To prove optimality, Seshadri et al. (1994) and Vickson and Fujimoto (1996) used a similar approach to that of Bergmans (1972) (both also corrected Bergmans' original proof). In her proof, Bengü (1995) followed a similar logic to that of Wong (1983).

Abdel-Malek and Tang (1994) present heuristics for locating items in a single carousel, and account for interdependence between item requests. They show that interactions within requests have a significant influence on the optimal sequencing solutions. The optimal solution for the interaction case does not seem to have a clear structure (e.g., an organ-pipe-like structure as in the no-interactions case).

Hamacher et al. (1998) and Kallrath (2005) consider the problem of sequencing trays that are input into a carousel system that allows the simultaneous operation of several layers. They apply their solution procedure to a case of distribution centre of a major German department store.

The storage-location problem in a system of several carousels has not received much attention until recently. Emerson and Schmatz (1981) use simple simulation models to study different storage schemes in a two-carousel setting. The storage schemes that they test are simple and include random storage, sequential storage, and storage in the carousel with the largest number of openings. They also consider the scheme that if an arriving item is already stored in the carousel, it would be stored in the same location. Their aim is to study the degree of carousel usage. They find that there is no significant difference between the carousel loads among all the six storage schemes. They do not treat the problem of optimally assigning items to carousel bins. Hassini and Vickson (2003) describe a problem of storing products in carousels that are grouped in pods of two. Each pod is served by one operator. The aim is to minimize the long-run average rotational time per retrieval operation. They formulate the problem as a new type of non-linear partitioning problem and discuss several heuristic solution procedures. Hassini (2002b) studies the problem of storage location in multiple carousels in more detail. He employs genetic algorithms to derive efficient solution procedures from several variants of the storage-location problem.

LeBaron and Hoffman (1998) present a simulation model that is used to assess two storage options in a double-stacked carousel system. In Option 1, incoming items are assigned to the closest empty bin in either the upper or lower carousel. In Option 2, items are stored in upper and lower carousels in an alternate fashion. Results of these two options are not supplied.

Expected Time Models: Ha and Hwang (1994) study class-based storage in a single-carousel system. They develop expected single- and dual-cycle times for the two class-based storage assignments and show that this policy is better than the randomized storage policy. Hwang and Ha (1991) develop cycle time expressions for single- and dual-command cycles in single- and double-stacked carousel systems, where a single storage/retrieval machine performs pick-up/discharge operations. Hwang et al. (1999) find cycle time expressions for a single carousel and a double-rotary carousel, where a double shuttle is used for storage retrieval. Hamacher et al. (1998) have looked at computing the expected duration of an I/O cycle in a carousel system that allows parallel operation of several layers.

Recently, in a series of papers, Litvak and colleagues have characterized the travel time for picking a set of items in a carousel under different picking strategies. Litvak and Adan (2001) and Litvak et al. (2001) consider the case when the carousel travels under the nearest-item heuristic. Litvak and Adan (2002) consider a carousel operating under an m-step strategy; the carousel goes in the shortest distance after collecting at most m items and reverses direction at most once. Litvak and van Zwet (2004) look at the case when the items are randomly located on the carousel according to independent uniform distributions. The case of nonuniform distributions is studies in Litvak (2006).

9.5.3 Storage-Space Allocation

9.5.3.1 General Systems

Retail Shelf-Space Allocation: Empirical findings that sales of style merchandise are proportional to displayed inventory have led the marketing and operations management communities to investigate ways to optimally allocate retail shelf space. Here we provide representative works of how these investigations evolved through time. Corstjens and Doyle (1981) were among the first to present an analytical model to optimally allocate retail space to merchandise. Bookbinder and Zarour (2001) integrate the direct product profitability technique into the optimization model presented in Corstjens and Doyle (1981). Borin et al. (1994) consider the simultaneous determination of product assortment and self-space allocation. Urban (1998) presents a model that combines inventory, product assortment, and shelf-space allocation decisions. A typical shelf-space allocation model includes an objective function (e.g., maximize profit or return on investment) and product supply limitations, store capacity constraint, and space bounds for each product.

Work-in-Process (WIP) Buffer Storage Allocation: Tang (1988) considers a general integer programming allocation problem, a special case of which is a WIP storage-space allocation problem. Components' storage space is allocated in a manner that would maximize the number of subassemblies made before a resupply of components occurs. Larson and Kusiak (1995) present a generalized transportation model to optimally allocate WIP space in a job shop setting. The objective of their model is to minimize material-handling costs and ensure a certain production level. Gershwin and Schor (2000) provide a set of algorithms to efficiently select buffer space in a flow line to achieve a specified production rate. They also present a detailed review of related studies on buffer space optimization in flow lines.

Warehouse Space Allocation: Kim and Kim (1999) develop models to optimally allocate storage space to containers in port terminals. Containers that arrive together are stored in the same area (a policy referred to as segregation). These containers are then put in randomly arriving trucks to be transported to their final destination. Due to the stacking away of some containers to reach a requested container, rehandling is sometimes unavoidable. The objective is to find the height to which containers will be stacked so that rehandling is minimized. Anily (1991) extends the economic order quantity (EOQ) model by incorporating a storage-space cost that is proportional to the maximum inventory held at a warehouse. She discusses the problem of determining optimal replenishment periods that would minimize total-setup, storage-space, and inventory-holding costs. Sagan and Bishir (1991) consider a scenario where only two products are stored in a storage facility. They study the problem of allocating storage space between the two products with the objective of maximizing the expected number of requests satisfied until one of the stocks is exhausted. They show that for small storage facilities it is not always the case that a product will be allocated a space that is proportional to its demand frequency.

9.5.3.2 Space Allocation in Carousels

There is only one major study that looked at storage-space allocation in carousels. Jacobs et al. (2000) consider the case where items are picked in groups, each consisting of a known number of items of each type. Items are stored in cases of fixed sizes. They study the problem of finding how many cases of each product should be stored to maximize the number of retrievals until replenishment is needed. They present a simple heuristic to solve the problem and show bounds on its performance. Hassini (2002a) has argued that the results that Jacobs et al. (2002) arrived at can be obtained in a more intuitive and trivial way. Yeh (2002) proposes a simpler procedure for finding a better approximate solution.

9.5.4 Interface of Storage Location and Order Management

Several researchers realized that the problems of storage location and order picking are interdependent and opted for solving them simultaneously. Using a Markov chain model, Stern (1986) studied the effect of order frequency on storage locations in a carousel. However, he did not discuss how to store items based on their request frequencies. Recently, Ruben and Jacobs (1999) develop batching heuristics for the three classical storage policies (randomized, class based, and dedicated). They conclude that order-batching and storage-location decisions significantly affect the order-picking efficiency. Guenov and Raeside (1992) develop expected travel times for multiple command order-picking cycles as a function of the number of picks and the area of rack used. Brynzér and Johansson (1996) describe methods for organizing stock in warehouses in a way that would lead to a more efficient material handling from the picker's point of view. Recently, Petersen and Schmenner (1999) study the interaction of routing and storage policies in volume-based storage systems and their effects on the efficiency of order picking. Liu (1999) discusses the problem of jointly grouping items and customers. He employs clustering techniques to form the groupings. His aim is to assign items in groups to gravity-flow racks and to sequence the picking lists by customers.

Recently, some researchers have developed sophisticated models to capture the stochastic dependence of storage requirements and order requests. Malmborg (1996) worked out the probability distribution of aggregate space requirements in a randomized storage system as a function of order-picking intensity. His results are useful in, for example, measuring the trade-offs between space requirements and retrieval efficiency among random and dedicated storage policies. Chew and Tang (1999) study a storage-location assignment and order picking in a rectangular warehouse system. They also derive the probability distribution of order-picker tour and use it to analyze order-batching and storage-location strategies. Simulation is used to validate their results. Tang and Chew (1997) investigate batching and storage-allocation strategies in a manual order-picking system of small parts, which processes a high volume of orders. Order arrival is assumed to follow a Poisson process and order quantities are assumed to be independently and identically distributed negative binomial variates. They use a two-stage queueing system to model the picking system.

Several researchers have used simulation to study the interactions between storage location and order picking. Badalamenti and Bao (1986) present several simulation modules that can be useful in studying different stocking and picking policies in carousels. van den Berg and Gademann (1999) also used simulation methods to investigate different storage and order-sequencing policies. Linn and Wysk (1987) used simulation to study different sequencing and storage policies in an AS/RS. They concluded that random storage is best for low utilization, while class-based storage is better at very high utilization. In addition, they found that the efficiency of control algorithms increases as the demand rate increases.

9.5.5 Related Applications

OPA was proved to be an optimal arrangement by several researchers in different times and fields such as Mathematics, Computer Sciences, and Operations Research. In an attempt to avoid such occurrences, in this section, we mention some relevant work in other fields that dealt with problems that are analogous (at least mathematically) to some of the problems we propose to study.

9.5.5.1 Computer Storage Systems

Yue and Wong (1973) and Wong (1983) argue for the optimality of the OPA policy in computer storage systems. They use the Schur function concept to prove their results. Grossman and Silverman (1973) use another approach to prove the same result. As was shown by Pratt (1972) and later noted by Grossman and Silverman (1973), and demonstrated again by Vickson and Lu (1998) for the carousel problem, an alternate way to show the optimality of OPA is to use a theorem of Hardy et al. (1926). Pratt (1972) presents a set of permutation problems, a special case of which would be an arrangement of items on a carousel. Burkard et al. (1998) give yet a broader generalization that includes Pratt's problem as a special case. These generalizations are presented in Hassini (2002b), where it is also shown how some of the carousel storage problems can be seen as special cases of these generalizations.

Bengü (1995) also used the same approach as Wong (1983) to prove the optimality of the OPA in single-carousel systems. Seshadri et al. (1994) study the problem of arranging cartridges and file-portioning schemes in carousel-type mass storage systems. They show that the OPA policy is optimal for both anticipatory and nonanticipatory carousels (using a proof that is based on Bergmans' [1972] proof). In addition, they show that this policy minimizes the mean queueing delay and time spent in the system when arrivals are arbitrary. Han and Diehr (1991) consider the problem of jointly selecting computer storage devices and assigning files to them.

Wong (1983) also provides some algorithms for solving some versions of two-dimensional storage-location problems in computer storage systems.

Coffman and Leighton (1989) present an efficient algorithm for the dynamic storage allocation of computer files. In dynamic storage, items (files or records) of different sizes enter and leave the storage device in a random manner. This randomness, coupled with the fact that items are stored wholly (cannot be fragmented), creates interior holes (wasted storage space) in the storage device. The goal is to minimize such wasted space. An analogy to the carousel storage systems can be as follows. Each shelf can store only one type of item. Each item type is stored in consecutive shelves. Replenishment (supply) and picking (demand) rates are random. This type of model would be useful in carousels that are used in WIP storage environments. Kipnis and Robert (1990) investigate the merits of compressing the storage system each time a hole is created.

9.5.5.2 Location of Servers

Researchers in the area of location sciences have extensively studied the problem of locating servers. Two interesting applications are the "strategic" location of emergency vehicles and idle elevator(s) in buildings.

Anderson and Fontenot (1992) study the problem of positioning service units along a coordinate line. Demand for the services occurs probabilistically at any point along the coordinate line. In particular, it is reported that it is optimal to locate a single server at the median of the demand distribution. Lu (1995) considers a class of stochastic location models on a line. He discusses the problems of locating idle servers as well as optimal probability arrangement of products under stochastic demand. Parts of his results for the optimal anticipatory position for an idle disk head are reported in Gerchak and Lu (1996). Vickson and Lu (1998) discussed the problem of simultaneously finding the optimal server base and product location in both linear and circular storage systems.

Surprisingly, not many Operations Researchers have studied the location and operation of elevators in buildings. Most studies in the literature on elevators have been about the optimal dispatching control of elevators (e.g., Levy et al., 1977; Alexandris et al., 1979; Pepyne and Cassandras, 1997) or about system performance evaluation using simulation (e.g., Hummet et al., 1978; Siikonen, 1993). Recently, Matsuzaki et al. (1999) consider finding the optimal number and location of elevators in the context of multi-floor layouts. They use simulated annealing to optimize the whole layout design and genetic algorithms to study the number and location of elevators. Newell (1998) discusses strategies for operating elevators at peak traffic in tall buildings. Both studies point to the lack of literature for this type of problem. The study of optimal locations for idle elevators will be even more interesting with the introduction of new elevator technology in which elevators would have the capability to move both horizontally and vertically (e.g., Lacob, 1997) for an account of new technology directions for elevators.

Similar works are found in the area of warehousing. Egbelu (1991) discussed optimal dwell-point selection for minimizing service response time in AS/RS. Later, Egbelu (1993) solves the same problem for the special case of a loop layout in an automated guided vehicle system. Vickson et al. (1995) derive necessary optimality conditions for the problem of anticipatory positioning of a head in mirrored (data is duplicated across two or more disk drives) disk storage systems.

In the Computer Science literature, King (1990) addresses the problem of optimally positioning an idle head under different scenarios for request location and demand rate.

9.5.5.3 Cellular Manufacturing

Foulds and Wilson (1993) consider a tool carousel in which some pockets (bins) are not used for storage and there is desirability for locating tools in pairs (adjacency rating). Their objective is to locate the tools to maximize the total adjacency ratings. Other similar problems are also discussed by Foulds and Wilson (1998) and Wilson (1987). Gaboune et al. (1994a,b) and Laporte et al. (1998) discuss the problem of partitioning tools in a tool carousel to minimize machine completion time.

9.5.5.4 Patrolling Repairman

A group of identical machines (e.g., winding machines) is attended by a single operator. The operator keeps patrolling the machines in a circular fashion and whenever he or she encounters a machine that is down he or she has to fix it. It is desired to study the system's performance under several patrolling schemes. This problem has been studied by Mack (1957), Mack and Murphy (1957), and Bunday and Mack (1973). As detailed earlier, their results were then applied by Koenigsberg and Mamer (1982), Koenigsberg (1986), Kim and Koenigsberg (1987), and Bunday and El-Badri (1988) to carousel systems.

9.5.5.5 Probability Arrangement

Bergmans (1972) developed sufficient and necessary conditions for optimal probability arrangements to minimize expected travel time on different geometrical patterns. In particular, he shows that OPA is optimal for the linear arrangement case (his proof was later shown to be incomplete in Fujimoto, 1991; Seshadri et al., 1994; Vickson and Fujimoto, 1996).

9.6 Open Problems

There are several problems that have not been studied so far. We cite the ones that are most relevant to the topics discussed in this chapter:

1. Would the OPA remain optimal if the picker has multiple access locations to a single carousel? For example, one expects that if there are two access points, a "bimodal" storage location may lead to better average carousel-rotational times.
2. Would the OPA remain optimal with the twin-bin carousel design? Several carousel manufacturers have introduced a new carousel design where it is possible for an operator to pick from two adjacent bins in a single carousel stop. Even if OPA is likely to be the optimal storage-location policy, the question of how to pair products in adjacent bins is interesting. The answer to this question would certainly involve looking at cross-product-demand correlations.
3. How robust would a storage plan (e.g., OPA) be to changes in product mix or demand rate changes? This problem is paramount in cases where carousels are used in WIP storage in a multiproduct flexible manufacturing environment. This brings in the issue of how often do we want to update our product-bin assignments to minimize carousel interruptions.
4. Almost all studies in carousel systems considered that the picking operation involves unidirectional movement—the rotation of the carousel. However, in real systems individual carousels can reach heights of up to 4 m and an operator might use a ladder to access upper shelves (e.g., Goetze, 1998). Multitiered carousels can reach up to 9 m. In such

cases, a lift platform or an automatic extractor is used that moves both horizontally and vertically (e.g., Rix, 1998). The preliminary work by Wong (1983) in the area of mass storage systems can be of help here.

5. In a multiple carousel system, should the picker be moving or fixed? In other words, given some target service level, what is the optimal number of operators in a multiple carousel system and where should they be located?
6. In a multiple carousel system, should individual item stocks be stored in a single carousel or in several carousels? This question is necessarily linked to the previous question. Intuitively, one expects that the more operators we have in the system, the more there will be a one-to-one assignment of products to carousels.
7. In a multiple carousel system, is it optimal to have the same number of bins in all carousels? In general, how does the number of bins in each carousel affect the efficiency of a carousel system?
8. What is the best storage-location policy for demand-correlated items? Except for the case of Abdel-Malek and Tang (1994), studies have assumed independent demand. In many real applications this could be an oversimplification of the carousel storage system. This question becomes more relevant particularly for multiple carousel systems, where correlation may play a role in not only how products are located within individual carousels, but also in how they are assigned to carousels.
9. In the literature, it has been implicitly assumed that there is ample storage capacity in the carousels. Assuming that the total carousel storage capacity is sufficient for storing the aggregated product requirements, in cases where an individual product requirement exceeds the capacity limit of a single bin, where should the excess stock be stored? In a single-carousel case, one expects that excess demand will be distributed uniformly throughout the carousel bins. For example, when a product with excess capacity has a total requirement that is less than twice the capacity of a bin, one expects the product to be stored in bins that are "diametrically" opposite.

9.7 Conclusions

The literature on carousels is concentrated on single-carousel models. Most studies consider developing performance measures for carousel operations, rather than how to effectively operate the carousels. Only a few optimality results have been developed, and most studies provide only heuristic and simulation-based results. Some results have been developed by several different researchers in different fields and times. This last fact stems from the fact that there have been no comprehensive research projects in this field.

The area of storage location and space allocation in carousel systems presents a host of challenging problems. Only a few of these problems have been dealt with in the literature. Previous studies have used a multitude of techniques to tackle these problems, such as simulation, meta-heuristics (genetic algorithms, tabu search, and simulated annealing), combinatorial and continuous optimization, cluster analysis, queueing theory, Markov chain models, and graph theory.

References

Abdel-Malek, L. and Tang, C. A heuristic for cyclic stochastic sequencing of tasks on a drum-like storage system. *Computers and Operations Research*, 21 (4), 385–396, 1994.

Ackerman, K.B. *Practical Handbook of Warehousing*. New York: International Thomson Publishing, p. 31, 1997.

Air Canada. Winning the space race at Air Canada. *Materials Management and Distribution*, 39 (3), 35–36, 1994.

Alexandris, N.A., Barney, G.C., and Harris, C.J. Multi-car lift system analysis and design. *Applied Mathematical Modeling*, 3, 269–274, 1979.

Allen, S.L. A selection guide to AS/R systems. *IIE Solutions*, 24 (3), 28–31, 1992.

Allnoch, A. Carousel system facilitates timely distribution operation. *IIE Solutions*, December, 56–57, 1997.

Anderson, L.R. and Fontenot, R.A. Optimal positioning of service units along a coordinate line. *Transportation Science*, 26 (4), 346–351, 1992.

Anily, S. Multi-item replenishment and storage problem (MIRSP): Heuristics and bounds. *Operations Research*, 39 (2), 233–243, 1991.

Anonymous. "Circulating stockroom"—Unique service to production. *Modern Materials Handling*, 31 (12), 43, 1976a.

Anonymous. Why IBM chose carousels for small parts storage. *Modern Materials Handling*, 31 (8), 44, 1976b.

Anonymous. A great new way to organize picking for production. *Modern Materials Handling*, 35 (4), 82, 1980.

Anonymous. A new level of automation in work-in-process handling. *Modern Materials Handling*, 37 (17), 48–51, 1982.

Anonymous. How to store and retrieve parts in an integrated carousel and AS/R system. *Material Handling Engineering*, 44 (6), 85–85, 1989.

Anonymous. Hospital discovers carousel cure: System speeds picking rates, supply-room replenishment. *Materials Management and Distribution*, 36 (11), 17 and 19, 1991.

Anonymous. It pays to know ergonomics equipment. *Material Handling Engineering*, 49 (4), 58–63, 1994.

Anonymous. Carousels drive hallmark's new order selection system. *Material Handling Engineering*, 50 (12), 37–39, 1995a.

Anonymous. ESKAY supplies benchmark hybrid distribution center. *Industrial Engineering*, 25 (4), 50, 1995b.

Anonymous. Automated tool dispenser cuts costs 20%. *Modern Materials Handling*, 51 (6), 45, 1996a.

Anonymous. Boeing uses horizontal carousels to handle emergency orders. *Material Handling Engineering*, 51 (4), 84, 1996b.

Anonymous. Carousels speed orders for seed company. *Material Handling Engineering*, 51 (2), 62, 1996c.

Anonymous. Order picking productivity takes good planning. *Modern Materials Handling*, 51 (7), 20–23, 1996d.

Anonymous. Pfizer's Rx for world-class distribution. *Modern Materials Handling*, 51 (11), 26–29, 1996e.

Anonymous. Black and Decker open a huge—and hugely successful—new DC. *Modern Materials Handling*, 52 (12), 36–39, 1997a.

Anonymous. Carousels cut pick errors in half. *Transportation and Distribution*, BG4, July, 1997b.

Anonymous. Carousels increase throughput, storage capacity. *Modern Materials Handling*, 52 (11), 51, 1997c.

Anonymous. Carousels, software double pick rates. *Modern Materials Handling*, 52 (11), 65, 1997d.

Anonymous. Meeting Panasonic's needs five years into the future. *Modern Materials Handling*, 52 (13), 38–41, 1997e.

Anonymous. Product focus: Small parts storage. *Modern Materials Handling*, 52 (13), 50–54, 1997f.

Anonymous. Automating the art of book distribution. *Modern Materials Handling*, 53 (3), 47–50, 1998a.

Anonymous. Top trends and developments of 1998. *Modern Materials Handling*, 53 (14), 42–48, 1998b.

Anonymous. Storage system improves efficiency, accuracy, flexibility. *Modern Materials Handling*, 54 (4), 101, 1999.

Auguston, K.A. flexible JIT plant for the 21st century. *Modern Materials Handling*, 50 (12), 42–45, 1995a.

Auguston, K. Giving customers what they want, when they want it. *Modern Materials Handling*, 50 (10), 44–46, 1995b.

Auguston, K. How we produce printers on a JIT basis (Part 2). *Modern Materials Handling*, 50 (1), 47–49, 1995c.

Auguston, K. How Lawson achieved phenomenal productivity. *Modern Materials Handling*, 51 (14), 30–32, 1996.

Badalamenti, J. and Bao, H. Simulation modeling of robotic assembly cell incorporating carousel. *Material Flow*, 3, 121–130, 1986.

Bengü, G. An optimal storage assignment for automated rotating carousels. *IIE Transactions*, 27, 105–107, 1995.

Berglund, L. Measuring vendor performance: A BC hospital looks at the supply chain for improvement in back orders and vendor errors. *Modern Purchasing*, 39 (6), 29–30, 1997.

Bergmans, P.P. Minimizing expected travel time on geometrical patterns by optimal probability rearrangements. *Information and Control*, 20, 331–350, 1972.

Bookbinder, J.H. and Zarour, F.H. Direct product profitability and retail shelf-space allocation models. *Journal of Business Logistics*, 22 (2), 183–208, 2001.

Borin, N., Farris, P.W., and Freeland, J.R. A model for determining retail product category assortment and shelf space allocation. *Decision Sciences*, 25 (3), 359–384, 1994.

Briggs, P. Shipping the next best seller: McGraw-Hill Ryerson is at the leading edge of the publishing industry when it comes to automated data collection in the warehouse. *Canadian Transportation Logistics*, 100 (6), 20–22, 1997.

Brynzér, H. and Johansson, M.I. Storage location assignment: Using the product structure to reduce order picking times. *International Journal of Production Economics*, 46–47, 595–603, 1996.

Bunday, B.D. and El-Badri, W.K. The efficiency of *M* groups of *N* machines served by a travelling robot: Comparison of two models. *International Journal of Production Research*, 26 (2), 299–308, 1988.

Bunday, B.D. and Mack, C. Efficiency of bi-directionally traversed machines. *Journal of Royal Statistics Society: C*, 22, 74–81, 1973.

Burkard, R.E., Ela, E.C., Bericht, R., and Woeginger, G.J. The quadratic assignment problem with an anti-monge matrix and a toeplitz matrix: Easy and hard cases. *Mathematical Programming*, 82, 125–158, 1998.

Calgary General Hospital. What the doctor ordered: Storage carousel system enhances picking productivity at Calgary General Hospital. *Materials Management and Distribution*, 37 (7), 12–14, 1992.

Canadian Tire Corp. Ltd. Distribution deluxe: Canadian tire's showcase distribution centre delivers the goods. *Materials Management and Distribution*, 38 (10), 59–60, 1993.

Canedy, D. Web-based grocers betting on time-starved consumers. *Herald Tribune*, no. 36, 243 (9), September 11–12, 1999.

Chew, E.P. and Tang, L.C. Travel time analysis for general item location assignment in a rectangular warehouse. *European Journal Operational Research*, 112 (3), 582–597, 1999.

Coffman, E.G., Jr. and Leighton, F.T. A provably efficient algorithm for dynamic storage allocation. *Journal of Computer and System Sciences*, 38 (2–35), 1989.

Cormier, G. and Gunn, E.A. A review of warehouse models. *European Journal of Operational Research*, 58, 3–13, 1992.

Corstjens, M. and Doyle, P. A model for optimizing retail space allocations. *Management Science*, 27 (7), 822–833, 1981.

Diamond Phoenix Systems. Roundabout: Advanced carousel system enables small warehouse operation to reap big rewards in logistics productivity. *Materials Management and Distribution*, 42 (11), 27–28, 1997.

Egbelu, P.J. Framework for dynamic positioning of storage/retrieval machines in an automated storage/retrieval system. *International Journal of Production Research*, 29 (1), 17–37, 1991.

Egbelu, P.J. Positioning of automated guided vehicles in a loop layout to improve response time. *European Journal Operational Research*, 71, 32–44, 1993.

Egbelu, P.J. and Wu, C. Relative positioning of a load extractor for a storage carousel. *IIE Transactions*, 30, 301–317, 1998.

Emerson, C.R. and Schmatz, D.S. Results of modeling an automated warehouse system. *Industrial Engineering*, 13, 28–33, 1981.

ESAB Welding and Cutting Products. Expansion allows ESAB to enhance services. *Metalworking Production and Purchasing*, 23 (4), 26, 1996.

Feare, T. Powered systems for small parts. *Modern Materials Handling*, 54 (9), 59–61, 1999a.

Feare, T. Tracking space-bound materials for earth assembly. *Modern Materials Handling*, 54 (5), 44–46, 1999b.

Field, K.A. A DC designed to handle 3,000 orders daily. *Modern Materials Handling*, 52 (3), 37–39, 1997.

Forger, G. Leading edge order fill system sets record. *Modern Materials Handling*, 54 (4), 36–38, 1999.

Forger, G. The bar has been raised. Modern Materials Handling, 56 (6), 3–5, 2001.

Foster, T.A. More growth, new applications. *Warehousing Management*, 6 (6), 44–52, 1999.

Foulds, L.R. and Wilson, J.M. Formulation and solution of problems of tool positioning on a single machine center. *International Journal of Production Research*, 31, 2479–2485, 1993.

Foulds, L.R. and Wilson, J.M. Tool carousel design via integer programming. *33rd Annual Conference of Operational Research Society*. New Zealand: University of Auckland, 1998.

Frazelle, E.H. and Sharp, G.P. Correlated assignments: A parts assignment strategy to reduce order pick times. Material Handling Center Report MHRC-OP-87–04. Atlanta, GA: Georgia Institute of Technology, 1987.

Frazelle, E.H. and Sharp, G.P. Correlated assignments strategy can improve any order-picking operation. *Industrial Engineering*, 21 (4), 33–37, 1989.

Fujimoto, A. *Models for Determining Storage Strategies for Carousel Storage and Retrieval Systems*. M.Sc. Thesis. Waterloo, Ontario, Canada: Department of Management Sciences, University of Waterloo, 1991.

Gaboune, B., Laporte, G., and Soumis, F. Optimal strip sequencing strategies for flexible manufacturing operations in two or three dimensions. *International Journal of Flexible Manufacturing Systems*, 6 (2), 123–135, 1994a.

Gaboune, B., Laporte, G., and Soumis, F. Optimal tool partitioning rules for numerically controlled punch press operations. *Recherche Operationnelle*, 28 (3), 209–220, 1994b.

Gerchak, Y. and Lu, X. Optimal anticipatory position of a disk arm for queries of random length and location. *INFOR*, 34 (4), 251–262, 1996.

Gershwin, S.B. and Schor, J.E. Efficient algorithms for buffer space allocation. *Annals of Operations Research*, 93 (1), 117–144, 2000.

Globe Electric. Bright ideas: Growing electrical and lighting manufacturer/distributor charges ahead to increase capacity and throughput. *Materials Management and Distribution*, 39 (5), 35–36, 1994.

Goetschalckx, M. and Ratliff, H.D. Shared storage policies based on the duration stay of unit loads. *Management Science*, 36 (9), 1120–1132, 1990.

Goetze, L.A. Teamwork plus technology boost productivity 165%. *Modern Material Handling*, 53 (5), 41–43, 1998.

Grand and Toy Ltd. Grand and Toy: Your partner in the supply chain. *Modern Purchasing*, 41 (2), S1–S34, 1999.

Graves, S.C., Hausman, W.H., and Schwarz, L.B. Storage-retrieval interleaving in automatic warehousing systems. *Management Science*, 23 (9), 935–945, 1977.

Gray, A.E., Karmarkar, U.S., and Seidmann, A. Design and operation of an order-consolidation warehouse: Models and application. *European Journal of Operational Research*, 58, 14–36, 1992.

Grossman, D.D. and Silverman, H.F. Placement of records on a secondary storage device to minimize access time. *Journal of the Association for Computing Machinery*, 20 (3), 429–438, 1973.

Gudehus, T. *Principles of Order Picking: Operations in Distribution and Ware-Housing Systems*. Essen, Germany: W. Girardet, 1973 (in German).

Guenov, M. and Raeside, R. Zone shapes in class based storage and multicommand order picking when storage/retrieval machines are used. *European Journal of Operational Research*, 58 (1), 37–47, 1992.

Gupta, N.K. Kitting matrix adds accuracy to small part picking system. *Industrial Engineering*, 14, 34–38, 1982.

Ha, J. and Hwang, H. Class-based storage assignment policy in carousel system. *Computers and Industrial Engineering*, 26 (3), 489–499, 1994.

Hackman, S.T. and Rosenblatt, M.J. Allocating items to an automated storage and retrieval system. *IIE Transactions*, 22 (1), 7–14, 1990.

Hamacher, H.W., Müller, M.C., and Nickel, S. Modelling ROTASTORE: A highly parallel, short term storage system. *Operations Research Proceedings 1998*, Berlin: Springer-Verlag, pp. 513–521, 1998.

Han, B.T. and Diehr, G. An algorithm for storage device selection and file assignment. *European Journal of Operational Research*, 61, 326–344, 1991.

Hardy, G.H., Littlewood, J.E., and Polya, G. The maximum of a certain bilinear function. *Proceedings of the London Mathematical Society*, 25, 265–282, 1926.

Harmatuck, D.J. A comparison of two approaches to stock location. *Logistics and Transportation Review*, 12, 282–284, 1976.

Hassini, E. Comments on a note by Jacobs, Peck and Davis. *Computers and Operations Research*, 29 (11), 1609–1610, 2002a.

Hassini, E. *Storage Location and Space Allocation in Carousel Systems*. Ph.D. Thesis. Waterloo, Ontario, Canada: Deparment of Management Sciences, University of Waterloo, 2002b.

Hassini, E. and Vickson, R.G. A two-carousel storage location problem. *Computers and Operations Research*, 30 (4), 527–539, 2003.

Hausman, W.H., Schwarz, L.B., and Graves, S.C. Optimal storage assignment in automatic warehousing systems. *Management Science*, 22 (6), 629–638, 1976.

Heskett, J.L. Cube-per-order index: A key to warehouse stock location. *Transportation and Distribution Management*, 3, 27–31, 1963.

Heskett, J.L. Putting the cube-per-order index to work in warehouse layout. *Transportation and Distribution Management*, 4, 23–30, 1964.

Hodgson, T.J. and Lowe, T.J. Production lot sizing with material-handling cost considerations. *IIE Transactions*, 14 (1), 44–51, 1982.

Homes, G.S. Carousel. http://www.answers.com/topic/carousel?cat=technology accessed on September 16, 2007.

Hummet, G.T., Moser, T.D., and Powell, B.A. Real time simulation of elevators. *Winter Simulation Conference*, Miami Beach, FL, pp. 393–402, 1978.

Hwang, H. and Ha, J. Cycle time models for single/double carousel system. *International Journal of Production Economics*, 25, 129–140, 1991.

Hwang, H., Kim, C., and Ho, K. Performance analysis of carousel systems with double shuttle. *Computers and Industrial Engineering*, 36, 473–485, 1999.

Jacobs, D.P., Peck, J.C., and Davis, J.S. A simple heurisitc for maximizing service of carousel storage. *Computers and Operations Research*, 27, 1351–1356, 2000.

Jarvis, J.M. and McDowell, E.D. Optimal product layout in an order picking warehouse. *IIE Transactions*, 23 (1), 93–102, 1991.

Kallina, K. and Lynn, J. Application of the cube-per-order index rule for stock location in a distribution warehouse. *Interfaces*, 7 (1), 37–46, 1976.

Kallrath, J. *Online Storage Systems and Transportation Problems with Applications*. Berlin: Springer, 2005.

Kardex System. Kardex carousel storage keeps British airways aircraft on the move. *Aircraft Engineering and Aerospace Technology*, 63, 4–6, 1991.

Kaylan, A. and Medeiros, D.J. Analysis of storage policies for mini-load AS/RS. *International Journal of Production Economiccs*, 13 (4), 311–318, 1988.

Kevin, R. Carousel system allows future expansion at GE. *Modern Materials Handling*, 40 (15), 58–61, 1985.

Kim, K.H. A joint determination of storage locations and space requirements for correlated items in a miniload storage–retrieval system. *International Journal of Production Research*, 31 (11), 2649–2659, 1993.

Kim, K.H. and Kim, H.B. Segregating space allocation models for container inventories in port container terminals. *International Journal of Production Economics*, 59, 415–423, 1999.

Kim, W.B. and Koenigsberg, E. The efficiency of two groups of *n* machines served by a single robot. *Journal of Operational Research Society*, 38 (6), 523–538, 1987.

King, R.P. Disk arm movement in anticipation of future requests. *ACM Transactions on Computer Systems*, 8 (3), 214–229, 1990.

Kipnis, C. and Robert, P.H. A dynamic storage process. *Stochastic Processes and Their Applications*, 34, 155–169, 1990.

Klitz, J.K. Simulation of an automated logistics and manufacturing system. *European Journal of Operational Research*, 14 (1), 36–39, 1983.

Kobuki, M. Piece picking using a carousel system. *Proceedings of the 8th International Conference on Automation in Warehousing*, San Francisco, CA, pp. 283–294, 1987.

Koenigsberg, E. Analysis of the efficiency of carousel and tote-stacker performance. *Proceedings of the 7th International Conference on Automation in Warehousing*, San Francisco, CA, pp. 173–183, 1986.

Koenigsberg, E. and Mamer, J. The analysis of production systems. *International Journal of Production Research*, 20 (1), 1–16, 1982.

Kulturel, S., Ozdemirel, N.E., Sepil, C., and Bozkurt, Z. Experimental investigation of shared storage assignment policies in automated storage/retrieval systems. *IIE Transactions*, 31, 739–749, 1999.

Kulwiec, R.A. *Materials Handling Handbook*. New York: Wiley, 1985.

Lacob, M. Elevators on the move. *Scientific American*, 277 (4), 136–137, 1997.

Landers, T.L., Beavers, M.K., Sadiq, M., and Stuart, D.E. Software for dynamic reconfiguration order picking systems. *Computers and Industrial Engineers*, 27 (1–4), 245–248, 1994.

Laporte, G., Lopes, L., and Soumis, F. Optimal sequencing rules for some large-scale flexible manufacturing problems under the Manhattan and Chebychev metrics. *International Journal of Flexible Manufacturing Systems*, 10 (1), 27–42, 1998.

Larson, N. and Kusiak, K. Work-in-process space allocation: A model and an industrial application. *IIE Transactions*, 27, 497–506, 1995.

LeBaron, T. and Hoffman, M.L. Using simulation to optimize a horizontal carousel storage system. In: D.J. Medeiros, E.F. Watson, J.S. Carson, and M.S. Manivannan (Eds.), *Proceedings of the 1998 Winter Simulation Conference*, pp. 1247–1253, 1998.

Lee, M. A storage assignment policy in a man-on-board automated storage/retrieval system. *International Journal of Production Research*, 30 (10), 2281–2292, 1992.

Lee, M. and Hwang, H. An approach in the design of a unit-load automated carousel storage system. *Engineering Optimization*, 13, 197–210, 1988.

Levy, D., Yadin, M., and Alexandrovitz, A. Optimal control of elevators. *International Journal of Systems Science*, 8 (3), 301–320, 1977.

Lim, W.K., Bartholdi, J.J. III, and Platzman, L.K. *Storage Schemes for Carousel Conveyors under Real Time Control*. Material Handling Center Report MHRC-TR-85–10. Atlanta, GA: Georgia Institute of Technology, 1985.

Linn, R.J. and Wysk, R.A. An analysis of control strategies for automated storage and retrieval systems. *INFOR*, 25 (1), 66–83, 1987.

Litvak, N. Optimal picking of large orders in carousel system. *Operations Research Letters*, 34, 219–227, 2006.

Litvak, N. and Adan, I. The travel time in carousel systems under the nearest item heuristic. *Journal of Applied Probability*, 38, 45–54, 2001.

Litvak, N. and Adan, I. On a class of order pick strategies in paternosters. *Operations Research Letters*, 20, 377–386, 2002.

Litvak, N. and van Zwet, W.R. On the optimal travel time needed to collect *n* items on a circle. *The Annals of Applied Probability*, 14 (2), 881–902, 2004.

Litvak, N., Adan, I.J.B.F., Wessels, J., and Zijm, W.H.M. Order picking in carousel systems under the nearest item heuristic. *Probability in the Engineering and Informational Sciences*, 15, 135–164, 2001.

Liu, C. Clustering techniques for stock location and order-picking in a distribution center. *Computers and Operations Research*, 26 (10–11), 989–1002, 1999.

Lu, X. *A Class of One Dimensional Stochastic Location Models and Their Application*. Ph.D. Thesis, Waterloo, Ontario, Canada: Department of Management Sciences, University of Waterloo, 1995.

Luton, D. Consider your basic options for broken-case picking. *Materials Management and Distribution*, 42 (5), 54, 1997.

Luton, D. Benefits of reducing your order checking function. *Materials Management and Distribution*, 43 (8), 95, 1998.

Mack, C. The efficiency of *N* machines uni-directionally patrolled by one operative when walking time is constant and repair times are variable. *Journal of the Royal Statistical Society: B*, 19, 173–179, 1957.

Mack, C. and Murphy, T. The efficiency of *N* machines uni-directionally patrolled by one operative when walking and repair times are constants. *Journal of the Royal Statistical Society: B*, 19, 166–172, 1957.

Malmborg, C.J. Storage assignment policy tradeoffs. *International Journal of Production Research*, 34 (2), 363–378, 1996.

Malmborg, C.J. and Bhaskaran, K. A revised proof of optimality for the cube-per-order index rule for stored item location. *Applied Mathematical Modeling*, 14, 87–95, 1990.

Malmborg, C.J. and Deutsch, S.J. A stock location model for dual address order picking systems. *IIE Transactions*, 20 (1), 44–52, 1988.

Malmborg, C.J. and Krishnakumar, B. Optimal storage assignment policies for multiaddress warehousing systems. *IEEE Transactions on Systems, Man, and Cybernetics*, 19 (1), 197–204, 1989.

Maloney, D. e-fulfillment: Two different views. *Modern Materials Handling*, 56 (6), 10–11, 2001.

Matson, J.O. and White, J.A. Operational research and material handling. *European Journal of Operational Research*, 11, 309–318, 1982.

Matsuzaki, K., Irohara, T., and Yoshimoto, K. Heuristic algorithm to solve the multi-floor layout problem with the consideration of elevator utilization. *Computers and Industrial Engineers*, 36, 487–502, 1999.

Mcgillivray, R. and Geiger, J. Bright future. *Materials Management and Distribution*, 42 (1), 18–20, 1997.

McGraw-Hill Ryerson Ltd. One for the books: Publisher's computer-controlled carousel system boosts pick efficiency, saves big on space. *Materials Management and Distribution*, 38 (3), 20–21, 1993.

Meller, R.D. and Klote, J.F. A throughput model for carousel/vlm pods. *IIE Transactions*, 36, 725–741, 2004.

Muralidharan, B., Linn, R.J., and Pandit, R. Shuffing heuristics for the storage location assignment in an AS/RS. *International Journal of Production Research*, 3 (6), 1661–1672, 1995.

Newell, G.F. Strategies for serving peak elevator traffic. *Transportation Research: B*, 32 (8), 583–588, 1998.

Paris, W.F., III. Automated warehouse system expedites medical resupply to hospital departments. *IIE Solutions*, 20 (10), 36–38, 1988.

Park, B.C. and Rhee, Y. Performance of carousel systems with "organ-pipe" storage. *International Journal of Production Research*, 43 (921), 4685–4695, 2005.

Park, B.C., Frazelle, E.D., and White, J.A. Buffer sizing models for end-of-aisle order picking systems. *IIE Transactions*, 31, 31–38, 1999.

Park, B.C., Park, J.Y., and Foley, R.D. Carousel system performance. *Journal of Applied Probability*, 40, 602–612, 2003.

Pepyne, D.L. and Cassandras, C.G. Optimal dispatching control for elevator systems during uppeak traffic. *IEEE Transactions on Control Systems Technology*, 5 (6), 1997.

Petersen, C.G., II. and Schmenner, R.W. An evaluation of routing and volume-based storage policies in an order picking operation. *Decision Sciences*, 30 (2), 481–501, 1999.

Pletsch, A. Technology invades the warehouse. *Plant*, 57 (9), 18, 1998.

Pratt, V.R. An $n \log n$ algorithm to distribute n records optimally in a sequential access file. In: R.E. Miller and J.W. Thatcher (Eds.), *Complexity of Computer Computations: Proceedings*, New York: Plenum Press, 1972.

Rix, R. Right blend: Integrated warehouse solution gets key parts to BMW dealers. *Materials Management and Distribution*, 43 (11), 20–21, 1998.

Robertson, R. Tall order: New equipment makes inventory picking easier. *Materials Management and Distribution*, 43 (8), 83–87, 1998.

Robertson, R. More Choice: Healthy logistics at London drugs. *Materials Management and Distribution*, 44 (5), 14–16, 1999a.

Robertson, R. Pain relief: Bayer Inc enjoys productivity gains with warehouse management system. *Materials Management and Distribution*, 42 (4), 14–18, 1999b.

Rosenblatt, M.J., Roll, Y., and Zyser, V. A combined optimization and simulation approach for designing automated storage/retrieval systems. *IIE Transactions*, 25 (1), 40–50, 1993.

Rouwenhorst, B., van den Berg, J.P., van Houtum, G.J., and Zijm, W.H.M. Performance analysis of a carousel system: Progress in material handling research—1996. Charlotte, NC: The Material Handling Institute, pp. 495–511, 1997.

Rouwenhorst, B., Mantel, R.J., Reuter, B., Stockrahm, V., van Houtum, G.J., and Zijm, W.H.M. Warehouse design and control: Framework and literature review. Working Paper 44. Enschede, The Netherlands: Beta Research Institute, 1998.

Ruben, R.A. and Jacobs, F.R. Batch construction heuristics and storage assignment strategies for walk/ride and pick systems. *Management Science*, 45 (4), 575–596, 1999.

Sadiq, M., Landers, T.L., and Taylor, G.D. An assignment algorithm for dynamic picking systems. *IIE Transactions*, 28, 607–616, 1996.

Sagan, H. and Bishir, J.W. Optimal allocation of storage space. *European Journal of Operational Research*, 55 (1), 82–90, 1991.

Sargent, T.A. and Kay, M.G. Implementation and utilization of a decentralized storage system costing model. *International Journal of Operations and Production Management*, 15 (9), 210–219, 1995.

Schulz, D.P. Consumer changes spur growing use of carousels in supermarket warehouses. *Stores*, 81 (2) Section 1, 40–43, 1999.

Schwarz, L.B., Graves, S.C., and Hausman, W.H. Scheduling policies for automatic warehousing systems: Simulation results. *AIIE Transactions*, 10 (3), 260–270, 1978.

Sciex Ltd. Power picking: Carousels play important kitting role for manufacturer of medical products. *Materials Management and Distribution*, 41 (11), 18, 1996.

Seshadri, S., Rotem, D., and Segev, A. Optimal arrangement of cartridges in carousel type mass storage systems. *The Computer Journal*, 37 (910), 873–887, 1994.

Siikonen, M. Elevator traffic simulation. *Simulation*, 61 (4), 257–267, 1993.

Stern, H.I. Parts location and optimal picking rules for a carousel conveyor automatic storage and retrieval system. *Seventh International Conference on Automation in Warehousing*, San Francisco, CA, pp. 185–193, 1986.

Su, C. Performance evaluations of carousel operation. *Production Planning and Control*, 9 (5), 477–488, 1998.

Subaru Auto Canada Ltd. Subaru Canada: All function, no forms. *Materials Management and Distribution*, 38 (11), 14–15, 1993.

Sung, C.S. and Han, Y.H. Determination of automated storage/retrieval system size. *Engineering Optimization*, 19, 269–286, 1992.

Tang, C.S. A max-min allocation problem: Its solution and applications. *Operations Research*, 36 (2), 359–367, 1988.

Tang, L.C. and Chew, E. Order picking systems: Batching and storage assignment strategies. *Computers and Industrial Engineering*, 33 (3–4), 817–820, 1997.

Thomas, J. A safe decision. *Warehousing Management*, 6 (3), 42–44, 1999.

Timex Canada Inc. Quick response time: How Timex sets the pace. *Materials Management and Distribution*, 41 (5), 16–17, 1996.

Torok, D.B. Paperless distribution: How pioneer-standard ensures customer satisfaction. *Material Handling Engineering*, 47 (6), 36–40, 1992.

Trevino, J., Liao, F., and Chopra, S. Cardes: Carousel design. *International Journal of Production Research*, 32 (5), 1013–1026, 1994.

Trunk, C. Powerful handling, control and ergonomics with carousels. *Material Handling Engineering*, 48 (7), 51–58, 1993.

Trunk, C. Horizontal carousels tackle the big jobs. *Material Handling Engineering*, 51 (7), 47–48, 1996a.

Trunk, C. Solve small parts problems with horizontal carousels. *Material Handling Engineering*, 51 (8), 53–55, 1996b.

Urban, T.L. An inventory-theoretic approach to product assortment and shelf-space allocation. *Journal of Retailing*, 74 (1), 15–35, 1998.

Valenti, M. Making merry on the Midway. *Mechanical Engineering*, 119 (5), 64–69, 1997.

van den Berg, J.P. Class-based storage allocation in a single-command warehouse with space requirement constraints. *International Journal of Industrial Engineering*, 3 (1), 21–28, 1996.

van den Berg, J.P. A literature survey on planning and control of warehousing systems. *IIE Transactions*, 31 (8), 751–762, 1999.

van den Berg, J.P. and Gademann, A.J.R.M. (Noud). Simulation study of an automated storage/retrieval system. *International Journal of Production Research*, 59 (1–3), 519–528, 1999.

van den Berg, J.P. and Zijm, W.H.M. Models for warehouse management: Classification and examples. *International Journal of Production Economics*, 59, 519–528, 1999.

van den Berg, J.P., Sharp, G.P., Gademann, A.J.R.M. (Noud), and Pochet, Y. Forward-reserve allocation in a warehouse with unit-load replenishment. *European Journal of Operational Research*, 111, 98–113, 1998.

van Oudheusden, D.L. and Zhu, W. Storage layout of AS/RS racks based on recurrent orders. *European Journal of Operational Research*, 58, 48–56, 1992.

Veterinary Purchasing. Pet project: Veterinary purchasing's inventory and distribution overhaul results in faster, more accurate deliveries of essential vet supplies. *Materials Management and Distribution*, 41 (3), 42 and 44, 1996.

Vickson, R.G. and Fujimoto, A. Optimal storage locations in a carousel storage and retrieval system. *Location Science*, 4 (4), 237–245, 1996.

Vickson, R.G., Gerchak, Y., and Rotem, D. Optimal positioning of read/write heads in mirrored disks. *Location Science*, 3 (2), 125–132, 1995.

Vickson, R.G. and Lu, X. Optimal product and server locations in one-dimensional storage racks. *European Journal of Operational Research*, 105, 18–28, 1998.

Vlasiou, M., Adan, I.J.B.F., and Wessels, J.A. Lindley-type equation arising from a carousel problem. *Journal of Applied Probability*, 41, 1171–1181, 2004.

Weiss, D.J. Computer controlled carousels. *Third International Conference on Automation in Warehousing*, Stratford-upon-Avon, UK, 1980.

Wilson, H.G. Order quantity, product popularity, and the location of stock in a warehouse. *AIIE Transactions*, 9 (3), 230–237, 1977.

Wilson, J.M. Formulation and solution of a set of sequencing problems for FMS. *Transactions of the Institute of Mechanical Engineering*, 20, 247–249, 1987.

Wilson, M.J. Computer-integrated flexible manufacturing. *Production*, 96 (6), 48–50, 1985.

Wisnia, S.E. Running to the rescue. *Industrial Distribution*, 86 (4), 120–123, 1997.

Witt, C.E. Automation brings fast, accurate service to General Motors' parts warehouse. *Material Handling Engineering*, 44 (3), 82–86, 1989.

Witt, C.E. Carousels handy in craft distribution center. *Material Handling Engineering*, 50 (2), 56–59, 1995.

Wong, C.K. *Algorithmic Studies in Mass Storage Systems*. New York: Springer-Verlag, 1983.

Yeh, D.A. Note on a simple heuristic for maximizing service of carousel storage. *Computers and Operations Research*, 29 (11), 1605–1608, 2002.

Yue, P.C. and Wong, C.K. On the optimality of the probability ranking scheme in storage applications. *Journal of the Association for Computing Machinery*, 20 (4), 624–633, 1973.

Chapter 10

Algebraic Deadlock Avoidance Policies for Sequential Resource Allocation Systems*

Spyros Reveliotis

CONTENTS

---

* Some of the material presented in this chapter originally appeared in J. Park and S. Reveliotis, Deadlock avoidance in sequential resource allocation systems with multiple resource acquistions and flexible routings. *IEEE Transactions On Automatic Control*, 46 (10), 1572–1583 (© [2001] IEEE), S. Reveliotis, Implicit siphon control and its role in the liveness enforcing supervision of sequential resource allocation systems. *IEEE Transactions on Systems, Man, and Cybernetics*—Part A, 37 (3), 319–328 (© [2007] IEEE), S. Reveliotis, *Real-Time Management of Resource Allocation Systems: A Discrete Event Systems Approach*, Springer, New York (© [2005] Springer), S. Reveliotis and J.Y. Choi, Designing reversibiilty-enforcing supervisors of polynomial complexity for bounded Petri nets through the Theory of Regions. Lecture Notes in Computer Science, 4024, 322–341 (© [2006] Springer).

Abstract As many contemporary technological applications move to operational modes of more extensive and flexible automation, there is a rising need to design and control the underlying resource allocation not only for efficiency, but also for logical correctness and internal consistency. The material presented in this chapter offers a unifying and comprehensive treatment of a class of policies that have been proposed as an effective and efficient solution to this emerging class of logical control problems.

10.1 Introduction

This chapter deals with the problem of managing the resource allocation that takes place in various contemporary technological applications, including flexibly automated production systems, automated railway or monorail transportation systems, electronic workflow management systems, and business transaction supporting systems. A distinguishing trait of all the aforementioned applications is that they seek to limit the role of the human element to remote high-level supervision, while placing the burden of the real-time monitoring and coordination of the ongoing activity upon a computerized control system. This development is justified by a number of technical, economic, and safety considerations, and it is facilitated by the advent of modern computing and sensing technologies. At the same time, the effective support of such an extensively automated operational mode poses new challenges to the designers and supervisors of these systems. A particularly challenging task in the emerging regime is the synthesis of the "control logic" that will manage the allocation of the resources of the aforementioned systems to the various running processes in a way that guarantees the orderly and expedient execution of all these processes, while preserving the operational "flexibility" sought by these environments.

The applications depicted in Figures 10.1 and 10.2 exemplify the aforementioned problem and they highlight the currently prevailing practice. Figure 10.1 depicts a small robotic cell* with three processing stations, W_1, W_2, and W_3. Each of these stations can accommodate only one part at a time, and collectively they support the production of two different part types, J_1 and J_2, whose processing routes are annotated in the figure. It should be clear to the reader that the state depicted in Figure 10.1 is a problematic state, because the two depicted jobs mutually block each other. Furthermore, this blockage will persist until it is realized and resolved, probably only through human intervention, by unloading one of the two jobs, a rather costly operation in the considered setting. Similarly, Figure 10.2 depicts a zone-controlled by automated guidance vehicle (AGV) system, where three vehicles block permanently the advancement of each other at a junction of the guide-path network.

* The depicted configuration is very similar in its basic topology to the "cluster tools" used extensively in the contemporary semiconductor manufacturing industry.

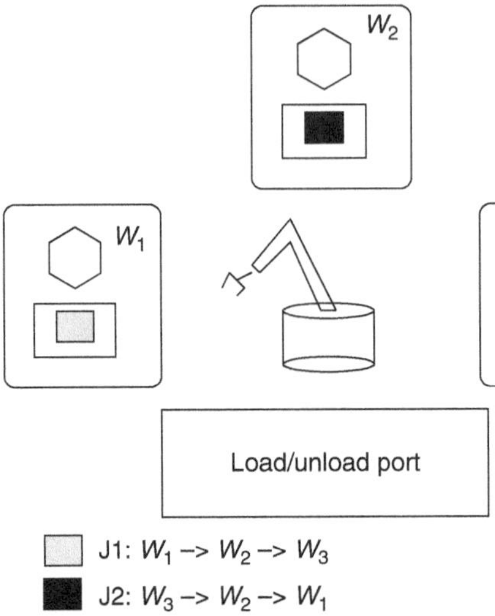

Figure 10.1 A manufacturing system deadlock.

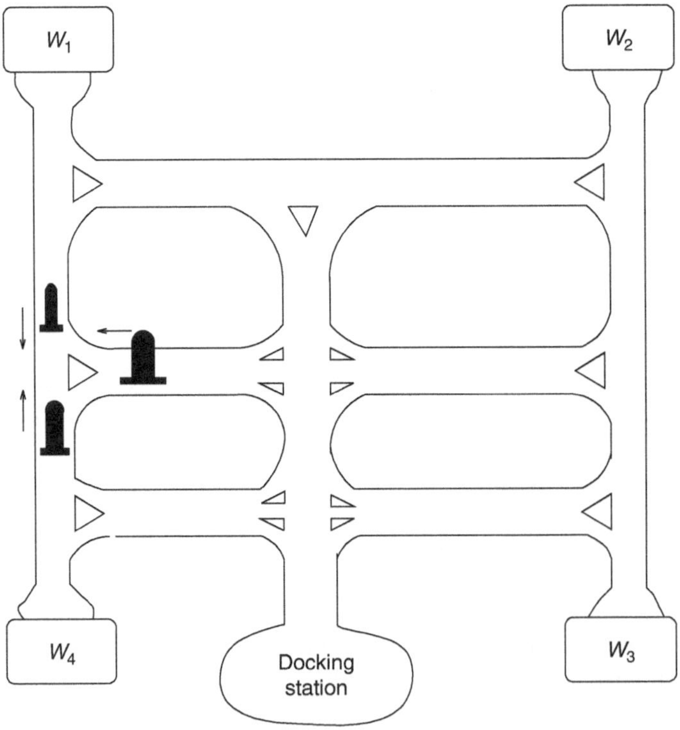

Figure 10.2 An AGV system deadlock.

The situations depicted in Figures 10.1 and 10.2 are known respectively as a manufacturing and an AGV "deadlock." In operational contexts relying on manual labor, such deadlock problems have been typically addressed through last-minute improvisation. However, in a fully automated context, the resolution of these problems must be part of the overall design process. In a lack of a systematic methodology to address these issues, past engineering practice has resorted to rather simplistic approaches that provide a robust solution to the problem, but only at the expense of the system operational flexibility, efficiency, and productivity. Hence, to prevent the occurrence of the manufacturing deadlock mentioned above, most contemporary cells are operated in a "batching" mode that separates the production of the supported part types. By preventing the concurrent production of the supported parts, the system will always be operated in a unidirectional flow that is free of any deadlocking problems. However, such a solution is a substantial departure from the notion of flexible automation and its advertised advantages. In a similar spirit, most contemporary AGV systems are designed according to the "tandem" configuration depicted in Figure 10.3, where the entire guide-path network is decomposed to a number of unidirectional loops, interfacing at a number of strategically preselected points. Although managing to avoid deadlock, tandem AGV systems have to experience expensive "hand-off" procedures at the loop junctions, and the vehicle filing at any single loop implies that the pacing in that loop is determined by the slowest vehicle.

The effective and systematic resolution of the aforementioned deadlock problems must be based on a detailed study of the event sequences that take place during the system operation. Hence, the analysis and resolution of these problems necessitate a methodological framework that places the

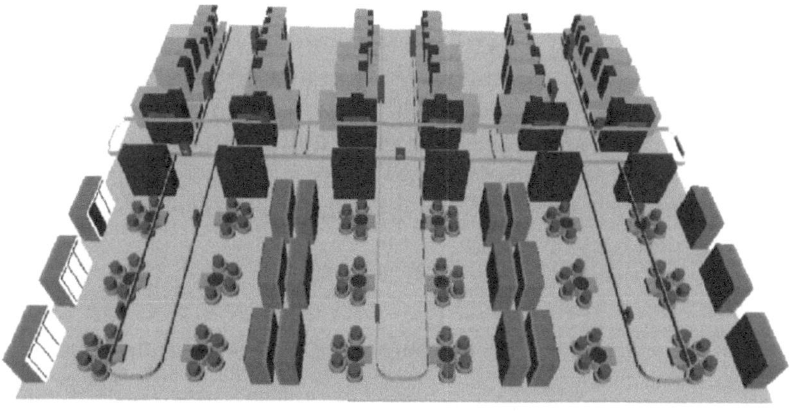

Figure 10.3 A tandem AGV system.

emphasis on the analysis and shaping of these event sequences, and it is substantially different from those that have been traditionally applied to performance-oriented control.* Such a methodological framework has been provided by an area of modern control theory known as qualitative or logical analysis and control of discrete event systems (DES) (Cassandras and Lafortune, 1999), and the last 15 years have seen the emergence of a substantial body of results on the aforementioned deadlock problems that are based on representations and analytical tools coming from this area. The fundamental abstraction that underlies the development of these results is the sequential resource allocation system (RAS), formally defined as follows.

Definition 1 (Reveliotis, 2005): A sequential RAS is defined as a 5-tuple $\Phi = <\mathcal{R}, C, \mathcal{P}, \mathcal{A}, \mathcal{T}>$, where:

1. $\mathcal{R} = \{R_1, \ldots, R_m\}$ is the set of the system resource types.
2. $C: \mathcal{R} \rightarrow Z^+$—the set of strictly positive integers†—is the system capacity function, characterizing the number of identical units from each resource type available in the system. Resources are considered to be reusable, that is, each allocation cycle does not affect their functional status or subsequent availability, and therefore, $C(R_i) \equiv C_i$ constitutes a system invariant for each i.
3. $\mathcal{P} = \{\Pi_1, \ldots, \Pi_n\}$ denotes the set of the system process types supported by the considered system configuration. Each process type Π_j is a composite element itself, in particular, $\Pi_j = <\mathcal{S}_j, \mathcal{G}_j>$, where:
   (a) $\mathcal{S}_j = \{\Xi_{j1}, \ldots, \Xi_{j,l(j)}\}$ denotes the set of processing stages involved in the definition of process type Π_j, and
   (b) $\mathcal{G}_j$ represents some data structure communicating some sequential logic that applies to the execution of any process instance of type Π_j.
4. $\mathcal{A}: \bigcup_{j=1}^{n} \mathcal{S}_j \rightarrow \prod_{i=1}^{m} \{0, \ldots, C_i\}$ is the resource allocation function associating every processing stage Ξ_{jk} with a resource allocation request $\mathcal{A}(j,k) \equiv A_{jk}$. More specifically, each A_{jk} is an m-dimensional vector, with its i-th component indicating the number of resource units of resource type R_i necessary to support the execution of stage Ξ_{jk}. Obviously, in a well-defined *RAS*, $A_{jk}(i) \leq C_i, \forall j,k,i$.

* While it is true that event timing can provide a mechanism for enforcing event sequences, such an approach will tend to be very brittle in the face of the stochasticity characterizing the operation of the considered applications.

† Also, in this document, Z_0^+ will denote the set of nonnegative integers, Z will denote the set of all integers, and $\Re$ will denote the set of reals.

5. $\mathcal{T} : \bigcup_{j=1}^{n} \mathcal{S}_j \rightarrow \mathcal{D}$ is the timing function, corresponding to each processing stage Ξ_{jk} a distribution D_{jk} that characterizes the statistics of the processing time t_{jk}, experienced during the execution of stage Ξ_{jk}.

The above characterization of the considered RAS is further qualified by the following conditions that detail their operation and facilitate the subsequent analysis:

Condition 1: Under expedient resource allocation, every activated process instance will terminate in a "finite" number of processing steps.

Condition 2: Every processing stage $\Xi_{jk} \in \mathcal{S}_j$ can be realized by at least one execution sequence supported by $\mathcal{G}_j$.

Condition 3: A process instance j_j advances from stage Ξ_{jk} to a successor stage $\Xi_{j,k+1}$ only upon being allocated the entire set of resources implied by the resource allocation request $A_{j,k+1}$. The allocation of all these resources takes place simultaneously, and it is only at this point that the process instance j_j releases the resources allocated to it for the execution of processing stage Ξ_{jk}.

Condition 4: The only way in which two distinct activated process instances can interact with each other is through their potential contest for some of the system resources.

Condition 1 excludes those pathological situations in which an executing process can entangle itself in an infinite loop. In well-designed applications, a process will not be allowed to run within the system indefinitely. From a representational standpoint, the satisfaction of this assumption allows the modeling of the process-defining logic through an "acyclic" data structure. Condition 2 essentially ensures that the process representation does not introduce redundant processing stages. Condition 3 introduces the "hold-while-waiting" effect in the considered resource allocation which is at the base of the considered deadlock problems.* Condition 4 applies primarily to complex process flows that involve parallelization, and implies that the logic coordinating the execution of the various process threads does not "confound" enacted subprocesses belonging to different process

* As demonstrated by the examples presented in Figures 10.1 and 10.2, this hold-while-waiting effect frequently results from the need to physically buffer the various process instances at any single point in time, that is, parts processed in a flexibly automated production system or vehicles in an AGV network are physical entities and they always need to be accommodated somewhere during their sojourn through the system. It must be noticed, however, that, while providing the necessary specificity for the underlying resource allocation dynamics, the aforestated assumptions do not

instantiations. Finally, to facilitate the subsequent discussion on the complexity of the posed problems and the proposed solutions, we also introduce the quantity $|\Phi| \equiv |\mathcal{R}| + |\bigcup_{j=1}^{n} \mathcal{S}_j| + \sum_{i=1}^{m} C_i$, which defines the size of the RAS Φ.

The problem of the real-time management of the resource allocation taking place in the considered RAS can be effectively addressed through the supervisory control (SC) framework depicted in Figure 10.4. As indicated in Figure 10.4, the proposed controller is event driven, that is, the control actions commanded to the underlying RAS can be perceived as the controller responses to the various events taking place in the RAS domain and

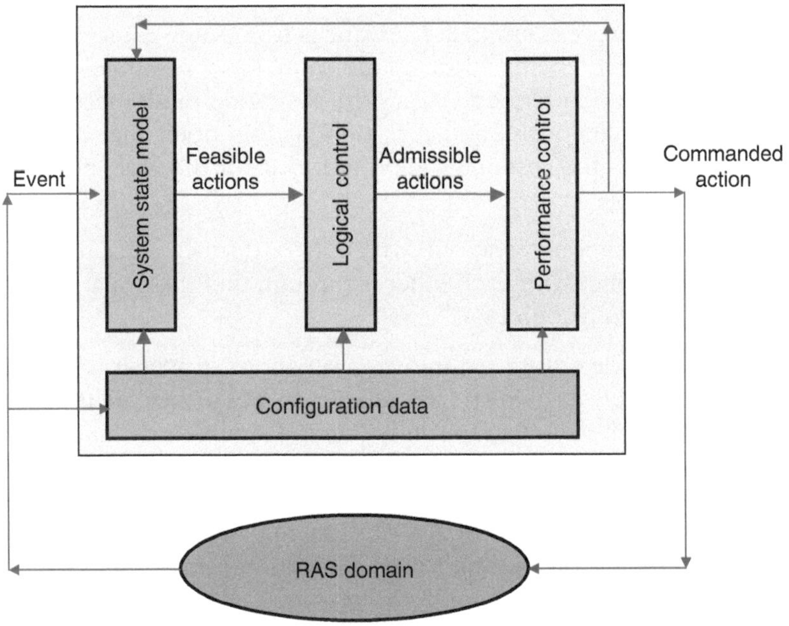

Figure 10.4 An event-driven framework for the RAS supervisory control problem.

compromise the modeling power of our framework, because one can capture any additional resource allocation dynamics by augmenting the specification of process Π_j. For example, one can model the fact that, at some particular process stage Ξ_{jk}, process Π_j might release (some of) its currently allocated resources before advancing to stage $\Xi_{j,k+1}$, by introducing to the process specification an intermediate process stage Ξ_{jq}, with resource allocation request A_{jq} equal to A_{jk} minus the deallocated resource set.

communicated to the controller through a monitoring function. Hence, the entire control function evolves in a number of cycles, with each cycle being triggered by a RAS event communicated to the controller. Conceptually, each cycle consists of three major phases: (1) In Phase I, the controller updates a representation of the RAS state so that it represents the RAS status after the occurrence of the communicated event. This representation, combined with the system knowledge about the running RAS configuration, encodes the entire set of feasible actions that could be executed by the RAS as a response to the occurring event. (2) In Phase II, the controller applies a logical control policy to filter out from the set of feasible actions identified in Phase I, the set of admissible actions,that is, this set of actions that satisfy some logical specification for the RAS behavior, like deadlock freedom. (3) Finally, in Phase III, the set of admissible actions is provided to the performance-oriented component of the RAS supervisor to select the one that will be communicated eventually to the RAS environment, in a way that observes some performance considerations. In addition to this basic functionality, the RAS controller should be able to respond to the various contingencies taking place in the RAS domain, by (1) appropriately updating the RAS configuration database and (2) revising the logical and performance-oriented control logic to apply in the emerging RAS configuration. This last function will be collectively characterized as (re-) configuration management.

A systematic exposition of most of the existing results concerning the design and deployment of the RAS control function depicted in Figure 10.4 can be found in Zhou and Fanti (2004) and Reveliotis (2005). The relevant theory provides (1) formal characterizations of the underlying RAS dynamics in the context of finite state automata (FSA) (Hopcroft and Ullman, 1979) and Petri net (PN) (Murata, 1989) modeling frameworks, as well as some more ad hoc representations; (2) a characterization of the optimal logical control problem for the considered RAS; (3) a rigorous study of the complexity of the aforementioned optimal control problem, that not only establishes its NP-hardness (Garey and Johnson, 1979) but also identifies important practical cases that admit optimal solutions of polynomial complexity with respect to the size $|\Phi|$ of the underlying RAS; (4) efficient solutions for the remaining cases that trade off optimality for computational tractability; (5) a formal characterization of the notions of robustness and re-configurability for the considered operational context; and (6) some preliminary results regarding the effective integration of the logical and the performance-oriented control.

The material presented in this chapter complements the aforementioned results by providing a unifying treatment of a particular class of RAS logical control policies known as "algebraic." More specifically, the chapter aggregates and systematizes a number of results regarding this class of policies that emerged during the last three years, partly in response to problems and thoughts generated during the development of Reveliotis (2005). Collectively, the presented results offer (1) a thorough characterization of the considered class of policies; (2) effective computational tools for their design and implementation on any given RAS instance, and (3) an insightful explanation of the mechanisms underlying the policy effectiveness and computational tractability. From an organizational standpoint, the chapter will evolve as follows: Section 10.2 provides a systematic characterization of the RAS dynamics by means of the PN modeling framework. Subsequently, Section 10.3 introduces the class of algebraic logical control policies, and it establishes that they also admit an effective representation within the PN modeling framework. Section 10.4 provides an analytical characterization of the entire set of algebraic logical control policies that can ensure the deadlock-free operation of any given RAS. Beyond its theoretical interest, this characterization enables also the introduction of a notion of optimality within the scope of the considered policies. In principle, such an optimized implementation is effectively computable through the presented developments. However, from a more practical standpoint, this computation is limited by a very high complexity. Hence, Section 10.5 offers an additional approach that can enable the synthesis of algebraic logical control policies for any given RAS while drastically mitigating the complexity problems arising from the previous approach. Section 10.6 offers some interesting and fundamental insights regarding the mechanism that facilitates the functionality of the considered policies. Finally, Section 10.7 concludes the chapter and suggests some directions for future developments. Throughout the following discussion, the emphasis is placed on the systematic and accessible presentation of the key results and their implications. Therefore, we have frequently omitted the detailed technical arguments underlying the relevant derivations; these arguments can be traced in the provided citations.

10.2 PN-Based Representation of the Considered RAS

As mentioned in Section 10.1, Petrinets (PN) (Murata, 1989) have been one of the primary modeling frameworks employed for the analysis and control of the RAS dynamics considered in this work. In this section, we define the PN subclass that characterizes the RAS behavior encompassed by Definition 1, presuming that the reader is already familiar with the basic PN concepts.* The subsequent discussion proceeds in three steps: (1) first we introduce a PN model that expresses the execution logic of any single process instance; (2) subsequently, this model is augmented with resource places to represent the dynamics of the associated resource allocation; and (3) finally, the complete RAS model is obtained by merging the various subnets developed in step (2) through their common resource places.

10.2.1 PN-Based Modeling of the RAS Process Types

In the PN modeling framework, the process type $\Pi_j = <S_j, G_j>$, introduced in item 3 of Definition 1, will be represented by the concept of the "process subnet," formally defined as follows.

Definition 2: A process subnet is an ordinary PN $\mathcal{N}_P = (P, T, W, M_0)$ such that:

i. $P = P_S \cup \{i, o\}$ *with* $P_S \neq \emptyset$;
ii. $T = T_S \cup \{t_I, t_F, t^*\}$;
iii. $i^\bullet = \{t_I\}$; ${}^\bullet i = \{t^*\}$;
iv. $o^\bullet = \{t^*\}$; ${}^\bullet o = \{t_F\}$;
v. $t_I^\bullet \subseteq P_S$; ${}^\bullet t_I = \{i\}$;
vi. $t_F^\bullet = \{o\}$; ${}^\bullet t_F \subseteq P_S$;
vii. $(t^*)^\bullet = \{i\}$; ${}^\bullet(t^*) = \{o\}$;
viii. the underlying digraph is strongly connected;
ix. $M_0(i) > 0 \wedge M_0(p) = 0, \forall p \in P \backslash \{i\}$;
x. $\forall M \in R(\mathcal{N}_P, M_0), M(i) + M(o) = M_0(i) \Rightarrow M(p) = 0, \forall p \in P_S$.

The PN-based process representation introduced by Definition 2 is depicted in Figure 10.5. Process instances waiting to initiate processing are represented by tokens in place i, while the initiation of a process instance is modeled by the firing of transition t_I. Similarly, tokens in place o represent completed process instances, while the event of a process completion is modeled by the firing of transition t_F. Transition t^* allows the token recirculation—that is, the token transfer from place o to place i—to model repetitive process execution. Finally, the part of the net between transitions t_I and t_F, that involves the process places P_S, models the sequential logic defining the considered process type. In particular, places $p \in P_S$ correspond to the various processing stages $\Xi_{jk} \in S_j$, while the net connectivity among these places concretizes component G_j of Π_j (c.f. item [3b] of Definition 1). As it can be seen in Definition 2, this part of the process

* A primer on the key PN concepts employed in this work is provided in the Appendix; for a more extensive discussion, the interested reader is referred to Murata, 1989.

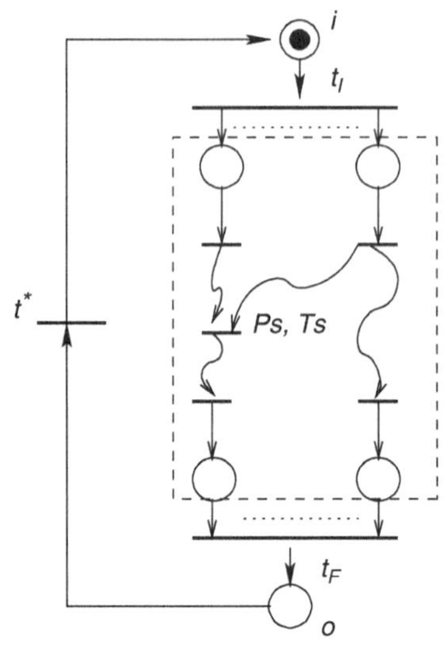

Figure 10.5 The process net structure of Definition 2.

subnet can be quite arbitrary. However, to capture the requirements posed by Conditions 1, 2, and 4 in Section 10.1, we further qualify the considered process subnets through the following three assumptions:

Assumption 1: The process subnets considered in this work are assumed to be acyclic, that is, the removal of transition t^* from them renders them acyclic digraphs.

Assumption 2: The process subnets considered in this work are assumed to be quasi-live for $M_0\,(i) = 1$.

Assumption 3: The process subnets considered in this work are assumed to be strongly reversible, that is, their initial marking M_0 can be reached from any marking $M \in R(\mathcal{N}_P, M_0)$, through a firing sequence that does not contain transition t_I.

Assumption 1 is introduced to satisfy Condition 1 of Section 10.1. Assumption 2 pertains to the satisfaction of Condition 2, by essentially stipulating that, in the considered process subnets, every transition models a meaningful event that can actually occur during the execution of some process instance, and therefore, it is not redundant. Assumption 3 pertains to the satisfaction of Condition 4, as it essentially stipulates that, at any point in time and under expedient resource allocation, all active process instances can advance to completion. As the main focus of this work is on the analysis and control of the resource allocation function taking place in the considered environments, we forego the further study of the process subnets themselves and the investigation of the structural and behavioral properties implied by Definition 2 and Assumptions 1–3. The interested reader can find some relevant discussion and results provided by Van der Aalst (1996, 1997), Jeng et al. (2002), Van der Aalst and Van Hee (2002), and Van Hee et al. (2003).

10.2.2 PN-Based Modeling of the Resource Allocation Function

The modeling of the resource allocation associated with the process stage Ξ_{jk} corresponding to any place $p \in P_S$, necessitates the augmentation of the process subnet $\mathcal{N}_P$, defined above, with a set of resource places $P_R = \{r_l, l = 1, \ldots, m\}$, of initial marking $M_0(r_l) = C_l$, $l = 1, \ldots, m$, and with the corresponding flow sub-matrix, Θ_{P_R}, expressing the allocation and de-allocation of the various resources to the process instances as they advance through their processing stages, according to the protocol stipulated by Condition 3 in Section 10.1. The resulting net will be called the "resource-augmented process subnet" and it will be denoted by $\overline{\mathcal{N}_P}$. Its basic structure is depicted in Figure 10.6. Notice that the characterization of transitions t^*, t_I, and t_F provided in the above discussion, implies that $(t^*)^\bullet \cap P_R = {}^\bullet(t^*) \cap P_R = (t_I)^\bullet \cap P_R = {}^\bullet(t_F) \cap P_R = \emptyset$. On the other hand, the reusable nature of the system resources presumed in this work, is modeled by the following assumption regarding the resource-augmented process net $\overline{\mathcal{N}_P}$:

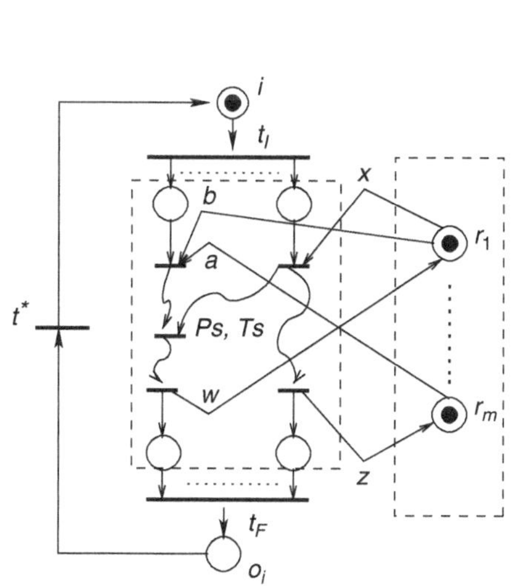

Figure 10.6 The resource-augmented process net.

Assumption 4: Let $\overline{\mathcal{N}_P} = (P_S \cup \{i,o\} \cup P_R, T, W, M_0)$ denote a resource-augmented process net. Then, $\forall l \in \{1,\ldots,|P_R|\}$, there exists a p-semiflow y_{r_l} such that: (i) $y_{r_l}(r_l)=1$; (ii) $y_{r_l}(r_j)=0, \forall j \neq l$; (iii) $y_{r_l}(i)=y_{r_l}(o)=0$; (iv) $\forall p \in P_S$, $y_{r_l}(p)=$ number of units from resource R_l required for the execution of the processing stage modeled by place p.

Furthermore, the following assumption extends the requirement for quasi-liveness of the process net $\mathcal{N}_P$, introduced by Assumption 2, to the resource-augmented process net $\overline{\mathcal{N}_P}$.

Assumption 5: The resource-augmented process subnets considered in this work are assumed to be quasi-live for $M_0\ (i)=1$ and $M_0\ (r_l)=C_l$, $\forall l \in \{1,\ldots,|P_R|\}$.

In general, assessing the quasi-liveness of a resource-augmented process net is an *NP*-hard problem (Roszkowska and Wojcik, 1993; Reveliotis, 2005). However, the main source of this complexity is the presence of synchronizing transitions in the underlying process net, and therefore, the problem remains polynomial in the quite frequent case that this process net is a state machine. In such a case, assessing the quasi-liveness of the resource-augmented process net is tantamount to validating that for every resource R_i, $C_i \geq \max_{j,k}\{A_{jk}(i)\}$, or equivalently in the PN formalism, that for all $r_l \in P_R$ and for all $p \in P_S$, $M_0(r_l) \geq y_{r_l}(p)$, where y_{r_l} are the p-semiflows introduced in Assumption 4. Some interesting and quite powerful computational tests for assessing quasi-liveness for the remaining cases can be found in the works of Jeng et al. (2002) and Reveliotis (2003b).

10.2.3 Complete RAS Model: Process-Resource Nets

The complete PN-based model, $\mathcal{N}=(P,T,W,M_0)$, of any given instance from the RAS class considered in this work is obtained by merging the resource-augmented process nets $\overline{\mathcal{N}_{P_j}} = (P_j, T_j, W_j, M_{0_j}), j = 1,\ldots,n$, modeling its constituent process types, through their common resource places. The resulting PN class is characterized as the class of process-resource nets with acyclic, quasi-live, and strongly reversible process subnets, and its basic structure is depicted in Figure 10.7. Let $P_S = \bigcup_j P_{S_j}$; $I=\bigcup_j \{i_j\}$; $O=\bigcup_j \{o_j\}$, and $P=\bigcup_j P_j = P_S \cup I \cup O \cup P_R$. Then, the reusable nature of the resource allocation taking place in the entire process-resource net is characterized by a p-semiflow y_{r_l} for each resource type R_l, $l=1,\ldots,m$, defined by: (i) $y_{r_l}(r_l)=1$; (ii) $y_{r_l}(r_j)=0$, $\forall j \neq l$; (iii) $y_{r_l}(i_j)=y_{r_l}(o_j)=0$, $\forall j$; (iv) $\forall p \in P_S$, $y_{r_l}(p) = y_{r_l}^{(j*)}(p)$, where $\overline{N_{P_{j*}}}$ denotes the resource-augmented process subnet containing place p, and $y_{r_l}^{(j*)}()$ denotes the corresponding p-semiflow for resource R_l. Furthermore, it is easy to see that Assumption 5, regarding the quasi-liveness of the constituent resource-augmented process subnets $\overline{\mathcal{N}_{P_j}}$, implies also the quasi-liveness of

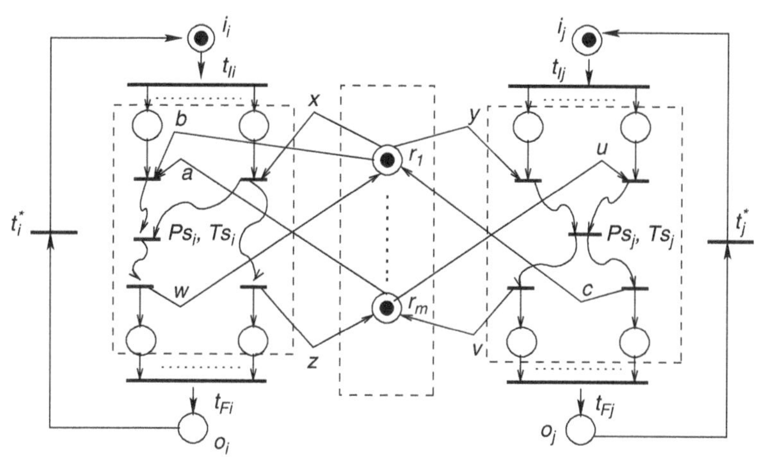

Figure 10.7 The process-resource net structure considered in this work.

the entire process-resource net $\mathcal{N}$. Finally, in the PN modeling framework, the size of a RAS Φ, modeled by a net $\mathcal{N} = (P_S \cup I \cup O \cup P_R, T, W, M_0)$, is defined as $|\Phi| = |\mathcal{N}| \equiv |P_R| + |P_S| + \sum_{r \in P_R} M_0(r)$.

Example 1: Figure 10.8 depicts the process-resource net modeling of the resource allocation taking place in the robotic cell of Figure 10.1. Each of the resource places $r_i, i = 1, 2, 3$, models the unit buffering capacity of the corresponding workstation in Figure 10.1, while the processing of an

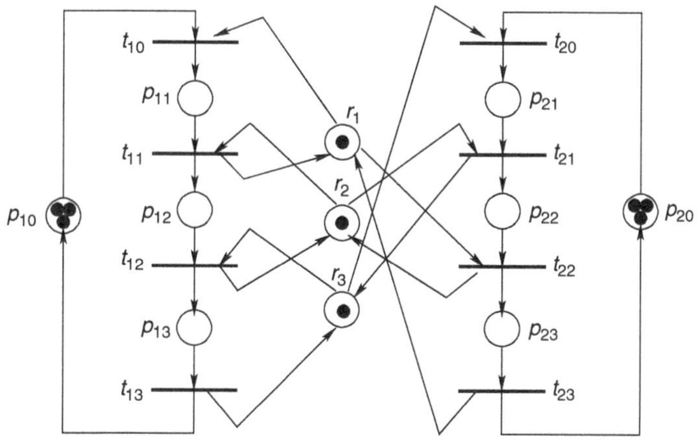

Figure 10.8 The process-resource net modeling the resource allocation taking place in the robotic cell of Figure 10.1.

instance of part type $J_j, j=1,2$ is modeled by the path $<t_{j0}, p_{j1}, t_{j1}, p_{j2}, t_{j2}, p_{j3}, t_{j3}>$. Furthermore, in the depicted net we have also adopted the common practice of aggregating the path $<i, t^*, o>$ of the underlying process nets to a single place p_0, that is called the process "idle place." Hence, the state depicted in Figure 10.8 corresponds to the initial RAS state, where the system is idle and empty of any processes. Finally, notice that we have set $M_0(i_j)=3$, $j=1,2$, so that these values do not constrain artificially the system loading. More generally, in the proposed PN-based RAS modeling, $M_0(i_j)$ must be set to a value that is an upper bound to the maximum number of process instances from process type Π_j that can be simultaneously loaded in the system. Such an upper bound will always exist due to the finiteness of the system resources.

10.2.4 RAS Deadlock and Deadlock Avoidance

In the PN-based modeling framework, the formation of RAS deadlock is expressed by the lack of reversibility of the corresponding process-resource net. Furthermore, in the underlying reachability space, $R(\mathcal{N}, M_0)$, this lack of reversibility is graphically represented by the formation of strongly connected components that are not co-reachable, that is, the initial marking M_0 is not reachable from them through any sequence of feasible transitions. As a case in point, Figure 10.9 depicts the reachability graph of the process-resource net of Figure 10.8, where it can be clearly seen that four reachable net markings are not co-accessible.

In the light of these characterizations, a correct "deadlock avoidance policy" (DAP), Δ, must try to restrict the system operation to a strongly connected component of the underlying reachability space, $R(\mathcal{N}, M_0)$, which contains the initial marking M_0. The RAS subspace that is reachable under—or admissible by—some DAP Δ will be denoted by $R_\Delta(\mathcal{N}, M_0)$. Given a RAS configuration, an applied DAP is characterized as "optimal," if the corresponding admissible subspace is the maximal strongly connected component of $R(\mathcal{N}, M_0)$ which contains the initial marking M_0. The set of markings admitted by the optimal DAP, Δ^*, is characterized as the (set of reachable) safe markings, and it will denoted by $R_S(\mathcal{N}, M_0)$. The complement of $R_S(\mathcal{N}, M_0)$ with respect to $R(\mathcal{N}, M_0)$ is denoted by $R_U(\mathcal{N}, M_0)$, and it constitutes the (reachable) unsafe markings.

The finiteness of the reachability space $R(\mathcal{N}, M_0)$ for the considered RAS class implies that the optimal DAP, Δ^*, is well defined, and it is effectively computable through a one-step look-ahead scheme that admits a tentative resource allocation iff the resulting marking is safe. However, the underlying safety problem is *NP*-complete, in general (Araki et al., 1977). In the light of this result, the research community has sought the development of

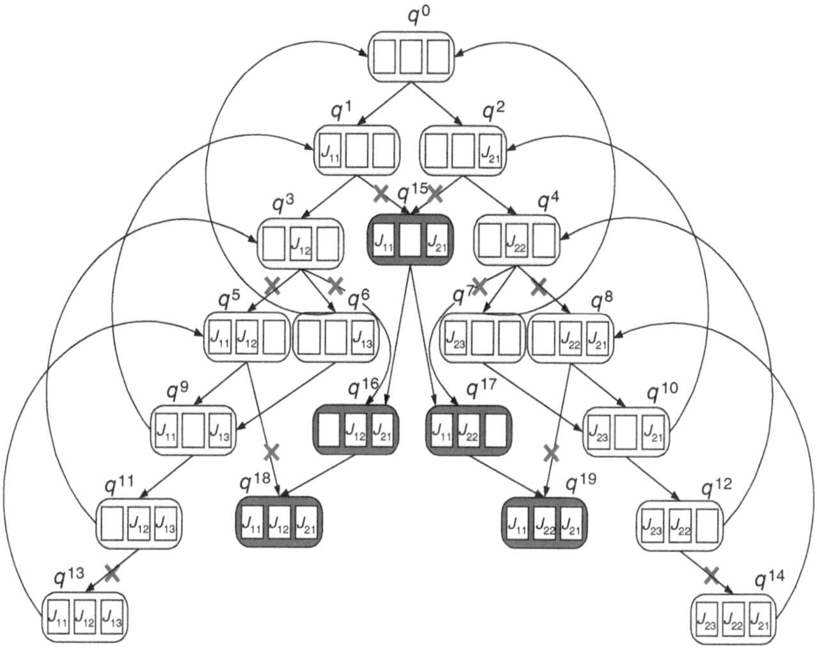

Figure 10.9 The reachability space of the process-resource net depicted in Figure 10.8; each depicted state indicates the occupancy of the workstation buffers in the original cell, while unsafe states are darkened.

suboptimal DAPs that are implementable in polynomial complexity with respect to the underlying RAS size, and yet, efficient, that is, they manage to admit a large part of $R_s(\mathcal{N}, M_0)$. This idea has been formalized by the concept of polynomial kernel (PK-) DAP (Reveliotis, 2005). From an implementational standpoint, a typical approach to the design of PK-DAPs is the identification of a property $\mathcal{H}(M), M \in R(\mathcal{N}, M_0)$, such that (i) the complexity of testing $\mathcal{H}()$ on the RAS markings is polynomial with respect to the RAS size; (ii) $\mathcal{H}(M_0) = \text{TRUE}$; and (iii) the subspace $\{M \in R(\mathcal{N}, M_0) : \mathcal{H}(M) = \text{TRUE}\}$ is strongly connected. The reader is referred to Zhou and Fanti (2004) and Reveliotis (2005) for a broad set of results regarding the design of PK-DAPs for various subclasses of the RAS class introduced in Definition 1, and also, for the identification of special RAS structure for which the optimal DAP Δ^* is implementable with polynomial complexity with respect to the corresponding RAS size, $|\Phi|$. In the rest of this chapter we focus on a particular subclass of PK-DAPs that is known as algebraic, and we present a series of results regarding the analysis and the design of these policies.

10.3 Algebraic DAPs and Their PN-Based Representation

10.3.1 Definition of Algebraic PK-DAPs

Algebraic PK-DAPs are defined as the particular class of PK-DAPs where the property $\mathcal{H}(M)$ constitutes a system of linear inequalities on the RAS marking M that is polynomially sized with respect to the RAS size $|\Phi|$. Furthermore, as the marking of the resource places $r \in P_R$ and of the process idle places, $p_{j0}, j=1,\ldots,n$, is determined by the marking of the remaining places $p \in P_S$, which expresses the state of the active process instances, the aforementioned inequalities will constrain explicitly only the restriction of M to its components corresponding to places $p \in P_S$, M_S. Hence, a typical algebraic PK-DAP would have the form:

$$A \cdot M_S \leq b \tag{10.1}$$

where A is a nonnegative integer matrix with K rows, b is a K-dimensional nonnegative integer vector, and K is polynomially related to the RAS size $|\Phi|$.

10.3.2 PN-Based Representation of Algebraic PK-DAPs

The constraints expressed by Equation 10.1 can be enforced in the PN-based representation of the RAS dynamics through the superimposition on the original process-resource net of a controlling subnet that is readily constructed through the theory of "control-place invariants" presented by Moody and Antsaklis (1998). According to Moody and Antsaklis (1998), each of the inequalities

$$A(k, \cdot) \cdot M_s \leq b(k), \quad k = 1, \ldots, K \tag{10.2}$$

can be imposed on the net behavior by superimposing on the original net structure a control place $p_c(k)$. The connectivity of place $p_c(k)$ to the rest of the network is determined by the flow matrix

$$\theta_{p_c(k)} = -A(k, \cdot) \cdot \Theta_S \tag{10.3}$$

where Θ_S denotes the flow sub-matrix of the uncontrolled network $\mathcal{N} = (P, T, W, M_0)$ corresponding to places $p \in P_S$. The initial marking of place $p_c(k)$ is set to

$$M_0(p_c(k)) = b(k) \tag{10.4}$$

The resulting controller imposes Constraint 2 on the original system behavior by establishing the place invariant

$$A(k, \cdot) \cdot M_S + M(p_c(k)) = b(k) \tag{10.5}$$

Example 2: Figure 10.10 depicts the implementation on the process-resource net of Figure 10.8 of an algebraic PK-DAP, Δ, that is expressed by the following set of inequalities on the marking M_S:

$$\begin{bmatrix} 1 & 0 & 0 & 0 & 0 & 1 \\ 1 & 1 & 0 & 1 & 1 & 0 \\ 0 & 0 & 1 & 1 & 0 & 0 \end{bmatrix} \cdot \begin{bmatrix} M(p_{11}) \\ M(p_{12}) \\ M(p_{13}) \\ M(p_{21}) \\ M(p_{22}) \\ M(p_{23}) \end{bmatrix} \leq \begin{bmatrix} 1 \\ 1 \\ 1 \end{bmatrix} \tag{10.6}$$

The reader can verify that the connectivity and the initial marking of the monitor places w_1, w_2, and w_3, that enforce each of the three constraints of Equation 10.6, satisfy the requirements of Equations 10.3 and 10.4, as well as Equation 10.5. The correctness of policy Δ for the considered process-resource net is manifested by Figure 10.11, that depicts the policy-admissible subspace, $R_\Delta(\mathcal{N}, M_0)$, and it is formally proven by Reveliotis and Ferreira (1996).

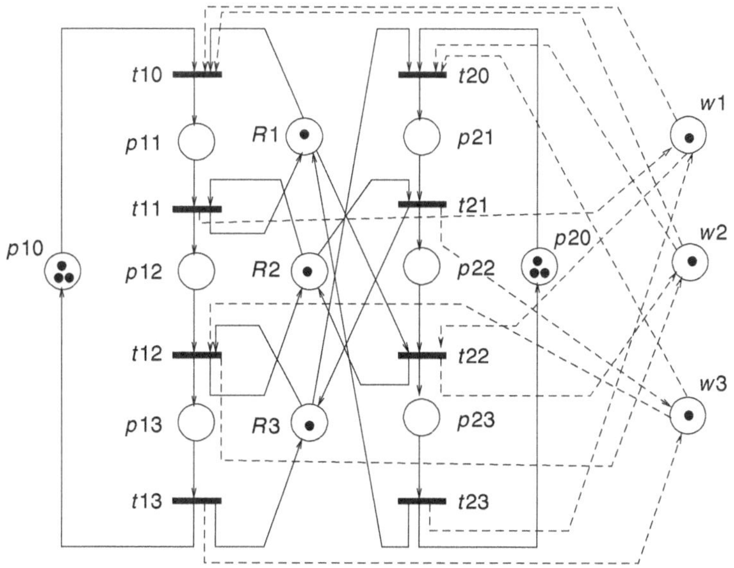

Figure 10.10 The process-resource net of Figure 10.8 under the control of the algebraic DAP of Equation 10.6.

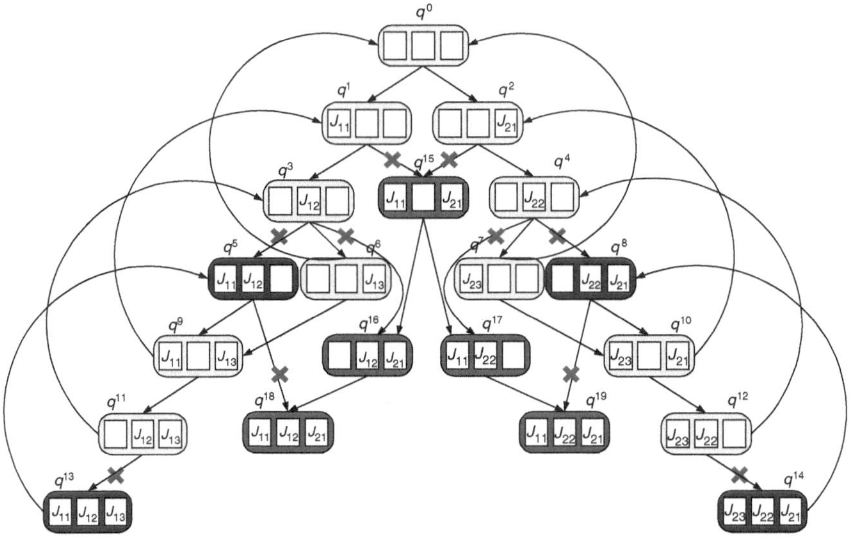

Figure 10.11 The policy-admissible space for the controlled net depicted in Figure 10.11: the inadmissible states are the darkened ones.

10.3.3 Monitor Places as Fictitious Resources

Equation 10.5, when interpreted in the light of Assumption 4 of Section 10.2, implies that the control places $p_c(k)$, implementing each of the constraints in the policy-defining Equation 10.1, essentially play the role of fictitious new resources in the dynamics of the net $\mathcal{N}^c$ that models the controlled system behavior.* As a result, the controlled net $\mathcal{N}^c$ remains in the class of process-resource nets that satisfy Assumptions 1 and 3. Let $P_C \equiv \cup_k \{p_c(k)\}$. If it can be shown that the net $\mathcal{N}^c$ satisfies also Assumption 5 with respect to the extended "resource" set $P_R \cup P_C$, it can be inferred that $\mathcal{N}^c$ belongs to the class of process-resource nets with acyclic, quasi-live, and strongly reversible process subnets, and therefore, all the analytical insights and results regarding the logical behavior of the uncontrolled RAS extend to their controlled counterparts. This remark will be especially useful in Section 10.5 where we derive correctness tests for algebraic PK-DAPs based on the structural properties of process-resource nets.

* This effect is manifested also in Figure 10.10.

10.4 Analytical Characterization of the Class of Algebraic DAPs

This section provides an analytical characterization for the entire set of algebraic PK-DAPs, Δ, that employ (up to) K constraints and can establish deadlock-free operation for some given process-resource net $\mathcal{N}$. Beyond its inherent theoretical interest, such a characterization enables also an optimal selection of the implemented policy Δ from the considered set of policies. The subsequent discussion addresses the aforementioned problem by considering the more general problem of characterizing the class of algebraic supervisors that employ (up to) K constraints and enforce the reversibility of some bounded PN $\mathcal{N}$; a solution to this extended problem can be easily customized to the notion of algebraic PK-DAPs for process-resource nets by restricting appropriately some elements of the policy-defining matrix A in the derived formulation. Hence, a formal statement of the problem considered in this section is as follows.

10.4.1 *Formal Statement of the Considered Problem*

Given a nonreversible, bounded PN $\mathcal{N}$ identify a set of constraints

$$A \cdot M \leq b \tag{10.7}$$

such that

i. when imposed on the plant net $\mathcal{N}$, they will incur the reversibility of the controlled system;
ii. the cardinality of the imposed constraint set must not exceed a prespecified parameter K;
iii. the matrix A and the vector b satisfy the constraint

$$\forall i,j, A(i,j) \in \{0, 1, ..., \overline{A}(i,j)\} \quad \text{and} \quad \forall i, b(i) \in \{0, 1, ..., \overline{b}(i)\}, \tag{10.8}$$

where $\overline{A}(i,j)$ and $\overline{b}(i)$ are finitely valued, externally provided parameters;
iv. assuming that every reachable marking $M_i \in R(\mathcal{N}, M_0)$ of $\mathcal{N}$ is associated with some value W_i, the developed supervisor will maximize the total value of the admissible markings over the set of supervisors satisfying the previous three requirements.

In the sequel, a PN supervisor that is defined by Equation 10.7 for some pricing of matrix A and vector b, will be referred to as the supervisor $\mathcal{S}(A,b)$.

10.4.2 Overview of the Proposed Solution

Next, we provide a mixed integer programming (MIP) formulation for the aforestated problem. The objective function of this formulation will express the optimality requirement stated in item (iv) above. Requirement (ii) will be captured by the structure of the decision variables of the presented formulation, while requirements (i) and (iii) will be explicitly encoded in its constraints. More specifically, given a pricing of the matrix A and the right-hand-side (rhs) vector b, the constraint set must check whether this pricing abides to requirement (iii) and it must also assess the ability of this pricing to satisfy requirement (i), that is, establish the reversibility of the controlled system. This last requirement further implies that all the markings $M \in R(\mathcal{N}, M_0)$ that remain reachable under the policy constraints are also co-reachable under these constraints. Hence, the constraint set of the proposed formulation must be able to assess the reachability and co-reachability of the markings $M \in R(\mathcal{N}, M_0)$ under the net supervision by any tentative constraint set, $A \cdot M \leq b$, and it must also be able to validate that all reachable markings are also co-reachable. The rest of this section proceeds to the detailed derivation of a formulation that possesses the aforementioned qualities.

10.4.3 Characterizing the Net Transition Firing under Supervision by $\mathcal{S}$(A,b)

To be able to assess the reachability and co-reachability of the various markings $M \in R(\mathcal{N}, M_0)$ under supervision by supervisor $\mathcal{S}(A,b)$, it is necessary to characterize how the various transitions, $t \in T$, of the plant net $\mathcal{N}$, retain their fireability in the controlled system. Next, we introduce a set of variables and constraints that will achieve this purpose. The main issue to be addressed is whether a transition t that was fireable in some marking $M_i \in R(\mathcal{N}, M_0)$, leading to another marking $M_j \in R(\mathcal{N}, M_0)$, will remain fireable under supervision by $\mathcal{S}(A,b)$. For this to be true, t must be enabled at M_i by all the monitor places, $p_c(k)$, $k = 1, \ldots, K$, that implement the supervisor $\mathcal{S}(A,b)$. Testing whether transition t is enabled at marking M_i by a monitor place $p_c(k)$ can be done through the employment of a binary variable z_{ij}^k, that will be priced to one, if this condition is true, and to zero, otherwise. A set of constraints that will enforce the pricing of z_{ij}^k according to the aforementioned scheme is the following:

$$M_0(p_c(k)) + \sum_{(u,v)\in\xi(i)} \Theta(p_c(k), t(u,v)) + \Theta(p_c(k), t(i,j)) + (z_{ij}^k - 1)L_{ij}^k \geq 0 \quad (10.9)$$

$$M_0(p_c(k)) + \sum_{(u,v)\in\xi(i)} \Theta(p_c(k), t(u,v)) + \Theta(p_c(k), t(i,j)) - z_{ij}^k U_{ij}^k \leq -1 \quad (10.10)$$

The parameter $\xi(i)$ appearing in Equations 10.9 and 10.10 denotes any path in $R(\mathcal{N}, M_0)$ leading from M_0 to M_i. (u,v) denotes an edge of $\xi(i)$ leading from node M_u to node M_v, and $t(u,v)$ denotes its labeling transition. L_{ij}^k denotes a lower bound for the quantity $M_0(p_c(k)) + \sum_{(u,v)\in\xi(i)} \Theta(p_c(k), t(u,v)) + \Theta(p_c(k), t(i,j))$, and U_{ij}^k denotes an upper bound for the quantity $M_0(p_c(k)) + \sum_{(u,v)\in\xi(i)} \Theta(p_c(k), t(u,v)) + \Theta(p_c(k), t(i,j)) + 1$. Then, it is clear that when $M_0(p_c(k)) + \sum_{(u,v)\in\xi(i)} \Theta(p_c(k), t(u,v)) + \Theta(p_c(k), t(i,j)) \geq 0$—that is, when transition $t(i,j)$ is enabled by monitor place $p_c(k)$ in marking M_i—the above set of constraints is satisfied by setting $z_{ij}^k = 1$. On the other hand, when $M_0(p_c(k)) + \sum_{(u,v)\in\xi(i)} \Theta(p_c(k), t(u,v)) + \Theta(p_c(k), t(i,j)) < 0$, the above constraint set is satisfied by setting $z_{ij}^k = 0$.

It remains to connect the variables $M_0\ (p_c(k))$ and $\Theta\ (p_c(k),\cdot)$ to the primary problem variables, A, b, and explain how to compute the bounds L_{ij}^k and U_{ij}^k employed in the above equations. Connecting $M_0(p_c(k))$ and $\Theta\ (p_c(k),\cdot)$ to the variables A, b can be done straightforwardly through Equations 10.3 and 10.4; the corresponding substitutions respectively transform Equations 10.9 and 10.10 to:

$$b(k) - \sum_{(u,v)\in\xi(i)} A(k,\cdot)\cdot\Theta(\cdot,t(u,v)) - A(k,\cdot)\cdot\Theta(\cdot,t(i,j)) + (z_{ij}^k - 1)L_{ij}^k \geq 0 \quad (10.11)$$

$$b(k) - \sum_{(u,v)\in\xi(i)} A(k,\cdot)\cdot\Theta(\cdot,t(u,v)) - A(k,\cdot)\cdot\Theta(\cdot,t(i,j)) - z_{ij}^k U_{ij}^k \leq -1 \quad (10.12)$$

Finally, it should be clear from the structure of Constraints 11 and 12 that the bound L_{ij}^k (resp., U_{ij}^k), defined above, can be obtained by minimizing (resp., maximizing) the quantity $b(k) - \sum_{(u,v)\in\xi(i)} A(k,\cdot)\cdot\Theta(\cdot,t(u,v)) - A(k,\cdot)\cdot\Theta(\cdot,t(i,j))$ over the space defined by the admissible ranges of the involved variables $A(k,\cdot)$ and $b(k)$ (c.f. item (iii) in the formal problem statement provided at the beginning of this section).

Once variables z_{ij}^k have been properly priced for all k, the feasibility of $M_i \xrightarrow{t(i,j)} M_j$ can be assessed by introducing another real variable, z_{ij}, that is priced according to the following constraints:

$$z_{ij} \leq z_{ij}^k, \ \forall k \in \{1, \ldots, K\} \quad (10.13)$$

$$z_{ij} \geq \sum_{k=1}^{K} z_{ij}^k - K + 1 \quad (10.14)$$

$$0 \leq z_{ij} \leq 1 \quad (10.15)$$

To understand the pricing logic behind Constraints 13–15, first notice that Constraint 15 restricts the variable z_{ij} within the interval [0,1]. Then, Constraint 13 sets it to zero, as long as any of the variables z_{ij}^k is priced to zero—and therefore, the corresponding monitor place $p_c(k)$ disables $t(i,j)$. On the other hand, when all variables z_{ij}^k are priced to one, Constraint 14 forces variable z_{ij} to its extreme value of one.

10.4.4 Characterizing the Reachability of the Markings $M_i \in R(\mathcal{N},M_0)$ under Supervision by $\mathcal{S}(A,b)$

The availability of the variables z_{ij}, defined above, subsequently enables the characterization of the reachability of the various markings $M_i \in R(\mathcal{N}, M_0)$ under supervision by the supervisor $\mathcal{S}(A,b)$. This can be done by introducing the real variables $y_i^l, 0 \leq i \leq |R(\mathcal{N}, M_0)|, 0 \leq l \leq \bar{l}$, and pricing them so that $y_i^l = 1$ indicates that marking M_i is reachable from the initial marking M_0 under supervision by $S(A,b)$ and the minimum length of any transition sequence leading from M_0 to M_i is l; if M_i is not reachable from M_0 under supervision by $\mathcal{S}(A,b)$, y_i^l should be set to zero for all l. Clearly, to satisfy this definition of y_i^l, $\bar{l}$ must be set to the length of the maximum path in $\mathcal{G}(N, M_0)$ that starts from M_0 and contains no cycles. Then, a set of constraints that achieve the pricing of y_i^l described above is as follows:

$$y_i^0 = \begin{cases} 1, & i = 0 \\ 0, & i \neq 0 \end{cases} \tag{10.16}$$

$$0 \leq y_i^l, \quad \forall i \in \{1, \ldots, |R(\mathcal{N}, M_0)|\}, l \in \{1, \ldots, \bar{l}\} \tag{10.17}$$

$$\sum_{l=0}^{\bar{l}} y_i^l \leq 1 \tag{10.18}$$

$$\delta_{ji}^l \leq y_j^{l-1}, \quad \forall j : (M_j, M_i) \in \mathcal{G}(\mathcal{N}, M_0) \tag{10.19}$$

$$\delta_{ji}^l \leq z_{ji}, \quad \forall j : (M_j, M_i) \in \mathcal{G}(\mathcal{N}, M_0) \tag{10.20}$$

$$y_i^l \leq \sum_j \delta_{ji}^l \tag{10.21}$$

$$y_i^l \geq y_j^{l-1} + z_{ji} - 1 - \sum_{q=0}^{l-1} y_i^q, \quad \forall j : (M_j, M_i) \in \mathcal{G}(\mathcal{N}, M_0) \tag{10.22}$$

Constraint 16 expresses the fact that marking M_0 is reachable from itself in zero steps, under supervision by $\mathcal{S}(A,b)$, and this is the only marking in $R(\mathcal{N},M_0)$ possessing this property. Constraint 17 states the nonnegative real nature of variables $y_i^l, i > 0, l > 0$, while Constraint 18 expresses the fact that, according to the pricing scheme discussed above, only one of the variables

$y_i^l, 0 \leq l \leq \bar{l}$, can be priced to one. Constraints 19, 20, and 21 express the fact that, under supervision by $\mathcal{S}(A,b)$, there is a minimal path from marking M_0 to marking M_i of length l, only if there is a minimal path of length $l-1$ from M_0 to some marking M_j such that (i) $(M_j, M_i) \in \mathcal{G}(\mathcal{N}, M_0)$ and (ii) this transition remains feasible under $\mathcal{S}(A,b)$. In particular, variables δ_{ji}^l are a set of auxiliary real variables that are used to force y_i^l to zero every time that the aforestated condition is violated for all the markings $M_j \in R(\mathcal{N},M_0)$ such that $(M_j, M_i) \in \mathcal{G}(\mathcal{N}, M_0)$. On the other hand, Constraint 22 tends to price variable y_i^l to one every time that there exists a marking M_j such that (i) $(M_j, M_i) \in \mathcal{G}(\mathcal{N},M_0)$; (ii) this transition remains feasible under $\mathcal{S}(A,b)$; and (iii) M_j is reachable from M_0 under supervision by $\mathcal{S}(A,b)$ through a minimal path of length $l-1$; however, this pricing is enforced only when the quantity $\sum_{q=0}^{l-1} y_i^q$ appearing in the rhs of this constraint is equal to zero—that is, only when the marking M_i cannot be reached from the initial marking M_0 through a path of smaller length.

10.4.5 Characterizing the Co-Reachability of the Markings $M_i \in R(\mathcal{N},M_0)$ under Supervision by $\mathcal{S}(A,b)$

Clearly, the co-reachability of a marking $M_i \in R(\mathcal{N},M_0)$ is equivalent to the reachability of the same marking in the graph $\mathcal{G}^R(\mathcal{N},M_0)$, obtained from $\mathcal{G}(N,M_0)$ by reversing all its arcs. In the light of this observation, the set of constraints characterizing the co-reachability of the markings $M_i \in R(\mathcal{N},M_0)$, under supervision by supervisor $\mathcal{S}(A,b)$, can be obtained through a straightforward modification of the Constraint set 16–22, characterizing the reachability of these markings. More specifically, let ψ_i^l be a real variable that will be priced to one, if $M_i \in R(\mathcal{N},M_0)$ is co-reachable under supervision by $\mathcal{S}(A,b)$, and a minimal transition sequence leading from M_i to M_0 has a length equal to l; otherwise, ψ_i^l should be priced to zero. By following a logic similar to that employed in the previous paragraph for the pricing of variables y_i^l, we obtain the following set of constraints for the pricing of variables ψ_i^l:

$$\psi_i^0 = \begin{cases} 1, & i = 0 \\ 0, & i \neq 0 \end{cases} \tag{10.23}$$

$$0 \leq \psi_i^l, \quad \forall i \in \{1, \ldots, |R(\mathcal{N}, M_0)|\}, \; l \in \{1, \ldots, \tilde{l}\} \tag{10.24}$$

$$\sum_{l=0}^{\tilde{l}} \psi_i^l \leq 1 \tag{10.25}$$

$$\eta_{ij}^l \leq \psi_j^{l-1}, \quad \forall j : (M_i, M_j) \in \mathcal{G}(\mathcal{N}, M_0) \tag{10.26}$$

$$\eta_{ij}^{l} \le z_{ij}, \quad \forall j : (M_i, M_j) \in \mathcal{G}(\mathcal{N}, M_0) \tag{10.27}$$

$$\psi_i^{l} \le \sum_{j} \eta_{ij}^{l} \tag{10.28}$$

$$\psi_i^{l} \ge \psi_j^{l-1} + z_{ij} - 1 - \sum_{q=0}^{l-1} \psi_i^{q}, \quad \forall j : (M_i, M_j) \in \mathcal{G}(\mathcal{N}, M_0) \tag{10.29}$$

The parameter $\tilde{l}$, appearing in Equations 10.24 and 10.25, denotes the length of the maximum path in $\mathcal{G}^R(\mathcal{N}, M_0)$ that leads from node M_0 to node M_i and contains no cycles, and the auxiliary variables η_{ij}^{l}, that appear in Constraints 26 and 27, play a role identical to that played by variables δ_{ji}^{l} in Constraints 19 and 20.

10.4.6 Characterizing the Closure of the Subspace that Is Reachable and Co-Reachable under Supervision by $\mathcal{S}$(A,b)

Let x_i be a real variable that will be priced to one when the marking $M_i \in R(\mathcal{N}, M_0)$ is reachable and co-reachable under supervision by $\mathcal{S}(A, b)$, and it will be priced to zero, otherwise. Then, in the light of the above characterizations of reachability and co-reachability, the desired pricing of x_i can be enforced by the following constraints:

$$x_i \le \sum_{l=0}^{\bar{l}} y_i^{l} \tag{10.30}$$

$$x_i \le \sum_{l=0}^{\tilde{l}} \psi_i^{l} \tag{10.31}$$

$$x_i \ge \sum_{l=0}^{\bar{l}} y_i^{l} + \sum_{l=0}^{\tilde{l}} \psi_i^{l} - 1 \tag{10.32}$$

$$0 \le x_i \le 1 \tag{10.33}$$

Constraint 33 restricts x_i in the interval [0,1]. Then, Constraints 30 and 31 force it to zero, when marking M_i is not reachable or co-reachable. On the other hand, if M_i is both reachable and co-reachable, Constraint 32 forces x_i to its extreme value of one.

Finally, the availability of variables x_i allows us to express the requirement for closure of the subspace of $R(\mathcal{N}, M_0)$ that is reachable and co-reachable under supervision by $\mathcal{S}(A, b)$, through the following constraint:

$$(1 - x_i) + x_j \ge z_{ij}, \quad \forall i, j : (M_i, M_j) \in \mathcal{G}(\mathcal{N}, M_0) \tag{10.34}$$

When $x_i = 1$ and $x_j = 0$—that is, when x_i belongs to the target space of markings that are reachable and co-reachable under supervision by $\mathcal{S}(A, b)$, but x_j does not belong to this set—Constraint 34 forces variable z_{ij} to zero—that is, it requires that the corresponding transition $M_i[t(i, j) > M_j$ is disabled by $\mathcal{S}(A, b)$. In any other case, the left-hand-side (lhs) of Constraint 34 is greater than or equal to one, and therefore, the constraint becomes inactive.

10.4.7 Objective Function of the Proposed Formulation

The objective function of the considered formulation is straightforwardly expressed as follows:

$$\max \sum_i w_i x_i \tag{10.35}$$

10.4.8 Correctness of the Proposed Formulation

The next theorem states the correctness of the derived formulation. A formal proof of this result can be based on concepts and arguments coming from the "theory of regions" (Badouel and Darondeau, 1998), a theory that concerns the design of PNs from a specification of their reachability space; the reader is referred to the work of Reveliotis and Choi (2006) for the relevant details.

Theorem 1: The formulation of Equations 10.8 and 10.11–10.35 returns an optimal solution to the problem stated at the beginning of Section 10.4, provided that such a solution exists; otherwise, this formulation will be infeasible.

10.4.9 Customizing the Derived Formulation to the Design of Algebraic PK-DAPs

We remind the reader that according to the definition of Section 10.3 algebraic PK-DAPs restrict explicitly only the projection M_S of the entire marking M of the corresponding process-resource net $\mathcal{N}$. This additional feature for the sought supervisors can be readily introduced in the formulation of Equations 10.8 and 10.11–10.35 by setting $\overline{A}(i, j) = 0$ for all the elements of matrix A corresponding to places $p \notin P_S$.* On the other hand,

* Although, from a computational standpoint, it is preferable to identify all these zero-valued variables and systematically remove them from the formulation.

the values for the remaining elements of matrix A and those of vector b should not be artificially restricted by the choice of the corresponding upper bounds, $\overline{A}(i,j)$ and $\overline{b}(i)$; this can be attained by maintaining these bounds to sufficiently large values. The resulting formulation will be always feasible, because it contains the trivial policy that confines the RAS to its initial state s_0.* Of course, such a result should be interpreted as lack of an effective DAP in the considered policy space. If we want such a negative result to be communicated as infeasibility by the proposed formulation, we can add the constraint

$$\sum_{i:i\neq 0} x_i \geq 1 \tag{10.36}$$

Furthermore, in most practical cases, one would like to enforce the existence of at least one policy-admissible process plan for each process type Π_j, $j=1,\ldots,n$. In such a case, letting $IN(j)$ denote the set of transitions corresponding to the initiation of an instance from process type Π_j, we can replace Constraint 36 with the following stronger requirement:

$$\forall j = 1,\ldots,n, \sum_{(i,q)\in IN(j)} z_{iq} \geq 1 \tag{10.37}$$

Finally, the typical objective for maximal permissiveness of the resulting policy can be communicated in the developed formulation by setting $w_i = 1, \forall i$.

Example 3: We demonstrate the efficacy of the design methodology developed in Section 10.5, by applying it to the design of an algebraic DAP for the PN depicted in Figure 10.12. This PN models a RAS consisting of three resource types, R_1, R_2, and R_3, with respective capacities $C_1 = C_3 = 1$, and $C_2 = 2$, and supporting two process types, Π_1 and Π_2, whose process plans are respectively modeled by the paths $<t_{10}p_{11}t_{11}p_{12}t_{12}p_{13}t_{13}>$ and $<t_{20}p_{21}t_{21}p_{22}t_{22}p_{23}t_{23}>$. The reachability space, $R(\mathcal{N},M_0)$, for the PN depicted in Figure 10.12 is provided in Figure 10.13, while the detailed characterization of the markings corresponding to the various nodes of the graph of Figure 10.13 can be found in Table 10.1.† Clearly, the considered net is not

* This policy is expressed by the single constraint $[11\cdots 1]\, M_s \leq 0$.

† Table 10.1 provides only M_s, that is, the markings of the places corresponding to the various processing stages, because the markings of the remaining places can be easily obtained from the net invariants corresponding to (i) the reusability of the system resources and (ii) the circuits established by the introduction of the process idle places.

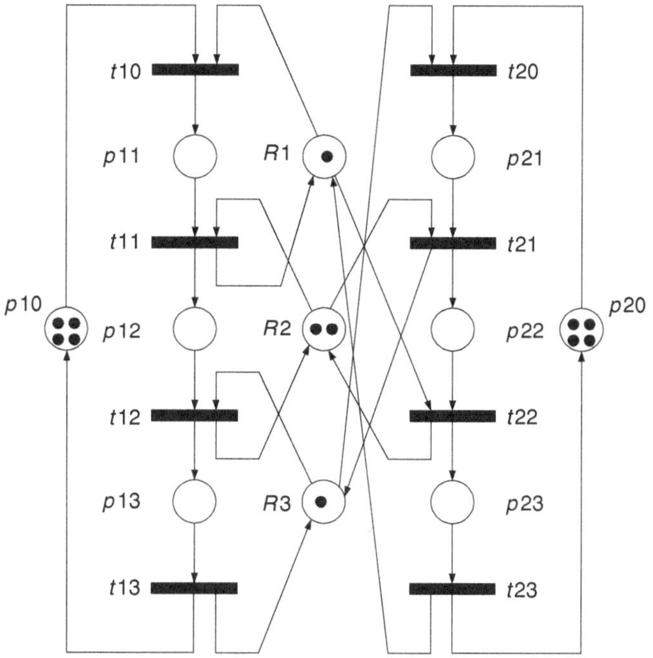

Figure 10.12 The process-resource net of Example 3.

reversible, because the states depicted by the darker-shaded nodes in Figure 10.13 are not co-reachable to M_0.

Two algebraic PK-DAPs for the process-resource net of Figure 10.12, originally developed by Park and Reveliotis (2000), are respectively expressed by the constraint sets:

$$\begin{bmatrix} 1 & & & 1 & 1 & 1 \\ 1 & 1 & & 1 & 1 & \\ 1 & 1 & 1 & 1 & & \end{bmatrix} \cdot M_S \leq \begin{bmatrix} 1 \\ 2 \\ 1 \end{bmatrix}. \tag{10.38}$$

$$\begin{bmatrix} 1 & & & 1 & 1 & 1 \\ 1 & 1 & & 1 & 1 & \\ 1 & 1 & 1 & 1 & & \end{bmatrix} \cdot M_S \leq \begin{bmatrix} 3 \\ 2 \\ 3 \end{bmatrix}. \tag{10.39}$$

It is interesting to notice that the Constraint set 39 is a relaxation of the Constraint set 38 as $A_1 = A_2$ and $b_1 \leq b_2$. Therefore, the supervisor established by the Constraint set 39 is expected to be more permissive than the supervisor established by the Constraint set 38, and this is indeed reflected in Figure 10.13 that also depicts the subspaces admitted by each of these

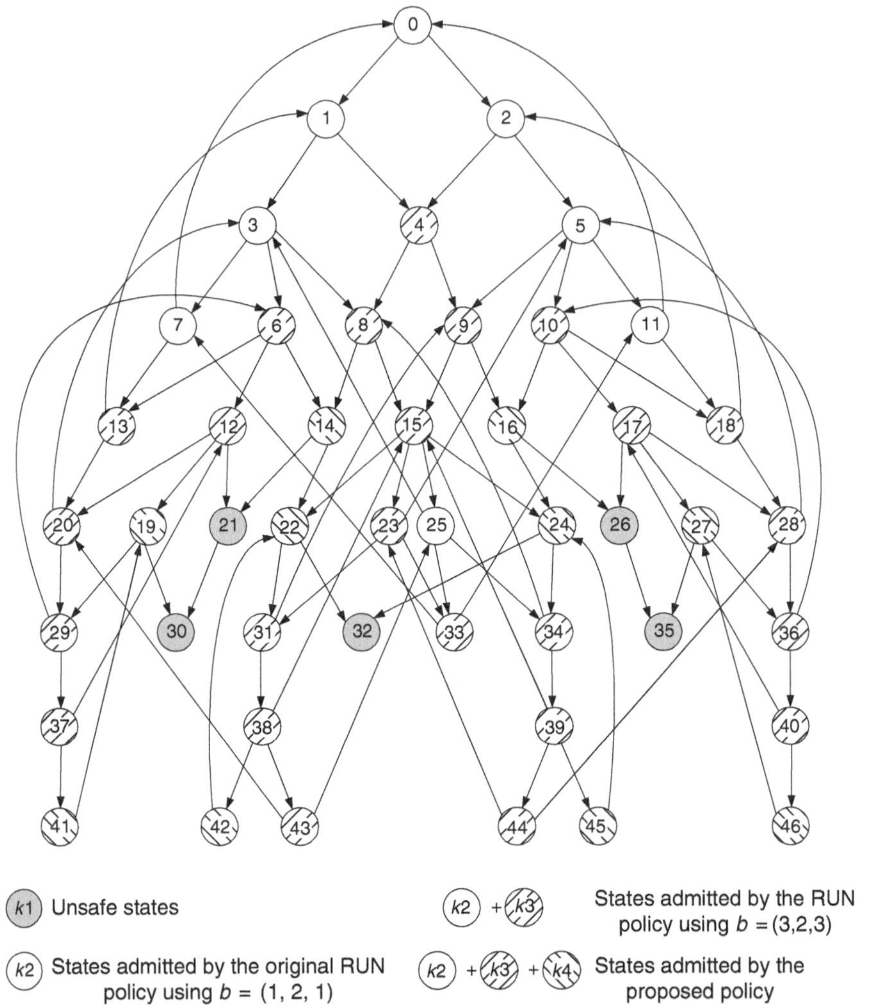

Figure 10.13 The reachability graph of the process-resource net of Figure 10.12, and a comparison of the subspaces admitted by the presented supervisors.

two supervisors. On the other hand, the application on the net of Figure 10.12 of the MIP formulation developed in the earlier parts of this section, with the number of policy constraints, K, set equal to 3, returned the following supervisor:

$$\begin{bmatrix} 1 & 0 & 0 & 0 & 3 & 0 \\ 0 & 1 & 0 & 2 & 0 & 0 \\ 2 & 2 & 0 & 2 & 3 & 0 \end{bmatrix} \cdot M_S \leq \begin{bmatrix} 6 \\ 3 \\ 8 \end{bmatrix} \tag{10.40}$$

Table 10.1 The Markings of Reachability Space Depicted in Figure 10.13

State	p_{11}	p_{12}	p_{13}	p_{21}	p_{22}	p_{23}	State	p_{11}	p_{12}	p_{13}	p_{21}	p_{22}	p_{23}
0	0	0	0	0	0	0	24	0	1	0	1	1	0
1	1	0	0	0	0	0	25	0	1	0	0	0	1
2	0	0	0	1	0	0	26	1	0	0	0	2	0
3	0	1	0	0	0	0	27	0	0	0	1	2	0
4	1	0	0	1	0	0	28	0	0	0	0	1	1
5	0	0	0	0	1	0	29	1	1	1	0	0	0
6	1	1	0	0	0	0	30	1	2	0	1	0	0
7	0	0	1	0	0	0	31	1	0	1	0	1	0
8	0	1	0	1	0	0	32	1	1	0	1	1	0
9	1	0	0	0	1	0	33	0	0	1	0	0	1
10	0	0	0	1	1	0	34	0	1	0	1	0	1
11	0	0	0	0	0	1	35	1	0	0	1	2	0
12	0	2	0	0	0	0	36	0	0	0	1	1	1
13	1	0	1	0	0	0	37	0	2	1	0	0	0
14	1	1	0	1	0	0	38	0	1	1	0	1	0
15	0	1	0	0	1	0	39	0	1	0	0	1	1
16	1	0	0	1	1	0	40	0	0	0	0	2	1
17	0	0	0	0	2	0	41	1	2	1	0	0	0
18	0	0	0	1	0	1	42	1	1	1	0	1	0
19	1	2	0	0	0	0	43	0	1	1	0	0	1
20	0	1	1	0	0	0	44	0	0	1	0	1	1
21	0	2	0	1	0	0	45	0	1	0	1	1	1
22	1	1	0	0	1	0	46	0	0	0	1	2	1
23	0	0	1	0	1	0							

The subspace admitted by the supervisor of Equation 10.40 is also depicted in Figure 10.13. As it can be seen in this figure, the obtained supervisor manages to recognize the entire safe space of the considered process-resource net, and therefore, it expresses the optimal DAP, Δ^*. Hence, this example corroborates the efficacy and analytical power of the proposed methodology.

10.4.10 Complexity Considerations

Yet, the practical applicability of the algebraic DAP design methodology developed in this section can be severely limited from the fact that it requires the explicit enumeration of the reachability space, $R(\mathcal{N}, M_0)$, of the underlying process-resource net, $\mathcal{N}$. It is well established in the relevant literature that the size of this state space is, in general, an exponential function of the size of the underlying net, $|\mathcal{N}|$, and it grows very fast. This situation is further complicated by the fact that the presented methodology eventually boils down to the solution of a MIP formulation with a number of variables and constraints that are determined by the size of the space $R(\mathcal{N}, M_0)$. A first approach to alleviate this problem is to restrict the target behavior for the controlled net to a subspace of $R(\mathcal{N}, M_0)$, so that the resulting formulation is computationally manageable. Another approach is to develop additional methodology that will provide correct algebraic DAPs for some given process-resource net $\mathcal{N}$, while avoiding the explicit enumeration of the reachability space $R(\mathcal{N}, M_0)$. We present such an alternative methodology in Section 10.5.

10.5 Analysis and Design of Algebraic DAPs through PN Structural Analysis

10.5.1 Structural Characterization for the Reversibility of Process-Resource Nets and the Correctness of Algebraic DAPs

The characterization of the RAS deadlock-freedom and the DAP correctness that was provided in the closing part of Section 10.2, and was behind the design methodology of Section 10.3, is "behavioral," because it engages patterns and structure that are traceable in the reachability space of the considered process-resource net. In this section we focus on an alternative characterization for these two concepts that is "structural," because it is based on the identification of special structure in the reachable markings of the underlying PNs. As it will be shown in the subsequent developments, the ability to confine the search for special structure in individual markings further enables the assessment of the deadlock-freedom of the underlying RAS or the correctness of any given algebraic DAP through an implicit enumeration of the underlying state space. At the basis of all the developments presented in this section is the following structural characterization for the reversibility of the process-resource nets defined in Section 10.2.

Theorem 2: Let $N=(P_S \cup I \cup O \cup P_R, T, W, M_0)$ be a process-resource net with acyclic, quasi-live, and strongly reversible processes. Then, the following hold true:

i. $\mathcal{N}$ is reversible iff it is live.
ii. $\mathcal{N}$ is live iff the space of modified reachable markings, $\overline{R(\mathcal{N}, M_0)}$, that is induced from $R(\mathcal{N}, M_0)$ through the projection

$$\overline{M}(p) = \begin{cases} M(p) & \text{if } p \notin I \cup O \\ 0 & \text{otherwise} \end{cases} \tag{10.41}$$

contains no deadly marked siphon S such that *(i)* $S \cap P_R \neq \emptyset$ and *(ii)* $\forall p \in S \cap P_R$, p is a disabling place at $\overline{M}$.

iii. In the particular case that $\mathcal{N}$ is *PT*-ordinary, $\mathcal{N}$ is live iff the space of reachable markings, $R(\mathcal{N}, M_0)$, contains no empty siphons.

A complete development of the results stated in Theorem 2 can be found in the work of Reveliotis (2003a, 2005). Here we give an intuitive explanation for the role of deadly marked and empty siphons in the interpretation of the RAS deadlock, under the PN representation introduced in Section 10.2. Hence, Figure 10.14 depicts the empty siphon that interprets the deadlock of the robotic cell depicted in Figure 10.1. As indicated in Figure 10.14, this siphon consists of the resources involved in the considered deadlock and the stages that involve the allocation of at least one of the deadlocked resources and are currently empty; these stages are known as "empty holders" in the relevant terminology. As a fundamental property of an empty siphon is that it will remain empty during the entire evolution of the net marking, the empty siphon of Figure 10.14 is a PN-based manifestation of the deadlock experienced in the underlying RAS. On the other hand, in the case of non-*PT*-ordinary PNs, the non-uniformity of the resource allocation requests for any single resource type, across the various processing stages, allows the existence of RAS states where certain process

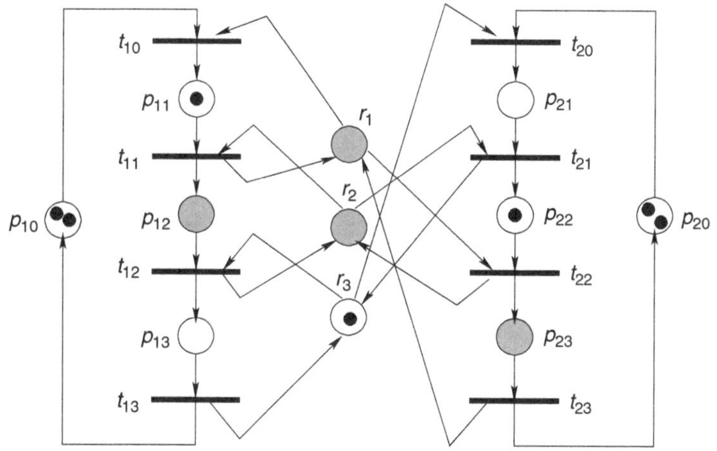

Figure 10.14 Interpreting the non-liveness of *PT*-ordinary process-resource nets with acyclic, quasi-live, and strongly reversible process subnets through empty siphons.

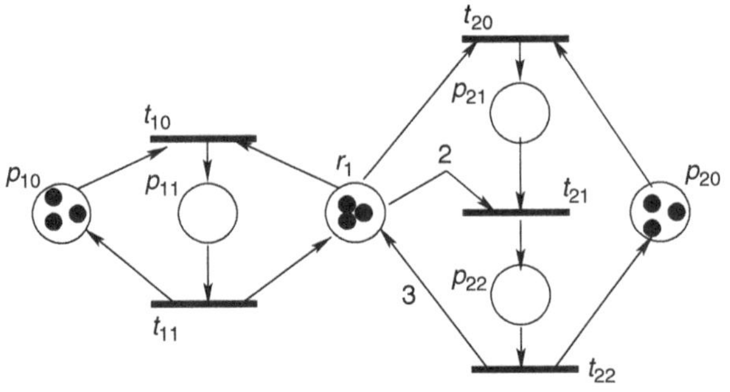

Figure 10.15 The considered process-resource net.

types can be executed repetitively, even though some of their supporting resource types are involved in a deadlock. The situation is depicted in Figures 10.15 through 10.17. In particular, Figure 10.16 depicts a reachable marking of the process-resource net depicted in Figure 10.15, where the active process instances corresponding to the tokens in place p_{21} are deadlocked, because their request for two extra units of resource R_1 cannot be met unless one of them releases its currently held resource. However, process instances executing stage p_{11} can still engage the remaining free unit of resource R_1 and successfully proceed to completion. But then, place p_{11} cannot be part of an empty siphon, even though it is an empty holder of resource R_1, that is involved in the depicted RAS deadlock. The aforementioned problem is remedied by the introduction of the concept of the "modified marking." The modified marking of the original marking depicted in Figure 10.16 is depicted in Figure 10.17. By removing all the tokens resident in places p_{j0}, $\forall j$, we construct a deadlock marking, in which the set S of all disabling places—depicted by shaded places in Figure 10.17—will be

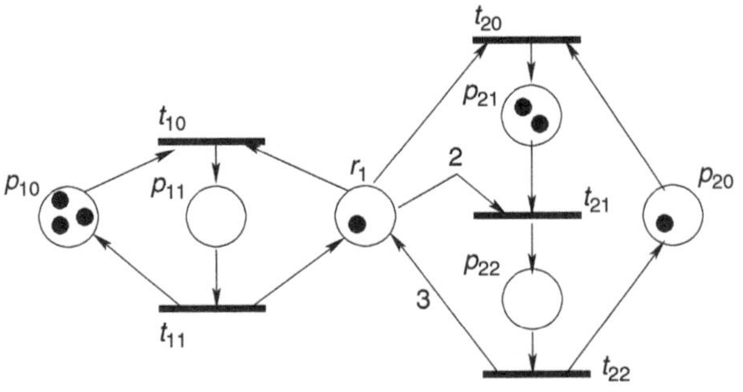

Figure 10.16 A net marking containing a RAS deadlock.

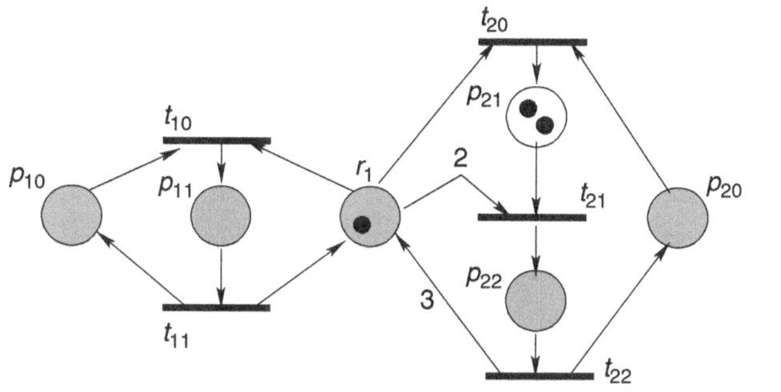

Figure 10.17 The resource-induced deadly marked siphon.

a deadly marked siphon.* It is interesting to notice that the constructed siphon S is deadly marked but not empty. On the other hand, the token removal implied by the definition of the modified marking $\overline{M}$ will also generate artificially empty siphons, especially, for those process types with no active process instances in the original marking M; in Figure 10.17, the place set $S' = \{p_{10}, p_{11}\}$ is such an artificially constructed empty siphon. Hence, to infer the net non-liveness, one must focus on the particular type of deadly marked siphon characterized in item 2 of Theorem 2; these siphons are called "resource-induced" deadly marked siphons in the relevant literature.

The liveness and reversibility criteria of Theorem 2 can also provide correctness criteria for any tentative DAP for a given process-resource net, provided that the controlled net remains in the class of process-resource nets with acyclic, quasi-live, and strongly reversible processes. In particular, as it was discussed in Section 10.3, the net $\mathcal{N}^c$ resulting from the imposition of some algebraic DAP on a given process-resource net $\mathcal{N}$ will possess this property as long as the superimposed monitor places do not affect the quasi-liveness of the resource-augmented process subnets. Although the assessment of the quasi-liveness of a resource-augmented process net can be a challenging problem in itself, there is a broad set of cases for which this problem is easily resolved,† and furthermore, as it will be demonstrated in the example provided in the closing part of this section, in certain cases the imposed policy can be defined in a way that it will ensure the sought quasi-liveness.

The remaining part of this section establishes that the criterion stated in item 2 of Theorem 2 can be effectively assessed through a mathematical programming formulation.

* c.f. Theorem 5 in the Appendix.

† c.f. The relevant discussion in Section 2.

10.5.2 Assessing the Liveness and Reversibility Criterion of Theorem 2 through a Mathematical Programming Formulation

The starting point for the development of the sought formulation is the realization that, given a PN $\mathcal{N} = (P, T, W, M_0)$ and a marking $M \in R(\mathcal{N}, M_0)$, the maximal deadly marked siphon S in M can be computed by the algorithm of Figure 10.18, originally developed by Park and Reveliotis (2001). In the case of "structurally bounded" nets, the algorithm of Figure 10.18 can be converted to a MIP formulation as follows: First, let $SB(p)$ denote a structural bound for the markings of place $p \in P$. Furthermore, let v_p, z_t, and f_{tp} be "binary indicator" variables respectively denoting the following conditions:

$v_p = 1 \Longleftrightarrow$ place p is removed by the algorithm, $\forall p \in P$
$z_t = 1 \Longleftrightarrow$ transition t is removed by the algorithm, $\forall t \in T$
$f_{pt} = 1 \Longleftrightarrow M(p) \geq W(p,t) \vee v_p = 1, \quad \forall W(p,t) > 0$

Then, we have the following theorem:

Theorem 3 (Park and Reveliotis, 2001; Reveliotis, 2005): Given a marking $M \in R(\mathcal{N}, M_0)$ of a structurally bounded PN $\mathcal{N} = (P, T, W, M_0)$, the maximal deadly marked siphon S contained in M is determined by

$$S = \{p \in P | v_p = 0\} \tag{10.42}$$

where v_p, $p \in P$, is obtained through the following IP formulation:

$$G(M) = \min \sum_{p \in P} v_p \tag{10.43}$$

Input: A PN $\mathcal{N} = (P, T, W, M_0)$ and a marking $M \in R(\mathcal{N}, M_0)$
output: The maximal deadly marked siphon in M, S

i. $S := P$; $N' := N$
ii. while $\exists t \in T$ such that t is fireable in the modified net N' do
(a) $S := S \setminus t^{\bullet}$
(b) Remove t from N'
(c) Remove $t^{\bullet}$ from N'
endwhile
iii. Return S

Figure 10.18 An algorithm for computing the maximal deadly marked siphon in a given marking *M*.

such that

$$f_{pt} \geq \frac{M(p) - W(p,t) + 1}{SB(p)}, \quad \forall W(p,t) > 0 \tag{10.44}$$

$$f_{pt} \geq v_p, \quad \forall W(p,t) > 0 \tag{10.45}$$

$$z_t \geq \sum\nolimits_{p \in {}^{\bullet}t} f_{pt} - |{}^{\bullet}t| + 1, \quad \forall t \in T \tag{10.46}$$

$$v_p \geq z_t, \quad \forall W(t,p) > 0 \tag{10.47}$$

$$v_p, z_t, f_{pt} \in \{0,1\}, \quad \forall p \in P, \quad \forall t \in T \tag{10.48}$$

To understand the result of Theorem 3, first notice that Equation 10.46 together with Equation 10.44 imply that all transitions z_t fireable in marking M will have $z_t = 1$. Furthermore, Equation 10.47 implies that all places $p \in t^{\bullet}$ for some t with $z_t = 1$ will have $v_p = 1$, which implements Step (2.b) in the algorithm of Figure 10.18. Similarly, Equation 10.45 combined with Equation 10.46 force $z_t = 1$ for all transitions t with $v_p = 1$, $\forall p \in {}^{\bullet}t$. Finally, the fact that no additional place p (resp., transition t) has $v_p = 1$ (resp., $z_t = 1$), is guaranteed by the specification of the objective function in the above formulation.

In the case that the net N is a process-resource net, the formulation of Theorem 3 can be restricted to the computation of the maximal resource-induced deadly marked siphon, through the introduction of the following two constraints:

$$\sum\nolimits_{r \in P_R} v_r \leq |P_R| - 1 \tag{10.49}$$

$$\sum\nolimits_{t \in r^{\bullet}} f_{rt} - |r^{\bullet}| + 1 \leq v_r, \quad \forall r \in P_R \tag{10.50}$$

Constraint 49 enforces that the identified siphon S must contain at least one resource place, while Constraint 50 requires that all resource places included in S must be disabling. The resulting necessary and sufficient condition for the nonexistence of resource-induced deadly marked siphons in a given marking M of a process-resource net is as follows:

Corollary 1 (Park and Reveliotis, 2001; Reveliotis, 2005): A given marking M of a process-resource net $\mathcal{N}$ contains no resource-induced deadly marked siphons, iff the corresponding formulation of Equations 10.43–10.50 is infeasible.

The test of Corollary 1 can be extended to a test for the nonexistence of resource-induced deadly marked siphons over the entire modified reachability space, $\overline{R(\mathcal{N}, M_0)}$, of a process-resource net $\mathcal{N} = (P, T, W, M_0)$, by

i. Substituting the marking M in the MIP formulation of Theorem 3 with the modified marking $\overline{M}$
ii. Introducing an additional set of variables, M, representing the net reachable markings
iii. Adding two sets of constraints, the first one linking variables M and $\overline{M}$ according to the logic of Equation 10.41, and the second one ensuring that the set of feasible values for the variable vector M is equivalent to the PN reachability space $R(\mathcal{N}, M_0)$

In the general case, the characterization of the set $R(\mathcal{N}, M_0)$ by a system of linear inequalities will involve a number of variables and constraints that is an exponential function of the net size $|\mathcal{N}|$ (Silva et al., 1998). However, in the case of the process-resource nets considered in this work, there is such a characterization of $R(\mathcal{N}, M_0)$ that is polynomially sized with respect to $|\mathcal{N}|$, and therefore, the plan outlined above remains a viable proposition; we refer to Reveliotis (2006) for the relevant details. Furthermore, a practical and frequently used implementation of the aforementioned plan substitutes the exact characterization of the reachability space $R(\mathcal{N}, M_0)$ by its superset that is provided by the state equation.* The resulting formulation provides a sufficient condition for the nonexistence of resource-induced deadly marked siphons S in the entire space $\overline{R(\mathcal{N}, M_0)}$ of a given process-resource net N, which in the light of Theorem 2, constitutes also a sufficient condition for the liveness and reversibility of process-resource nets with acyclic, quasi-live, and strongly reversible process subnets. The following corollary summarizes this discussion:

Corollary 2: Let $\mathcal{N} = (P, T, W, M_0)$ be a process-resource net with acyclic, quasi-live, and strongly reversible process subnets. If the MIP defined by (i) Equations 10.43–10.50, where vector variable M is replaced by vector variable $\overline{M}$, (ii) Equations 10.72–10.73, and (iii) Equation 10.41, is infeasible, then $\mathcal{N}$ is live and reversible.

Concluding this discussion, we notice that for the case of *PT*-ordinary process-resource nets with acyclic, quasi-live, and strongly reversible subnets, a similar but simpler liveness and reversibility sufficiency test can be obtained by focusing on the presence of empty siphons in the original net reachability space, $R(\mathcal{N}, M_0)$; refer to Chu and Xie (1997) and Park and Reveliotis (2000) for a detailed discussion of this formulation.

* This is Equation 10.7 in the Appendix.

10.5.3 Design of Efficient Algebraic DAPs through the Criterion of Corollary 2

As it was discussed in the opening part of this section, the correctness of an algebraic DAP can be established by the criterion of Corollary 2 as long as the superimposition of the relevant monitor places maintains the quasi-liveness of the underlying resource-augmented process subnets. Hence, the formulation of Corollary 2 can be embedded in a search process seeking a pair (A, b) that will render this formulation infeasible.* This search can be further assisted by additional insights regarding the dynamics of the underlying process-resource net. A particularly effective use of the aforementioned criterion has sought the systematic relaxation of the rhs vector, b, in algebraic DAPs that have been developed through alternative approaches; we refer the reader to Park and Reveliotis (2000, 2001) for some relevant examples. Next, we demonstrate the application of the criterion of Corollary 2 by focusing on a particular class of algebraic DAPs known as "process-release" policies. This class of policies seeks only to restrict the number of process instances that are loaded simultaneously into the system, rather than their access to particular segments of their process routes. Hence, they can be expressed by a single linear inequality

$$a \cdot M_S \leq b \tag{10.51}$$

where b defines the ceiling on the process concurrency imposed by the considered supervisor, and the elements of the row vector a are provided by the p-semiflows that characterize the flow logic of the various process types. The DAP of Equation 10.51 is superimposed to the original process-resource net $\mathcal{N}$ through the introduction of a single control place p_c with $p_c^{\bullet} \subseteq \bigcup_j \{t_{I_j}\}$. Hence, it should be obvious to the reader that the resulting controlled net $\mathcal{N}^c$ preserves the quasi-liveness of the original net $\mathcal{N}$, as long as $M_0(p_c) \geq 1$. The following example demonstrates the DAP synthesis method discussed in this section, and also, the particular concept of process-release control.

Example 4: Consider the process-resource net depicted in Figure 10.19. As it can be seen in the figure, the underlying RAS consists of two process types, Π_1 and Π_2, and five resource types, $R_1, \ldots, R_5$. Process type Π_1 has a flow represented by an acyclic marked graph, and it involves six processing stages, $\Xi_{11}, \ldots, \Xi_{16}$, with corresponding resource requirements: $(1,0,0,0,0)^T$, $(0,1,0,0,0)^T$, $(0,0,1,0,0)^T$, $(0,0,1,0,0)^T$, $(0,1,0,0,0)^T$, and

* For a more concrete experience, the reader is invited to apply the criterion of Corollary 2 on the algebraic DAPs that were presented in Examples 2 and 3.

$(0,0,0,0,1)^T$. Process type Π_2 has a flow represented by an acyclic state machine, and it involves four stages, $\Xi_{21}, \ldots, \Xi_{24}$, with corresponding resource requirements: $(0,1,0,0,0)^T$, $(1,1,0,0,0)^T$, $(0,1,1,0,0)^T$, and $(0,0,0,1,0)^T$. A closer inspection of the stage resource requirements for these two processes reveals that the only resources that could be entangled in a deadlock are R_1,R_2, and R_3. Therefore, the critical sections for Π_1 and Π_2 are respectively defined by the stage sets $\{\Xi_{11}, \Xi_{12}, \Xi_{13}, \Xi_{14}, \Xi_{15}\}$ and $\{\Xi_{21}, \Xi_{22}, \Xi_{23}\}$.

Our intention is to develop an algebraic DAP that will establish the liveness and reversibility of the controlled net by restricting the number of process instances that can execute simultaneously in their critical sections identified above. Hence, the proposed supervisor constitutes a more refined implementation of the general process-release control scheme, to the particular process-resource net of Figure 10.19. This supervisor is superimposed to the original process-resource net of Figure 10.19 by introducing the control place w, connected to the original process-resource net through the flow structure depicted by dotted lines in Figure 10.19.

Next we seek to determine the maximal initial marking for place w that leads to live and reversible behavior for the controlled net of Figure 10.19, using the siphon-based analysis that was developed in this section. For this, first we determine an upper bound to the maximal number of processes that can be executed simultaneously by the considered RAS. The reader can convince herself that, based on the resource capacities and the process flows annotated in Figure 10.19, an upper bound for the system concurrency with respect to process type Π_1 (resp., Π_2) is 7 (resp., 5) process

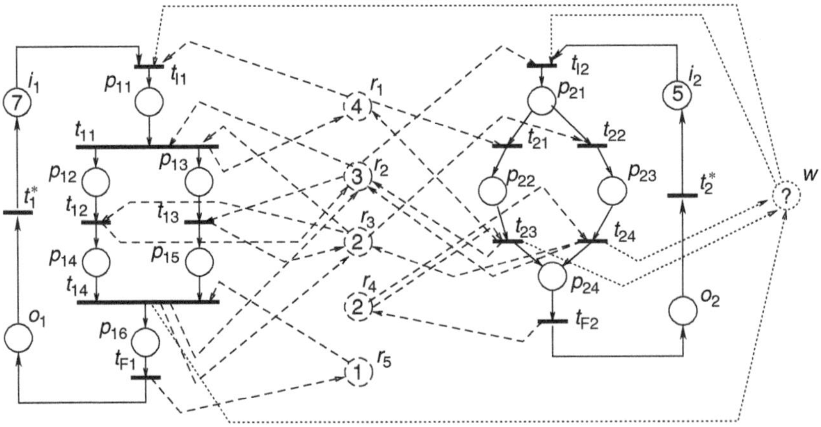

Figure 10.19 The process-resource net and the imposed process-release control policy for Example 4.

instances. Then, application of the MIP formulation of Corollary 2 in a binary search over the integer set {1, ... ,12} reveals that the maximal initial marking for control place w leading to a correct algebraic DAP—or equivalently, the maximal number of processes that can be simultaneously loaded and let to execute uncontrollably through the system without the possibility of running into any deadlocking problems—is 6. For completeness, we mention that the deadlock marking identified by the computerized solver when the MIP formulation of Corollary 2 was solved with $M_0(w) = 7$, is: $M(i_1) = 1$; $M(p_{11}) = 4$; $M(p_{12}) = M(p_{13}) = 2$; $M(i_2) = 4$; $M(p_{21}) = 1$; $M(r_4) = 2$; $M(r_5) = 1$; and zero for every other place.

Closing the discussion of this example, we want to point out that, although in the case of process-release control policies the satisfaction of the criterion of Corollary 2 presents a monotonicity with respect to the initial marking of the control place, p_c, that enables the search for the maximally permissive implementation through binary search, this property will not be true for algebraic DAPs implementing more involved control schemes. In those cases, the identification of a maximal marking leading to live and reversible behavior will necessitate a more careful search mechanism. Furthermore, in the more general case, the "structural liveness" of the controlled net with respect to markings $M_0(p_c(k))$, $k = 1, \ldots, \dim(b)$—that is, the existence of some marking $M_0(p_c(k))$, $k = 1, \ldots, \dim(b)$, that satisfies the liveness and reversibility condition of Corollary 2 for the resulting net $\mathcal{N}^c$—cannot be guaranteed a priori. Section 10.6 offers some further insights on these issues by providing an analytical characterization of the basic mechanism that enables the control of the net siphons through a limited number of monitor places.

10.6 Explaining the Functionality of Algebraic DAPs

10.6.1 Implicit Siphon Control

This section offers an analytical interpretation of the basic mechanism that enables the algebraic DAPs to control the marking of the entire set of siphons of a process-resource net with only a limited number of control places. Our discussion epitomizes the key insights and results of Reveliotis (2007), that constitutes a more formal and extensive reference for the subsequent developments. Furthermore, to enhance the clarity of the presentation, in the following section we shall confine our attention to *PT*-ordinary process-resource nets. Then, because of the last item of Theorem 2, we are able to focus on the empty siphons of the underlying process-resource net instead of the more elusive set of resource-induced deadly marked siphons. In particular, we shall say that a net siphon is "controlled" if it remains nonempty during the entire evolution of the net marking.

The following series of definitions introduce a number of concepts that are instrumental for the linkage of the structure of the algebraic DAPs to the control of the net siphons.

Definition 3: Consider a marked PN $\mathcal{N} = (P, T, W, M_0)$ and a vector $v \in \Re^{|P|}$, where $\Re$ denotes the set of reals. Then, for any marking $M \in R(\mathcal{N}, M_0)$, the generalized compound marking generated by v, is defined by

$$GCM(M,v) = \sum_{p \in P} v(p)M(p) = v^T M \tag{10.52}$$

The vector v will be called the "generator" of $GCM(M,v)$ and the set of places corresponding to nonzero elements of v will be denoted by P^v. Finally, in the particular case that $v(p) \in \{0,1\}$, $\forall p \in P$, a $GCM(M,v)$ reduces to the compound marking of the place subset P^v.

Definition 4: Consider a pure marked PN $\mathcal{N} = (P, T, W, M_0)$ and a GCM generator $v \in \Re^{|P|}$. Then, the net flow (vector) of v is defined by

$$NF(v) = v^T \Theta \tag{10.53}$$

where Θ denotes the flow matrix of $\mathcal{N}$.

Notice that $NF(v)$ is a $|T|$-dimensional row vector. Furthermore, the components of $NF(v)$ have the following very intuitive interpretation: For every transition $t \in T$, $NF(v,t)$ denotes the net change of $GCM(M,v)$ resulting by the firing of transition t at M.* Finally, the next definition connects the GCM and NF concepts to the concept of siphon.

Definition 5: Consider a siphon S of a pure marked PN $\mathcal{N} = (P, T, W, M_0)$. The characteristic vector of S is a $|P|$-dimensional binary vector λ_S such that

$$\forall p \in P, \quad \lambda_S(p) = 1 \Longleftrightarrow p \in S \tag{10.54}$$

Hence, the characteristic vector, λ_S, of any given siphon S, can be considered as a GCM generator with $GCM(M,\lambda_S)$ being equal to the token content of siphon S at marking M. Furthermore, the components of the corresponding net flow vector $NF(\lambda_S)$ express the net change incurred to the siphon marking by the firing of any single transition $t \in T$.

The next theorem, which constitutes the main result of this section, establishes the connection between the siphon control and the concepts introduced in the above definitions.

* This becomes obvious in the light of Equation 10.7 in the Appendix.

Theorem 4 (Reveliotis (2007)): Let S denote a siphon of a pure marked PN $\mathcal{N} = (P, T, W, M_0)$ such that

$$NF(\lambda_S) = \sum_{i=1}^{n} a_i NF(v^i) \tag{10.55}$$

where v^i, $i = 1, \ldots, n$, are *GCM* generators of $\mathcal{N}$, and $a_i \in \Re$, $\forall i$. Then,

$$S \text{ is controlled in } \mathcal{N} \Longleftrightarrow \lambda_S^T M_0 + G^* > 0 \tag{10.56}$$

where

$$G^* = \min_{M \in R(\mathcal{N}, M_0)} (M - M_0)^T \sum_{i=1}^{n} a_i v^i \tag{10.57}$$

To see the validity of this theorem, consider a marking $M \in R(\mathcal{N}, M_0)$. Then, there exists a vector $z \in (Z_0^+)^{|T|}$ such that $M = M_0 + \Theta z$ (c.f. Equations 10.2 and 10.3 in the Appendix). Therefore,

$$\begin{aligned}
M(S) &= \sum_{p \in S} M(p) \\
&= \lambda_S^T M \\
&= \lambda_S^T M_0 + \lambda_S^T \Theta z \\
&= \lambda_S^T M_0 + NF(\lambda_S) z \\
&= \lambda_S^T M_0 + \left[\sum_i a_i NF(v^i)\right] z \\
&= \lambda_S^T M_0 + \left[\sum_i a_i (v^i)^T \Theta\right] z \\
&= \lambda_S^T M_0 + \left[\sum_i a_i v^i\right]^T \Theta z \\
&= \lambda_S^T M_0 + \left[\sum_i a_i v^i\right]^T (M - M_0) \\
&= \lambda_S^T M_0 + (M - M_0)^T \sum_i a_i v^i
\end{aligned} \tag{10.58}$$

Clearly, the rhs of Equation 10.58 is minimized over $R(\mathcal{N}, M_0)$ by G^*, and therefore, S will be controlled iff the criterion of Equation 10.56 holds.

A siphon S controlled by means of the criterion of Theorem 4 will be characterized as an "implicitly" controlled siphon. The corresponding generator vectors v^i, $i = 1, \ldots, n$, of Equation 10.55, will be called the "controlling generators" of S. To operationalize the criterion of Theorem 4, we must provide an analytic characterization of the constraint $M \in R(\mathcal{N}, M_0)$. This can be done effectively using the theory presented by Reveliotis (2006). Alternatively, one can compromise for a sufficiency test by relaxing the requirement $M \in R(\mathcal{N}, M_0)$ in Equation 10.57 to that expressed by the state Equations 10.2 and 10.3, in the Appendix. We state the resulting criterion as a corollary.

Corollary 3: Let S denote a siphon of a pure marked PN $\mathcal{N} = (P, T, W, M_0)$ such that

$$NF(\lambda_S) = \sum_{i=1}^{n} a_i NF(v^i) \tag{10.59}$$

where v^i, $i = 1, \ldots, n$, are *GCM* generators of $\mathcal{N}$, and $a_i \in \Re$, $\forall i$. Also, let

$$G' = \min_{(M, z)} (M - M_0)^T \sum_{i=1}^{n} a_i v^i \tag{10.60}$$

such that

$$M = M_0 + \Theta z \tag{10.61}$$

$$M \geq 0, \quad z \in (Z_0^+)^{|T|} \tag{10.62}$$

Then,

$$\lambda_S^T M_0 + G' > 0 \Rightarrow S \text{ is controlled in } \mathcal{N} \tag{10.63}$$

Notice that the mathematical programming (MP) formulation involved in the criterion of Corollary 3 is a MIP, and therefore, it can be easily addressed through commercial solvers.* Next, we present another criterion that is weaker than the criterion of Corollary 3, but it connects the presented results to those originally derived by Li and Zhou (2004). Furthermore, this new criterion can be simpler, from a computational standpoint.

Corollary 4: Let S denote a siphon of a pure marked PN $\mathcal{N} = (P, T, W, M_0)$ such that

$$NF(\lambda_S) = \sum_{i=1}^{n} a_i NF(v^i) \tag{10.64}$$

where v^i, $i = 1, \ldots, n$, are *GCM* generators of $\mathcal{N}$, and $a_i \in \Re$, $\forall i$. Also, for every $i \in \{1, \ldots, n\}$, let $\underline{GCM}(v^i)$ and $\overline{GCM}(v^i)$ respectively denote a lower and an upper bound of $GCM(M, v^i)$, for all M such that

---

* In fact, the integrality requirement for z can be further relaxed to $z \geq 0$, providing a test that is computationally easier, but also with diminished resolution power, compared to the test of Corollary 3.

$$M = M_0 + \Theta z \tag{10.65}$$

$$M \geq 0, z \in (Z_0^+)^{|T|} \tag{10.66}$$

Finally, let

$$G'' = \sum_{i:a_i>0} a_i[\underline{GCM}(v^i) - GCM(M_0,v^i)] + \sum_{i:a_i<0} a_i[\overline{GCM}(v^i) - GCM(M_0,v^i)] \tag{10.67}$$

Then,

$$\lambda_S^T M_0 + G'' > 0 \Rightarrow S \text{ is controlled in } \mathcal{N} \tag{10.68}$$

The validity of Corollary 4 can be easily established by noticing that

$$(M - M_0)^T \sum_i a_i v^i = \sum_i a_i(M^T v^i - M_0^T v^i) = \sum_i a_i[GCM(M,v^i) - GCM(M_0,v^i)] \tag{10.69}$$

Then, the definitions of $\underline{GCM}(v^i)$ and $\overline{GCM}(v^i)$, when combined with Equations 10.60–10.62, 10.65–10.67, and 10.69, imply that $G' \geq G''$ and the validity of Corollary 4 follows from Corollary 3.

Beyond providing a sufficiency test for assessing whether a given siphon S is implicitly controlled by a set of GCM generator vectors $\{v^i: i=1,\ldots,n\}$, the result of Corollary 4 can also provide the basis for deploying a control mechanism that will actively enforce the implicit control of siphon S by some generator set $\{v^i: i=1,\ldots,n\}$. Under this approach, the upper and lower bounds $\overline{GCM}(v^i)$ and $\underline{GCM}(v^i)$, $i=1,\ldots,n$, are "design parameters," and their values are chosen such that they guarantee the condition of Equation 10.68. The selected bounds can be subsequently enforced on the behavior of the original net by the addition of appropriate "monitor places," according to the theory developed by Moody and Antsaklis (1998). Finally, it is also known that:

Proposition 1 (Li and Zhou, 2004): Given a pure marked PN $\mathcal{N}=(P,T,W,M_0)$, the rank of the space of net flow vectors $NF(\lambda_S)$, corresponding to the net siphons S, is bounded from above by min $\{|P|,|T|\}$.

Hence, the entire set of siphons, S, of a pure marked PN $\mathcal{N}=(P,T,W,M_0)$, can be potentially controlled by a set of generators, and the resultant monitor places, that is linearly sized with respect to the net size $|N|$. The following example demonstrates this capability and connects the above discussion to the context of process-resource nets and algebraic DAPs.

Example 5: Consider the net N depicted by solid lines in Figure 10.20, under the supervision of the algebraic DAP expressed by the following constraints:

$$\begin{bmatrix} 1 & & & 1 & 1 & 1 \\ 1 & 1 & & 1 & 1 & \\ 1 & 1 & 1 & 1 & & \end{bmatrix} \begin{bmatrix} M(p_{11}) \\ M(p_{12}) \\ M(p_{13}) \\ M(p_{21}) \\ M(p_{22}) \\ M(p_{23}) \end{bmatrix} \leq \begin{bmatrix} 1 \\ 2 \\ 1 \end{bmatrix} \tag{10.70}$$

The control subnet enforcing the constraints of Equation 10.70 on N is also depicted in Figure 10.20, through dashed lines. The resulting controlled net, $\mathcal{N}^c$, has been shown to be live and reversible by Park and Reveliotis (2000). Here we reestablish the liveness of net $\mathcal{N}^c$, and the correctness of the DAP expressed by Equation 10.70, by applying the siphon control criterion of Corollary 4.

The characteristic vectors of the minimal siphons in the controlled net $\mathcal{N}^c$ of Figure 10.20 are tabulated in Table 10.2. Siphons S_1–S_8 correspond to the support of p-semiflows, and therefore, they are already controlled. The net flows $NF(\lambda_{S_k})$ of the remaining uncontrolled siphons $S_k, k=9,10,11$, can be expressed as linear combinations of the net flows $NF(v^l)$ corresponding to

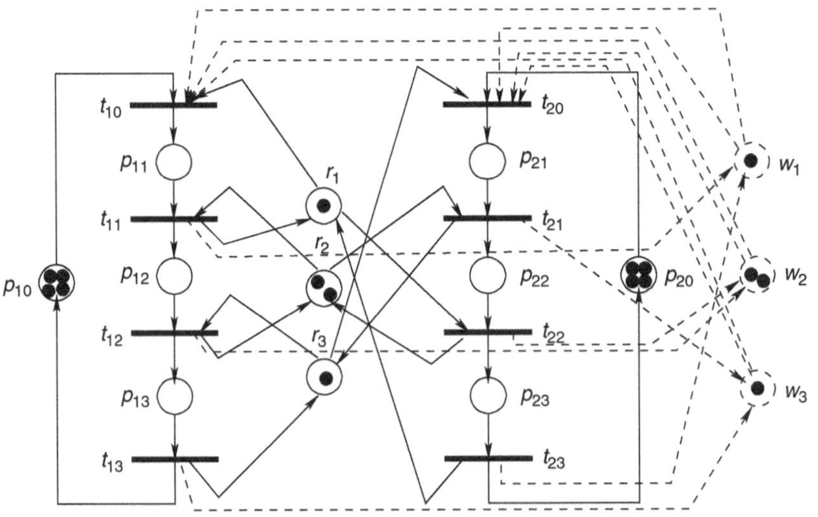

Figure 10.20 The nets $\mathcal{N}$ and $\mathcal{N}^c$ of Example 5.

Table 10.2 The Minimal Siphons of the Controlled Net $\mathcal{N}^c$

Siphon	p_{10}	p_{11}	p_{12}	p_{13}	p_{20}	p_{21}	p_{22}	p_{23}	r_1	r_2	r_3	w_1	w_2	w_3
S_1	1	1	1	1	0	0	0	0	0	0	0	0	0	0
S_2	0	0	0	0	1	1	1	1	0	0	0	0	0	0
S_3	0	1	0	0	0	0	0	1	1	0	0	0	0	0
S_4	0	0	1	0	0	0	1	0	0	1	0	0	0	0
S_5	0	0	0	1	0	1	0	0	0	0	1	0	0	0
S_6	0	1	0	0	0	1	1	1	0	0	0	1	0	0
S_7	0	1	1	0	0	1	1	0	0	0	0	0	1	0
S_8	0	1	1	1	0	1	0	0	0	0	0	0	0	1
S_9	0	0	1	0	0	0	0	1	1	1	0	0	0	0
S_{10}	0	0	0	1	0	0	1	0	0	1	1	0	0	0
S_{11}	0	0	0	1	0	0	0	1	1	1	1	0	0	0

the *GCM* generator vectors $v^l, l=1,\ldots,6$, presented in Table 10.3; Table 10.4 provides the relevant coefficients a^l_k, $l=1,\ldots,6$, $k=9,10,11$, for these expansions. Notice that the vector set $\{v^l, l=1,\ldots,6\}$ contains the *GCM* generator set $\{v^i, i=1,2,3\}$, that is induced by the DAP constraints of Equation 10.70, and an additional vector set $\{v^j, j=4,5,6\}$, selected in a way that facilitates the aforementioned expansion of the vector set $\{NF(\lambda_{S_k})\}$, $k=9,10,11$.

Table 10.3 also provides the bounds $\underline{GCM}(v^l)$ and $\overline{GCM}(v^l)$ used in the evaluation of G'', during the application of the criterion of Corollary 4 to the siphons S_k, $k=9,10,11$. The values of $\underline{GCM}(v^l)$ are obtained immediately by noticing that (i) $M^T \cdot v^l \geq 0$, $\forall l$, (ii) $v^l(p) > 0 \Rightarrow p \in P_S$, $\forall p \in P$, and (iii) $M_0(p)=0$, $\forall p \in P_S$. The values of $\overline{GCM}(v^l)$ were obtained by solving the following MIP for each $l \in \{1,\ldots,6\}$:

Table 10.3 Example 5: The *GCM* Generators, v^l, Employed for the Expansion of the Net Flows Vectors $NF(\lambda_{S_k})$, $k = 9,10,11$, and the Associated Bounds Used in the Evaluation of the Criterion of Corollary 4

Gen.	p_{10}	p_{11}	p_{12}	p_{13}	p_{20}	p_{21}	p_{22}	p_{23}	r_1	r_2	r_3	w_1	w_2	w_3	$\underline{GCM}$	$\overline{GCM}$
v^1	0	1	0	0	0	1	1	1	0	0	0	0	0	0	0	1
v^2	0	1	1	0	0	1	1	0	0	0	0	0	0	0	0	2
v^3	0	1	1	1	0	1	0	0	0	0	0	0	0	0	0	1
v^4	0	1	0	0	0	0	0	0	0	0	0	0	0	0	0	1
v^5	0	0	1	0	0	0	0	0	0	0	0	0	0	0	0	1
v^6	0	0	0	0	0	1	0	0	0	0	0	0	0	0	0	1

Table 10.4 The Coordinates for the Expansion of $NF(\lambda_{S_k})$, $k = 9,10,11$, as Linear Combinations of $NF(v^l)$, $l = 1,\ldots,6$, and the Obtained Value for the Test of Corollary 4

Siphon	a^1	a^2	a^3	a^4	a^5	a^6	$\lambda_S^T M_0 + G''(S)$
S_9	0.0	−1.0	0.0	0.0	1.0	1.0	3−2=1
S_{10}	0.0	0.0	0.0	0.0	−1.0	−1.0	3−2=1
S_{11}	0.0	−1.0	0.0	0.0	0.0	0.0	4−2=2

$$\overline{GCM}(v^l) = \max_{(M,z)} M^T \cdot v^l$$

s.t.

Equations 10.65 and 10.66

Finally, Table 10.4 provides also the values obtained for the lhs of the inequality that is employed by the test of Corollary 4 (c.f. Equation 10.68), based on the aforementioned expansions and bounds. As all the obtained values are strictly greater than zero, it is concluded that the net $\mathcal{N}^c$ is live, and the DAP of Equation 10.70 is a correct DAP for the original net $\mathcal{N}$.

The work of Park and Reveliotis (2000) has also established that the DAP obtained by replacing the rhs of Equation 10.70 with the vector $[2\ 4\ 2]^T$, is another correct DAP for net $\mathcal{N}$. Interestingly, the application of the test of Corollary 4, based on the *GCM* generator set $\{v^l\}$ of Table 10.3, fails to recognize the ability of this new DAP to control the siphons S_9 and S_{10} of Table 10.2. On the other hand, this effect is successfully recognized by the more powerful test of Corollary 3. We leave the relevant computational details to the reader.

10.7 Conclusions

This chapter started with the observation that the flexible automation pursued in the context of many contemporary technological applications necessitates the explicit logical analysis and control of these environments with respect to the underlying resource allocation, and subsequently it offered a unified and comprehensive treatment of the theory of algebraic deadlock avoidance policies, that provide an effective and efficient solution to the emerging logical control problems. The presented developments characterized the state of the art in the relevant research area, and, hopefully, they have also revealed its maturity and vigor. At the same time, these results can function as the starting point for additional developments in the field in terms of, both, theory and application.

On the theoretical side, a novel research direction was recently developed by Reveliotis et al. (2007), which introduced the notion of generalized algebraic DAP, based on the notion of "committee machine" that was borrowed from pattern recognition and machine learning. The key property of generalized algebraic DAPs is that, when viewed as pattern classifiers in the underlying state space, they can recognize nonconvex state subsets, something that is not possible with the linear structure of algebraic DAPs. The work of Reveliotis et al. (2007) extends the design methodology presented in Section 10.4 to this new class of policies, but currently, we lack a complete understanding and characterization of the properties of these policies in the spirit of Sections 10.5 and 10.6. An even more prominent open problem on the theoretical side is the development of the necessary theory for the effective and systematic integration of logical and performance-oriented control. Some preliminary thoughts and results along these lines are reported by Reveliotis (2005) and Choi and Reveliotis (2005).

From an application standpoint, the ultimate objective of the research program underlying the results presented in this work is the integration of the developed theory to a control architecture that will function as the next-generation "operating system," able to support robust, yet highly flexible and efficient operation of the target technological applications. Although this effort can be initiated and led by the relevant research community, a profound understanding of, and extensive interaction with, the target industries is of paramount importance for the successful implementation and the eventual acceptance of the final product.

Appendix: Petri Nets—Basic Concepts and Definitions

A formal definition of the Petri net (PN) model is as follows:

Definition 6 (Murata, 1989): A PN is defined by a quadruple $\mathcal{N} = (P, T, W, M_0)$, where

- P is the set of places,
- T is the set of transitions,
- W: $(P \times T) \cup (T \times P) \rightarrow Z_0^+$ is the flow relation, and
- M_0: $P \rightarrow Z_0^+$ is the net initial marking, assigning to each place $p \in P$, $M_0(p)$ tokens.

The first three items in Definition 6 essentially define a "weighted bipartite digraph" representing the system "structure" that governs its underlying dynamics. The last item defines the system initial state. A conventional graphical representation of the net structure and its marking depicts nodes corresponding to places by empty circles, nodes corresponding to transitions by bars, and the tokens located at the various places by small filled circles. The flow relation W is depicted by directed edges that link every nodal pair for which the corresponding W-value is nonzero. These edges point from the first node of the corresponding pair to the second, and they are also labeled—or, "weighed"—by the corresponding W-value. By convention, absence of a label for any edge implies that the corresponding W-value is equal to unity.

PN Structure-Related Concepts and Properties

Given a transition $t \in T$, the set of places p for which $(p,t) > 0$ (resp., $(t,p) > 0$) is known as the set of "input" (resp., "output") places of t. Similarly, given a place $p \in P$, the set of transitions t for which $(t,p) > 0$ (resp., $(p,t) > 0$) is known as the set of input (resp., output) transitions of p. It is customary in the PN literature to denote the set of input (resp., output) transitions of a place p by ${}^{\bullet}p$ (resp., $p^{\bullet}$). Similarly, the set of input (resp., output) places of a transition t is denoted by ${}^{\bullet}t$ (resp., $t^{\bullet}$). This notation is also generalized to any set of places or transitions, X, for example, ${}^{\bullet}X = \bigcup_{x \in X} {}^{\bullet}x$.

The ordered set $X = <x_1 \cdots x_n> \in (P \cup T)^*$ is a path, iff $x_{i+1} \in x_i^{\bullet}$, $i = 1, \ldots, n-1$. Furthermore, a path X is characterized as a circuit iff $x_1 \equiv x_n$.

A PN with a flow relation W mapping onto $\{0,1\}$ is said to be "ordinary." If only the restriction of W to $(P \times T)$ maps on $\{0,1\}$, the PN is said to be *PT-ordinary*. An ordinary PN such that $\forall t \in T$, $|t^{\bullet}| = |{}^{\bullet}t| = 1$, is characterized as a "state machine," while an ordinary PN such that $\forall p \in p$, $|p^{\bullet}| = |{}^{\bullet}p| = 1$, is characterized as a "marked graph."

A PN is said to be "pure" if $\forall (x,y) \in (P \times T) \cup (T \times P), W(x,y) > 0 \Rightarrow W(y, x) = 0$. The flow relation of pure PNs can be represented by the flow matrix $\Theta = \Theta^+ - \Theta^-$ where $\Theta^+(p,t) = W(t,p)$ and $\Theta^-(p,t) = W(p,t)$.

PN Dynamics-Related Concepts and Properties

In the PN modeling framework, the system state is represented by the net marking M, that is, a function from P to Z_0^+ that assigns a token content to the various net places. The net marking M is initialized to marking M_0, introduced in Definition 6, and it subsequently evolves through a set of rules summarized in the concept of "transition firing." A concise characterization of this concept is as follows: Given a marking M, a transition t is enabled iff for every place $p \in {}^{\bullet}t$, $M(p) \geq W(p,t)$, and this is denoted by $M[t>$. $t \in T$ is said to be disabled by a place $p \in {}^{\bullet}t$ at M iff $M(p) < W(p,t)$. Furthermore, a place $p \in P$ for which there exists $t \in p^{\bullet}$ such that $M(p) < W(p,t)$ is said to be a disabling place at M. Given a marking M, a transition t can be fired only if it is enabled in M. The enabling of t in M is denoted by $M[t>$, and its firing results in a new marking M, which is obtained from M by removing $W(p,t)$ tokens from each place $p \in {}^{\bullet}t$, and placing $W(t,p')$ tokens in each place $p' \in t^{\bullet}$. For pure PNs, the marking evolution incurred by the firing of a transition t can be concisely expressed by the "state equation":

$$M' = M + \Theta \cdot 1_t \tag{10.71}$$

where 1_t denotes the unit vector of dimensionality $|T|$ and with the unit element located at the component corresponding to transition t.

The set of markings reachable from the initial marking M_0 through any fireable sequence of transitions is denoted by $R(\mathcal{N},M_0)$ and it is referred to as the net "reachability space." In the case of pure PNs, a necessary condition for $M \in R(\mathcal{N}, M_0)$ is that the following system of equations is feasible in z:

$$M = M_0 + \Theta z \tag{10.72}$$

$$M \geq 0, \quad z \in Z_0^+ \tag{10.73}$$

A PN $\mathcal{N} = (P, T, W, M_0)$ is said to be "bounded" iff all markings $M \in R(\mathcal{N}, M_0)$ are bounded. $\mathcal{N}$ is said to be "structurally bounded" iff it is bounded for any initial marking M_0. $\mathcal{N}$ is said to be "reversible" iff $M_0 \in R(\mathcal{N},M)$, for all $M \in R(\mathcal{N}, M_0)$. A transition $t \in T$ is said to be "live" iff for all $M \in R(\mathcal{N}, M_0)$, there exists $M' \in R(\mathcal{N}, M)$ such that $M'[t>$; non-live transitions are said to be dead at those markings $M \in R(\mathcal{N}, M_0)$ for which there is no $M' \in R(\mathcal{N}, M)$ such that $M'[t>$. PN $\mathcal{N}$ is "quasi-live" iff for all $t \in T$, there exists $M \in R(\mathcal{N}, M_0)$ such that $M[t>$; it is "weakly live" iff for all $M \in R(\mathcal{N}, M_0)$, there exists $t \in T$ such that $M[t>$; and it is "live" iff for all $t \in T$, t is live. A marking $M \in R(\mathcal{N}, M_0)$ is a (total) "deadlock" iff every $t \in T$ is dead at M.*

Siphons and Their Role in the Interpretation of the PN Deadlock

Of particular interest for the liveness analysis of the PNs considered in this chapter is a structural element known as "siphon," that is, a set of places $S \subseteq P$ such that ${}^{\bullet}S \subseteq S^{\bullet}$. A siphon S is "minimal" iff there exists no other siphon S' such that $S' \subset S$. A siphon S is said to be empty at marking M iff $M(S) \equiv \sum_{p \in S} M(p) = 0$. S is said to be "deadly marked" at marking M, iff every transition $t \in {}^{\bullet}S$ is disabled by some place $p \in S$. Clearly, empty siphons are deadly marked siphons. It is easy to see that, if S is a deadly marked siphon at some marking M, (i) $\forall t \in {}^{\bullet}S$, t is a dead transition in M, and (ii) $\forall M' \in R(\mathcal{N}, M)$, S is deadly marked. Furthermore, the next theorem establishes a strong relationship between the notion of deadly marked siphon and that of the PN deadlock.

* Notice that the concept of deadlock in the PN framework is different from the usage of this term in the RAS context. Some further elaboration on this issue is provided in Section 10.2.

Theorem 5: (Reveliotis, 2005) Given a deadlock marking M of a PN $\mathcal{N}=(P, T, W, M_0)$, the set of disabling places $S \subseteq P$ in M constitutes a deadly marked siphon.

PN Semiflows

PN semiflows provide an analytical characterization of various concepts of "invariance" underlying the net dynamics. Generally, there are two types, p- and t-semiflows, with a p-semiflow formally defined as a $|P|$-dimensional vector y satisfying $y^T\Theta=0$ and $y \geq 0$, and a t-semiflow formally defined as a $|T|$-dimensional vector x satisfying $\Theta x=0$ and $x \geq 0$. In the light of Equation 10.72, the invariance property expressed by a p-semiflow y is that $y^T M = y^T M_0$, for all $M \in R(\mathcal{N}, M_0)$. Similarly, Equation 10.72 implies that for any t-semiflow x, $M=M_0+\Theta x=M_0$.

Given a p-semiflow y (resp., t-semiflow x) its support is defined as $||y|| = \{p \in P \mid y(p)>0\}$ (resp., $||x||=\{t \in T \mid x(t)>0\}$). A p-semiflow y (resp., t-semiflow x) is said to be minimal iff there is no p-semiflow y' (resp., t-semiflow x') such that $||y'|| \subset ||y||$ (resp., $||x'|| \subset ||x||$).

PN Merging

We conclude this introductory discussion on the PN concepts and properties by defining a merging operation for two PNs: Given two PNs $\mathcal{N}_1=(P_1,T_1, W_1, M_{01})$ and $\mathcal{N}_2=(P_2,T_2,W_2,M_{02})$ with $T_1 \cap T_2=\emptyset$ and $P_1 \cap P_2 = Q \neq \emptyset$ such that for all $p \in Q$, $M_{01}(p) = M_{02}(p)$, the PN $\mathcal{N}$ resulting from the merging of the nets $\mathcal{N}_1$ and $\mathcal{N}_2$ through the place set Q, is defined by $\mathcal{N} = (P_1 \cup P_2,\ T_1 \cup T_2,\ W_1 \cup W_2,\ M_0)$ with $M_0(p) = M_{01}(p),\ \forall p \in P_1 \backslash P_2$; $M_0(p) = M_{02}(p),\ \forall p \in P_2 \backslash P_1$; $M_0(p) = M_{01}(p) = M_{02}(p),\ \forall p \in P_1 \cap P_2$.

References

Araki, T., Sugiyama, Y., and Kasami, T. Complexity of the deadlock avoidance problem. In *2nd IBM Symp. on Mathematical Foundations of Computer Science*, pp. 229–257. IBM, 1977.

Badouel, E. and Darondeau, P. Theory of regions. In W. Reisig and G. Rozenberg (eds.), *LNCS 1491—Advances in Petri Nets: Basic Models*, pp. 529–586. Springer, New York, 1998.

Cassandras, C.G. and Lafortune, S. *Introduction to Discrete Event Systems*. Kluwer Academic, Boston, MA, 1999.

Choi, J.Y. and Reveliotis, S.A. Relative value function approximation for the capacitated re-entrant line scheduling problem. *IEEE Trans. on Automation Science and Engineering*, 2: 285–299, 2005.

Chu, F. and Xie, X.-L. Deadlock analysis of petri nets using siphons and mathematical programming. *IEEE Transactions on Robotics & Automation*, 13: 793–804, 1997.

Garey, M.R. and Johnson, D.S. *Computers and Intractability: A Guide to the Theory of NP-Completeness*. W.H. Freeman, New York, 1979.

Hopcroft, J.E. and Ullman, J.D. *Introduction to Automata Theory, Languages and Computation*. Addison-Wesley, Reading, MA, 1979.

Jeng, M., Xie, X., and Peng, M.Y. Process nets with resources for manufacturing modeling and their analysis. *IEEE Transactions. on Robotics & Automation*, 18: 875–889, 2002.

Li, Z.W. and Zhou, M.C. Elementary siphons of Petri nets and their application to deadlock prevention in flexible manufacturing systems. *IEEE Transactions on Systems, Man, and Cybernetics—Part A*, 34: 38–51, 2004.

Moody, J.O. and Antsaklis, P.J. *Supervisory Control of Discrete Event Systems using Petri Nets*. Kluwer Academic, Boston, MA, 1998.

Murata, T. Petri nets: Properties, analysis and applications. *Proceedings of the IEEE*, 77: 541–580, 1989.

Park, J. and Reveliotis, S. Algebraic synthesis of efficient deadlock avoidance policies for sequential resource allocation systems. *IEEE Transactions on Robotics & Automation*, 16: 190–195, 2000.

Park, J. and Reveliotis, S.A. Deadlock avoidance in sequential resource allocation systems with multiple resource acquisitions and flexible routings. *IEEE Transactions on Automatic Control*, 46: 1572–1583, 2001.

Reveliotis, S.A. Implicit siphon control and its role in the liveness enforcing supervision of sequential resource allocation systems. *IEEE Trans. on SMC-Part A*, 37(3): 319–328, 2007.

Reveliotis, S.A. On the siphon-based characterization of liveness in sequential resource allocation systems. Proceedings of ATPN 2003, Eindhoven, Netherlands, Springer LCNS 2679, pp. 241–255, 2003a.

Reveliotis, S.A. Structural analysis of assembly/disassembly resource allocation systems. In *Proceedings of ICRA 2003*. Institute of Electrical and Electronics Engineers, New York, 2003b.

Reveliotis, S.A. *Real-time Management of Resource Allocation Systems: A Discrete Event Systems Approach*. Springer, New York, 2005.

Reveliotis, S.A. A necessary and sufficient condition for the liveness and reversibility of process-resource nets with acyclic, quasi-live, serialisable and reversible process subnets. *IEEE Transactions on Automation Science and Engineering*, 3: 462–468, 2006.

Reveliotis, S.A. and Choi, J.Y. Designing reversibility-enforcing supervisors of polynomial complexity for bounded Petri nets through the theory of regions. In *Proceedings of ICATPN 2006*, Turku, Finland, Springer LCNS 4024, pp. 322–341, 2006.

Reveliotis, S.A. and Ferreira, P.M. Deadlock avoidance policies for automated manufacturing cells. *IEEE Transactions on Robotics & Automation*, 12: 845–857, 1996.

Reveliotis, S.A., Roszkowska, E., and Choi, J.Y. Generalized algebraic deadlock avoidance policies for sequential resource allocation systems. In *2007 IEEE International Conference on Robotics & Automation*, pp. 991–996. Institute of Electrical and Electronics Engineers, New York, 2007.

Roszkowska, E. and Wojcik, R. Problems of process flow feasibility in fas. In *CIM in Process and Manufacturing Industry*, pp. 115–120. Pergamon Press, Oxford, 1993.

Silva, M., Teruel, E., and Colom, J.M. Linear algebraic and linear programming techniques for the analysis of place / transition net systems. In W. Reisig and G. Rozenberg (eds.), *Lecture Notes in Computer Science*, Vol. 1491, pp. 309–373. Springer, New York, 1998.

Van der Aalst, W. Structural characterizations of sound workflow nets. *Technical Report Computing Science Reports* 96/23, Eindhoven University of Technology, Eindhoven, The Netherlands, 1996.

Van der Aalst, W. Verification of workflow nets. In P. Azema and G. Balbo (eds.), *Lecture Notes in Computer Science*, Vol. 1248, pp. 407–426. Springer, New York, 1997.

Van der Aalst, W. and Van Hee, K. *Workflow Management: Models, Methods and Systems*. The MIT Press, Cambridge, MA, 2002.

Van Hee, K., Sidorova, N., and Voorhoeve, M. Soundness and separability of workflow nets in the stepwise refinement approach. In W. Van der Aalst and E. Best (eds.), *Lecture Notes in Computer Science*, Vol. 2679, pp. 337–356. Springer, New York, 2003.

Zhou, M. and Fanti, M.P. (eds). *Deadlock Resolution in Computer-Integrated Systems*. Marcel Dekker, Singapore, 2004.

Chapter 11

Teaching Warehousing Concepts through Interactive Animations and 3-D Models

Gürdal Ertek

CONTENTS

Abstract A significant challenge in teaching warehousing and facility logistics is conveying the important managerial and operational concepts to students effectively. These concepts always involve the presence and interactions of real-world objects such as facilities, vehicles, materials, storage and retrieval equipment, and most importantly humans. Meanwhile, knowledge in the mentioned fields does not only consist of concepts and principles but also include algorithms that solve a variety of problems encountered in practice. Trying to communicate this body of knowledge in words with the help of photos and drawings frequently proves insufficient. In teaching his warehousing courses the author has realized that a set of educational media composed of interactive animations and virtual three-dimensional (3-D) models indeed facilitates the challenges described earlier.

Extensive educational media has been developed at Sabanci University in a coordinated effort of undergraduate students under the supervision of the author, and with the support of a multimedia expert. The media content has been selected and organized such that it can support and enrich classes and courses that would be based on two extremely useful books, one on implementing world-class practices in warehousing, and the other on strategies for saving space in the warehouse. The developed materials are freely available on the Internet to everyone, and are expected to contribute to the awareness, recognition, and growth of the fields of warehousing and facility logistics. In this chapter the educational materials developed, the rationale for the technology selections, and the teaching methods in class are explained. Meanwhile, resources for warehousing education are reviewed extensively and potential new technologies and approaches are highlighted.

11.1 Introduction

In a supply chain, physical products are delivered from their origin points (sources) to their destination points (sinks) through transshipment points. Warehouses are critical components within supply chains, because they may exist in all these three sets of points and significantly impact financial and service performance of supply chains. As of 1995, there were an estimated 550,000 warehouses in the United States alone (Andel, 1995). According to the Material Handling Industry of America (MHIA)*, an independent organization for warehousing professionals, "the consumption of material handling and logistics equipment and systems in America exceeds $125 billion per year, and producers employ in excess of 700,000 workers." Thus the best education of warehousing professionals and the adequate design, implementation, and operation of warehouses may have a large positive impact on the economy.

The main goal of this chapter is to provide readers with new perspectives on teaching warehousing and material handling. For this purpose, firstly, existing resources for warehousing education are introduced and discussed. Next, a project carried out at Sabanci University, Istanbul, Turkey is presented. This continuing project has so far received contributions from 11 undergraduate students at Sabanci University, and aims at developing educational materials for warehousing. Until now, the project has involved the development of 31 animations that explain warehousing concepts, and 33 three-dimensional (3-D) models of warehousing equipment. Finally, other alternative technologies and approaches that can be used in warehousing education are proposed.

11.2 Resources for Warehousing Education

In this section, the various warehousing education resources available in-print and online are reviewed. These resources have been developed or their development has been supervised by professional organizations, publishers, software companies, or university professors. The resources are sorted, based on a subjective combined assessment of their quality, accessibility, online availability, cost, applicability, and relevancy to the work described in this chapter, as perceived by the author.

The book used as the textbook for the "MS 420: Storage and Distribution Systems" course at Sabanci University is *World-Class Warehousing and Material Handling* by Frazelle (2001). It is suggested that the book *How to Save Warehouse Space: 153 Tested Techniques*, published by Distribution Group† also be used in warehousing and supply-chain courses. The educational materials described in this chapter are designed to support these two books.

* http://www.mhia.org/

† http://www.distributiongroup.com/

MHIA* is a leading professional organization that serves warehousing and supply-chain community. The MHIA case studies Web page† contains brief case studies that summarize implementations of automated storage/retrieval systems (AS/RS) at companies and institutions in very diverse industries. Articles in the Material Handling Classics and Material Handling Perspectives sections within the MHIA Web site cover practically every aspect of warehousing and material handling. Material Handling Institute Bookstore‡ hosts an impressive collection of online books and articles, which can be downloaded free of charge through registration.

John J. Bartholdi at Georgia Institute of Technology$ has made available on his Web site a broad collection of educational materials for warehousing. The primary resource at Bartholdi's Web site is the freely downloadable online book *Warehouse & Distribution Science* by J.J. Bartholdi and S. Hackman. The book's Web site also contains abundant supplementary materials to support the book. These include software tools, past class projects with all their relevant materials (such as problem descriptions, plant layouts, and datasets), virtual tours of warehouses through photos and text explanations, and a compiled list of logistics news. One of the computational tools is the Java applet "Bird's Eye View" that visualizes a given location-based statistic within a given warehouse. The applet reads a Microsoft Excel spreadsheet that contains a map of the warehouse, with every spreadsheet cell representing a section of rack/shelving or empty space. The cells that represent physical storage locations should be labeled uniquely. The applet also reads a text file that contains the values for the statistic of interest for every storage location. Once the color scheme is selected, the applet displays a colored bird's eye view of a warehouse, where the storage locations are painted based on their values for the given statistic. Bartholdi's Web site also contains extensive information on bucket brigades, a method of organizing workers on an assembly line that results in self-balancing of the line.

College-Industry Council on Material Handling Education (CICMHE)‖ is a professional organization which "prepares and provides information, teaching materials and various events in support of material handling education and research." Educational resources available at the council's Web site include 16 case studies, 13 lecture resources, classroom modules, a list of textbooks, and a list of specialty sites with tools for warehousing/material-handling educators. The council also hosts the Material Handling

* http://www.mhia.org/

† http://www.mhia.org/et/et_case_studies.cfm

‡ http://www.mhiastore.org/

$ http://www2.isye.gatech.edu/people/faculty/John_Bartholdi/

‖ http://www.mhia.org/et/ET_MHI_CICMHE_Home.cfm

Taxonomy[*] Web pages that have been developed by Micheal G. Kay at North Carolina State University. These Web pages are available under Kay's own Web site,[†] as well. This very wide collection of Web pages provides taxonomy of material-handling equipment, accompanied with brief information for and pictures of every equipment.

Material Handling Multimedia Bank[‡] is developed by Benoit Montreuil, Richard Legare, and Jonathan Bouchard at the Université Laval, Canada. The multimedia bank contains a rich collection of pictures and videos, besides application guides and information on vendor companies.

Another resource available on the Internet is the Interactive Warehouse[$] a Java applet created by Kees Jan Roodbergen, a professor at Erasmus University. The Interactive Warehouse allows users to learn about algorithms for order-picker routing. The applet is based on an article by Roodbergen and De Koster (2001). The applet walks the user through the steps of setting a warehouse layout, creating an order, manually creating a route, running one of the known routing algorithms, and viewing the results.

Modern Materials Handling[||] and *Logistics Management*[#] are two influential and valuable trade magazines that can be freely subscribed online (following an evaluation process). Once subscribed, one can read the full electronic versions of the magazine issues and can even download the issues in Adobe Acrobat (.pdf) format. Both magazines offer a wealth of information on warehousing, material handling, and supply-chain management.

Modern Materials Handling magazine has also made available online[**] several guideline documents that contain warehouse design plans and ideas. These guidelines have the following themes: picking strategies, value-added services, world-class warehouse productivity, facility layouts for third-party logistics, and world-class facility layouts for e-commerce.

Distribution Group[††] is a provider of useful practical information and news for distribution professionals around the world. Section 11.6 describes the educational materials developed to support the book *How to Save Warehouse Space: 153 Tested Techniques* (Kuchta and the staff of Gross & Associates 2004[‡‡]).

---

* http://www.mhia.org/et/mhe_tax.htm
† http://www.ise.ncsu.edu/kay/mhetax/
‡ http://www.centor.ulaval.ca/mhmultimediabank/
$ http://www.roodbergen.com/warehouse/
|| http://www.mmh.com/
http://www.logisticsmgmt.com/
** http://www.contentconvergence.com/2001web/tocmain.htm
†† http://www.distributiongroup.com/
‡‡ http://www.distributiongroup.com/htsws.php

Dosch Design* is a software company that provides computer graphics products, including 3-D models, textures, visualizations, images, and movie clips for "license free" use by companies. The "Industrial Objects V2" product† created and marketed by the company contains 3-D models of industrial objects and equipment from workshop, factory, and warehouse environments. The 3-D models come in a variety of 3-D graphic formats and have very high quality. Many of the objects and equipment available in the mentioned product were not modeled within the scope of our project. Therefore, the usage of this product is suggested to support the educational materials described here. The company also offers the "Utility Vehicles" product,‡ which includes 3-D models of a truck, a trailer, and a forklift.

Jeroen van den Berg Consulting,$ is a consulting company in the Netherlands that specializes in warehousing and logistical information systems. The company's Web site contains several publications including management outlook reports and journal articles. Under the Research link at the Web site one can find WOLF, an online program for computing the suitability of ~50 warehouse management systems (WMS) for a given company. The system takes as input, answers to several questions, including region, number of warehousing staff, warehouse surface area, and order lines per day. Then a computational engine is run and the WMS products available in the WOLF database are ranked from most suitable to the least, accompanied with scores that denote their suitability. This is the only tool that was found on the Internet that provides such a decision-support capability. Many consulting companies sell research reports that provide similar information for a price in the scale of hundreds of U.S. dollars. Thus the WOLF tool is a highly useful example of decision support for technology and software selection.

The Progress Group|| provides logistics and especially warehousing consultancy, and is based in Georgia, United States. The Publications link on the company's Web site leads to a compilation of insightful white papers and articles.

Tompkins Associates# is a warehousing and supply-chain consulting firm based in North Carolina, United States. The company Web site has a variety of publications available for download under the Publications link.

---

* http://www.doschdesign.com/

† http://www.doschdesign.com/products/3d/Industrial_Objects_V2.html

‡ http://www.doschdesign.com/products/3d/D3D_Utility_Vehicles.html

$ http://www.jvdbconsulting.com/

|| http://www.theprogressgroup.com/

\# http://www.tompkinsinc.com/

These include "supply chain edge articles," monographs, and white papers. The Tompkins Press books are also listed under the Publications link.

Armstrong & Associates Inc.* is a supply-chain market research and consulting firm based in Wisconsin, United States. The company has a listing of its guides and research reports on its Web site.[†]

Keck Virtual Factory Lab at Georgia Institute of Technology[‡] hosts two interactive tutorials on material-handling systems, one on AS/RS design and the other on "from-to-chart" analysis for product routings. The tutorials have been developed under the supervision of Gunter Sharp at Georgia Institute of Technology and Bala Ram at North Carolina A&T University.

Hkplanet.net Learning Center,[$] a portal created by HK Systems,[||] is a provider of material-handling solutions for warehousing and manufacturing systems. This portal hosts a selection of industry articles and presentations from Material-Handling & Logistics Conferences. The Virtual Warehouse[#] that can be downloaded from this Web site can be interactively browsed through a joystick or mouse, and is a wonderful example of how virtual 3-D environments can be used for warehousing education. The HK Systems Web site also contains various resources, including white papers[**] and videos,[††] which can be used in teaching warehousing.

Heragu et al. (2003) report the development of a multimedia system for warehousing/material-handling classes. The system teaches ten principles of materials handling and three equipment categories. The system also depicts the industrial applications of some of the selected equipment in each of the three equipment categories and teaches quantitative methods for solving facility design and technology selection problems. The authors also present results of a formal evaluation of the system and its modules through surveys and interviews. They report that "an overwhelming number of students perceived their interaction with the modules to be a valuable experience." The educational system is distributed on a compact disc with the title "10 Principles of Materials Handling" and can be purchased via the MHIA Store.[‡‡]

---

* http://www.3plogistics.com/

† http://www.3plogistics.com/shopsite/index.html

‡ http://factory.isye.gatech.edu/mhs/

$ http://www.hkplanet.net/learning_center/

|| http://www.hksystems.com/

\# http://www.hkplanet.net/learning_center/virtual_warehouse.cfm

** http://www.hksystems.com/RESOURCES/whitepapers.cfm?m=4&s=1

†† http://www.hksystems.com/RESOURCES/multimedia.cfm?m=5&s=1

‡‡ http://www.mhiastore.org/

11.3 Selected Technologies and Teaching Approach

In this section, the selected technologies for creating the interactive animations and the virtual equipment models are explained. Then the teaching approach followed in class with the teaching materials is outlined.

11.3.1 Interactive Animations

Flash technology by Adobe* and the Adobe Flash authoring software have been selected in creating the interactive animations. Certain significant advantages of the Flash technology and Adobe Flash software played role in our selection include:

- The Flash animations (with file extension of .swf) that are created using Adobe Flash can be played from within almost any operating system. These animations can also be accessed over the Internet and be played within almost any Web browser that has a Flash Player add-in.
- A great percentage of Internet users already have Flash Player installed on their computers.
- The Flash animation files (with the .swf extension) have very small sizes, due to the use of vector graphics.
- The Adobe Corporation has a very strong presence on the Internet and a great influence in the software industry. The well-known Adobe Acrobat format (.pdf files) is the almost-universal document format in the academic and business world.
- The "action script" programming language within Macromedia Flash enables developers to create easy-to-use graphical user interfaces (GUI) and offers many of the capabilities of general-purpose programming languages, such as C++.

In designing and implementing the animations, special attention was given to make sure that all the animations share the same user-interface elements: The same set of buttons, same set of font sizes, and the same height-to-weight ratios were used in all the animations. It is believed that this approach helps the users to perceive the animations developed by different people as components of an integrated whole.

11.3.2 3-D Models

The SolidWorks software was selected for developing the 3-D models of warehousing equipment. The main criteria that determined this selection

* http://www.adobe.com

were the ease of learning and the popularity of the software. The fact that SolidWorks software had been already well known by some of the students in the project also played an important role in our final decision.

11.3.3 Teaching Approach

The material described in this chapter has been used while teaching the "MS 420: Storage and Distribution Systems" course at Sabanci University in Spring 2006 semester. The course is offered to senior and junior students and has no prerequisites. The students have been exposed to the media related with Module 1 (World-Class Warehousing and Material Handling) twice throughout the class. The animations and 3-D models have been shown to the students by the instructor (the author of this chapter) as part of the lectures. Then, in a separate two-hour session, the students have been requested to bring their laptops to class with the Adobe Flash Player and SolidWorks Viewer software installed and the animations and the 3-D models downloaded. In this second exposure, the students interactively explored each of the animations and 3-D models in detail, while the instructor explained each of the concepts and equipment once again. The author believes that such an approach is much more robust in teaching the concepts and animations to students at a university where English is not the native language of the students. However, this hypothesis has to be tested through formal studies.

The animations in Module 2 (How to Save Warehouse Space) have been shown to students only once, in another two-hour hands-on session where the students interactively explored each of the Module 2 animations while listening to the instructor's explanations.

11.4 Related Work

One can find papers in engineering-education literature that present impressive animation-based educational materials in various fields of engineering. Leung et al. (2001) report the development and classroom use of an animated-simulation package to teach electromagnetic theory. Ong and Mannan (2002) describe a Web-based courseware to teach concepts and principles in metalworking, focusing on metallurgy aspects. Their system is developed using Adobe Flash (as in our work) together with Microsoft Front Page Web authoring system. Another system is developed by Ettouney et al. (2000) for the design and simulation of thermal desalination process. Their system has been used extensively in graduate, undergraduate, and training classes with great success.

One can also find examples of academic work where virtual 3-D models have been used: Ou et al. (2003) describe the implementation of a Web-based virtual reality system that facilitates teaching of computer-aided design (CAD).

11.5 Teaching World-Class Warehousing Concepts (Module 1)

For Module 1 of the educational materials, interactive animations and 3-D models have been developed for teaching warehousing concepts and material-handling equipment. This module is intended to support the book *World-Class Warehousing and Material Handling*.

11.5.1 Sample Animation 1: Order-Picking Schemes

This animation summarizes the picking schemes in a warehouse, as explained in Frazelle (2001). Freeform picking takes place when there is no zoning in the warehouse. In this scenario, two possibilities are single-order picking and batch picking (Figure 11.1a). In single-order picking, the order picker travels on a new route to pick each new order (Figure 11.1b). In batch picking, the order picker picks two or more orders on the same route.

In zone picking, each order picker is allocated to a particular zone, and is responsible of picking the orders in his or her zone. Because the items within the same order may be picked at different zones simultaneously, there is a need to combine these picks into a single whole. Two strategies to achieve this are progressive assembly and downstream sortation. In progressive assembly, the first order picker picks the items in an order that reside in his or her zone, and puts the tote or box containing the incomplete order to the origin point for the second zone (Figure 11.1c). The next order picker seizes what the preceding order picker has accumulated and continues picking the items in the order that reside in his or her region.

In downstream-sortation strategy, the order picker in each zone leaves the incomplete order, that he or she picked, into a conveyor, and the conveyor carries the totes or boxes to a sortation point (Figure 11.1d).

11.5.2 Sample Animation 2: Pick-to-Light System

Order picking in most warehouses involves minimal technology. A typical order-picking tour starts with receipt of a picking list, which displays

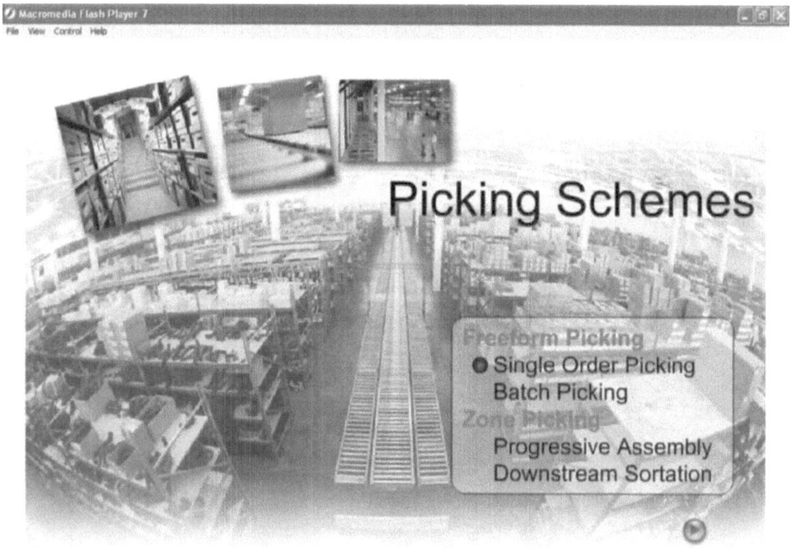

(a)

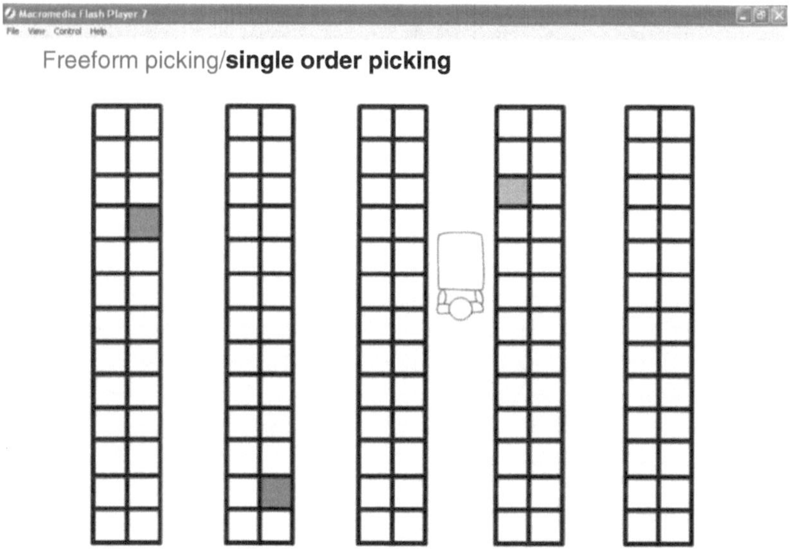

(b)

Figure 11.1 Animation for illustrating picking schemes in a warehouse.

(continued)

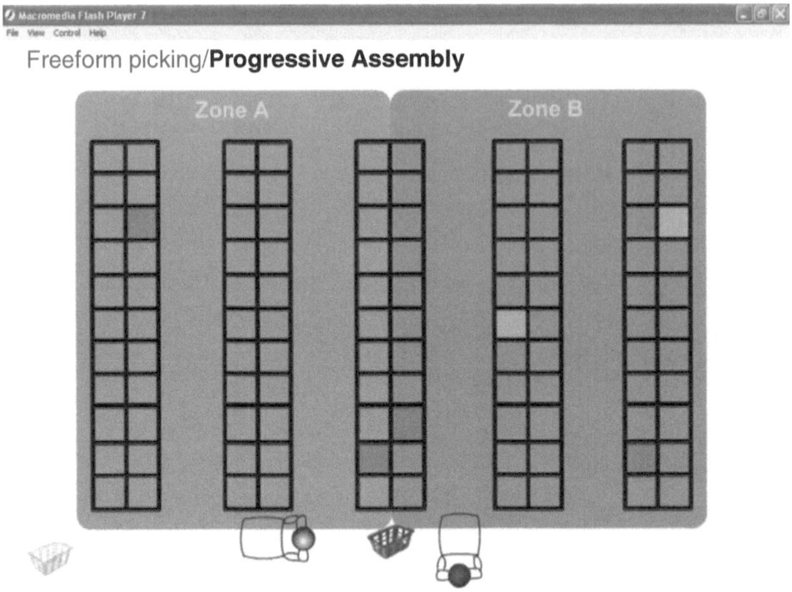

(c)

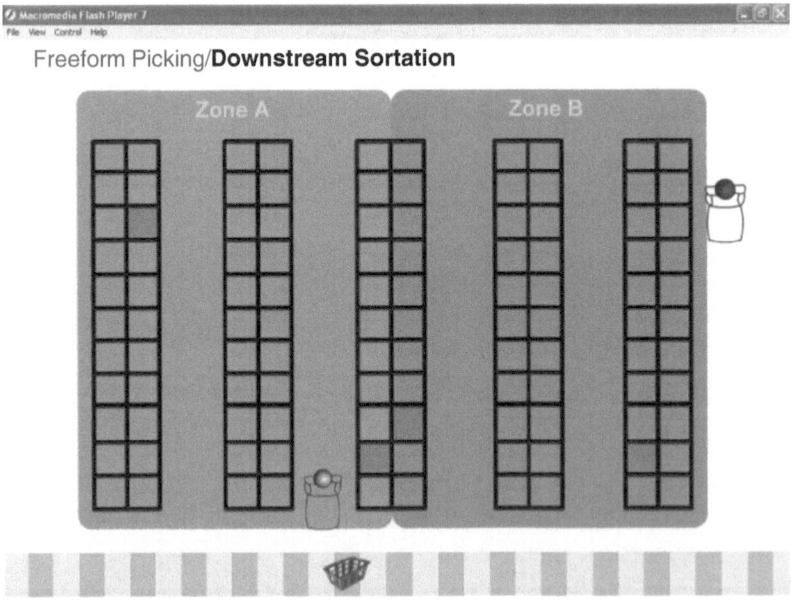

(d)

Figure 11.1 (continued)

the item number (also referred to as stock-keeping unit (SKU) number), the item location, and the number of units to pick. The order picker sorts the items to be picked in his or her mind, and then visits each item's location to make the picks (Figure 11.2a). Tasks that consume time during this classical mode of order picking are traveling, searching, extracting, and documenting.

One problem encountered in most order-picking operations is the erroneous picking of items. A pick-to-light system is a solution that technology offers to reduce the times for the mentioned tasks and to reduce order-picking errors (Figure 11.2b). In a pick-to-light system, an indicator light and an electronic numeric display inform the order picker on where to pick from and in what quantity. The order picker is relieved from the burden of searching the item locations and—to a degree—relieved from other tasks. Picking errors are also observed to decrease when pick-to-light systems are implemented. Thus the pick-to-light system is a viable option that can reduce costs and increase order-picking accuracy in a warehouse.

In this animation, the user is expected to pick two separate orders, first from storage locations without any pick-to-light systems and second from

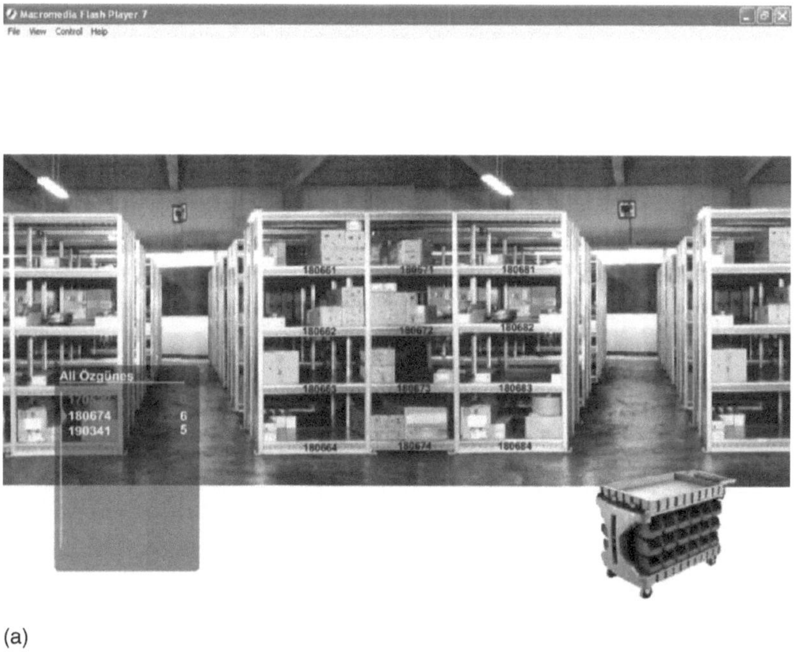

(a)

Figure 11.2 Pick-to-light system animation.

(continued)

(b)

Figure 11.2 (continued)

storage locations with pick-to-light systems. Even though the user is given pick lists in both scenarios, order picking is greatly facilitated by the pick-to-light systems in the second scenario. Thus the user experiences firsthand the benefits of pick-to-light systems.

11.5.3 Sample 3-D Model

Figure 11.3 illustrates the 3-D model for automated item-dispensing machine, as viewed from within a Web browser. It is required to install the SolidWorks Viewer software beforehand to be able to view the 3-D models.

The automated item-dispensing machine is also referred to as A-frame, because it is composed of two rows of dispensers positioned in the form of the letter A, with a belt conveyor running underneath. The information regarding the items to be picked (dispensed) is communicated to the A-frame and the dispensers automatically kick the bottommost packages of the items to be picked onto the conveyor. Workers continuously feed the dispensers with items picked from other systems such as the gravity

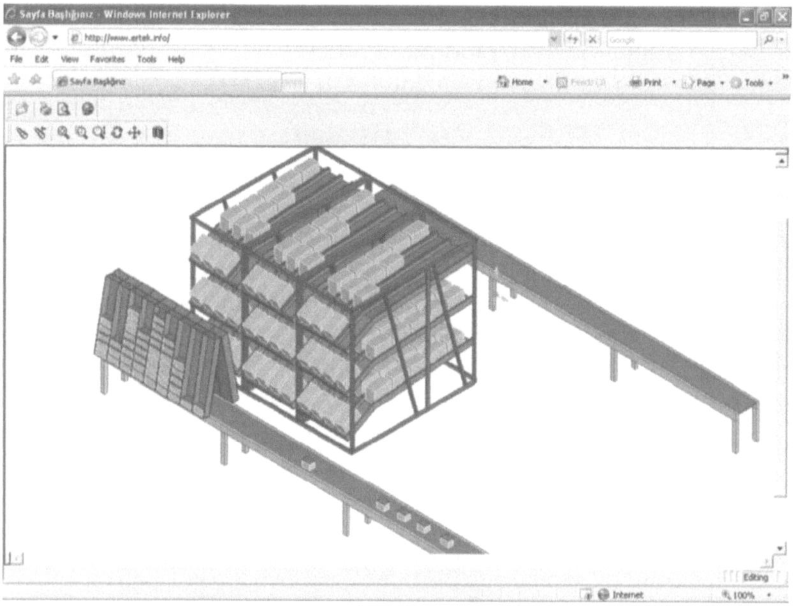

Figure 11.3 3-D model for automated item-dispensing machine (A-frame).

flow rack (as in Figure 11.3). Pick (dispensing) rates with the A-frame can reach up to 2000 picks per hour. The A-frame is especially applicable in industries which require "high throughput of small items with uniform size and shape" (Frazelle, 2001). One can find this equipment being used in warehouses from which cosmetics, wholesale drugs, compact discs, and publications are distributed.

Once the 3-D model is opened from within a browser, the user is able to interactively explore the equipment model. Zooming, panning, and rotating are some of the ways in which the user can interact with the model. These actions enable the user to focus on details, focus at specific sections of the equipment, and to observe the model from the best viewing angle.

11.6 Teaching How to Save Warehouse Space (Module 2)

In this section, a sample animation is explained from Module 2, "How to Save Warehouse Space." The animation illustrates Idea 72 suggested by

Kuchta and the staff of Gross & Associates (2004). It should be noted that some of the animations in Module 2 reflect the ideas presented in the *Modern Materials Handling* magazine (MMH, 2004).

11.6.1 Sample Animation 3

Figure 11.4 illustrates how to save warehouse space through providing bridges over cross-aisles. Figure 11.4a shows a storage block without any cross-aisles, which frequently requires the order picker to detour around the storage block. This causes wasted time during order picking. The classic solution to remedy this problem is to establish a cross-aisle that provides a quick pass way from one side of the storage block to the other side. However, cross-aisles are frequently implemented so as to consume the whole vertical space, resulting in loss of storage space. The solution is to provide bridges over cross-aisles as in Figure 11.4b, which enables the reclamation of otherwise-lost vertical space.

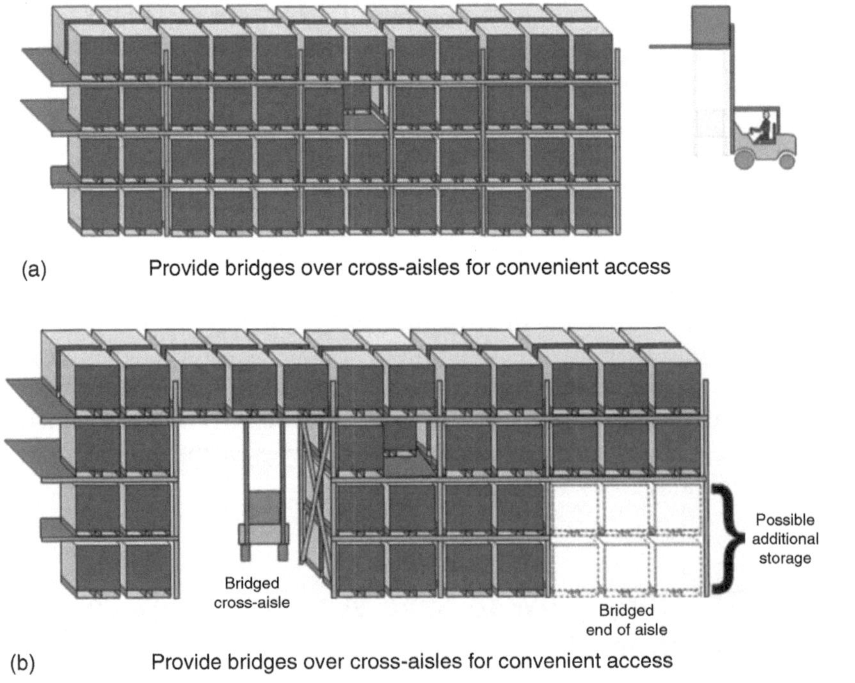

Figure 11.4 Animation showing how to save warehouse space through providing bridges over cross-aisles.

11.7 Other Possible Technologies and Approaches

In this section some of the other possible technologies and approaches, which can be applied in developing warehousing course materials, are discussed.

11.7.1 3-D Virtual Warehouses

Instructors can develop or contribute to the development of 3-D virtual environments that simulate real-world warehouses. These virtual environments can be created using simulation software with 3-D visualization capabilities, such as Automod.* Simulation models of warehousing systems have been used to teach simulation (Standridge, 2000), and can as well be used for teaching warehousing concepts. One can also build the virtual warehouses using 3-D visualization and animation software libraries or game engines.[†,‡] Developers of such learning environments can learn a great deal from practices of high-profile companies that provide learning solutions with interactive 3-D graphics. One such company is 3-Dsolve.[$] The company provides software platforms for collaborative simulation-based training, including implementations through the open-source Croquet operating system.[||]

Another way to construct virtual warehouses is to take 3-D photos or film 3-D videos[#] and show them to the students as they wear 3-D glasses. These pictures and videos, seen through 3-D glasses, give the feeling of observing the real world. Many Hollywood films, including titles such as *Shrek* and *Spy Kids 3D Game Over*, are available as 3-D movies and come with 3-D glasses for home entertainment. In the mean time, many technology companies including Philips, Mitsubishi Electric, and IBM have already developed prototypes for three-dimensional television (3-D TV), which will revolutionize the electronics, media, and entertainment industries. These televisions enable viewing of high-resolution 3-D content with bare eyes, without the need for any special apparatus such as 3-D glasses. Creating 3-D warehousing-education content for these televisions may greatly contribute to the recognition and advancement of the field of warehousing.

---

* http://www.brookssoftware.com/
† http://en.wikipedia.org/wiki/Game_engine
‡ http://www.devmaster.net/engines/
$ http://www.3dsolve.com/
|| http://www.opencroquet.org/
\# http://www.whurl.net/pages/3dgallery.php

11.7.2 360° Panoramic Scenes

Virtual warehouses can also be created through immersive 360° panoramic scenes. A panoramic scene is constructed through stitching together a series of images covering the 360° surrounding view using specialized software.[*,†,‡] Some panorama software products enable establishing hyperlinks to other scenes through hot spots within a given scene. In using this technology one can create a multitude of 360° scenes within a warehouse and can thus provide the users a virtual tour of the facility.

11.7.3 Media-Enhanced Case Studies

Supply-chain management classes typically involve discussion and assignment of business case studies. Implementation of these case studies can significantly be enhanced by describing the problems using animations and 3-D virtual worlds. Case studies covering supply-chain management and warehousing can be found in various textbooks, two of which are by Dornier et al. (1998) and by Simchi-Levi et al. (2002). Some of the cases in these books are from the Harvard Business School (HBS) Cases collection,[$] the richest source of business cases in the world. Two case studies from the HBS Cases collection that are especially suitable for adopting in warehousing classes are the following:

- "Amazon.com's *European Distribution Strategy*," by Janice Hammond, HBS Cases no: 9-605-002
- "*Velky Potraviny: Prague*," by William Coyle and Jay Rao, HBS Cases no: BAB013

It is proposed that the discussion of the above two cases and other relevant cases can be enhanced through educational media described earlier.

11.7.4 Software Demonstration Videos

Discussion and teaching of software tools are major contributors to the quality and usefulness of a warehousing course. It is suggested that instructors

---

* http://www.ulead.com/cool360/

† http://www.360dof.com/

‡ http://www.easypano.com

$ http://www.hbsp.harvard.edu/b01/en/cases/cases_home.jhtml

develop, contribute to the development of, or at least use software demonstration videos for teaching various software tools that are used in warehouse planning, design, and management. Software packages that can be taught include WMS, decision-support software, spreadsheets, databases, and modeling environments for simulation and optimization of warehousing systems. The software demonstration videos are created by first capturing the on-screen actions and recording any audio input as the user performs actions, editing the captured screens and audio, and then finally publishing the resulting videos in popular multimedia or video formats, such as .avi and .swf (Adobe Flash animation format), respectively. Once a software demonstration is recorded, it is also possible to enhance it by adding captions, images, buttons, and highlights. One highly capable commercial package that enables creation of software demonstration videos is Adobe Captivate.* DemoCharge† is an alternative commercial package for creating such videos: DemoCharge is available at a lower price and has the advantage of creating videos in various formats, including Java applets. However it falls behind Adobe Captivate with respect to editing of captured screen actions and incorporating other media (such as Adobe Flash animations) into the software demonstrations. The author has found this video-based method to be much more superior in teaching software applications when compared to other approaches, such as static tutorial documents.

11.8 Conclusions

This chapter presented existing resources, some newly developed resources used at Sabanci University, and alternative technologies and approaches for teaching warehousing and material handling.

The educational materials described in this chapter have been made freely available on the Internet through the MIT License.‡ They can be accessed through the Free Information Fountain link at http://www.ertek.info or http://people.sabanciuniv.edu/ertekg. The MIT License gives complete freedom to anyone to use the material in any way he or she likes, subject to the condition that the authors will not be held liable in any way from any unfavorable situations that might arise. This license was selected because it was found to be the most "freedom-granting" software license among those approved by the Open Source Initiative.$

* http://www.adobe.com/products/captivate/

† http://www.yessoftware.com/

‡ http://opensource.org/licenses/mit-license.html

$ http://www.opensource.org/licenses/

11.9 Future Work

One future path to improve the educational material described here is to convert the developed 3-D models to other file formats so that they can be embedded into simulation software products with 3-D animation capability, such as Automod, Arena, Promodel, Quest, and Taylor II. For example, the Automod software* supports importing of the VRML (.wrl), Open Inventor (.iv), 3D Studio (.3ds), AutoCAD (.dxf), and LightWave (.lwo) 3-D graphic formats (*Automod 11.0 User's Guide*, 2003). However the models developed in our project are currently available in SolidWorks graphic format and have to be converted to one of the mentioned formats and tested. This may be a very time-consuming task, because our trials to convert the models to VRML format from within SolidWorks resulted in unsatisfactory output.

Another future work is converting the models into 3-D graphic file formats which enable them to be embedded directly into Web pages, without having to install special viewer programs. One notable company that provides the technology for such functionality is Demicron,† which provides the software that enables Web-based interaction with 3-D models from within Java applets.

Enhancing the interactive animations through real-world videos, better graphics, and better user interfaces, and implementing the technologies discussed in Section 11.7 are yet other possibilities for future work.

One important consideration for those building e-learning environments is to conform to well-established specifications and standards, such as shareable content object reference model (SCORM).‡ Unfortunately, the e-learning materials reported in this chapter were not developed in conformance with any such standards. Revising the developed materials and the general structure of the learning modules according to one of the well-accepted standards is hence another possible area for future work.

Acknowledgments

The author is thankful to the following current and past students of Sabanci University for contributing to the project described in this chapter: Halil Keskin, Ömer Gürarslan, Yiğit Hüseyin Güler, Hasan Şahin, Şehmuz Cacina, Ali Can Akkaş, Eren Maden, Kivanç Kolukisa, Emir Cevheri, Emir Ceyhun, Güliz Menkü, and Giray Haci Turan. The author thanks Dr. Kutluk Özgüven, now at Doğuş University, for his insightful suggestions regarding

* http://www.brookssoftware.com/

† http://www.demicron.com/

‡ http://www.adlnet.gov/scorm/

the design of the interactive animations. And finally, the author thanks Sabanci University undergraduate students Hakan Kalelioğlu and Umut Karaarslan for their guidance in selecting the most "freedom-granting" license, namely the MIT license, before making the educational materials available on the Internet.

References

Andel, T. (1995) Get fit for the future. *Transportation & Distribution*, 36(1), 83–97.

Automod 11.0 User's Guide (2003) Brooks Automation Inc., Chelmsford, Massachussets.

Dornier, P.-P., Ernst, R., Fender, M., and Kouvelis, P. (1998) *Global Operations and Logistics: Text and Cases*. John Wiley, New York.

Ettouney, H., El-Dessouky, H., Al-Fulaij, H., and Al-Ansary, A. (2000) Computer package for design/rating of thermal desalination processes, *Computer Application in Engineering Education*, 8(2), 80–103.

Frazelle, E.H. (2001) *World-Class Warehousing and Material Handling*. McGraw-Hill, New York.

Heragu, S.S., Graves, R.J., Malmborg, C.J., Jennings, J., and Newman, D.L. (2003) Multimedia tools for use in materials handling classes. *European Journal of Engineering Education*, 28(3), 375–393.

Kuchta, J. (2004) *How to Save Warehouse Space: 153 Tested Techniques*. Distribution Group, New York.

Leung, W., Chen, Y., and Lee, Y. (2001) EM Wave simulation: An animated electromagnetic wave teaching package, *Computer Applications in Engineering Education*, 9(4), 208–219.

MMH (2004) 20 Space Saving Ideas. *Modern Materials Handling*, January, pp. 26–27.

Ong, S.K. and Mannan, M.A. (2002) Development of an interactive multimedia teaching package for a course on metalworking, *Computer Applications in Engineering Education*, 10(4), 215–228.

Ou, S., Sung, W., Hsiao, S., and Fan, K. (2003) Interactive web-based training tool for CAD in a virtual environment. *Computer Applications in Engineering Education*, 10, 182–193.

Roodbergen, K.J. and De Koster, R. (2001) Routing methods for warehouses with multiple cross aisles. *International Journal of Production Research*, 39, 1865–1883.

Simchi-Levi, D., Kaminsky, P., and Simchi-Levi, E. (2002) *Designing Managing the Supply Chain*. McGraw-Hill, New York.

Standridge, C.R. (2000) Teaching simulation using case studies, in J.A. Joines, R.R. Barton, K. Kang, and P.A. Fishwick (Eds.), *Proceedings of the 2000 Winter Simulation Conference*. Institute of Electrical and Electronics Engineers, Piscataway, New Jersey, pp. 1630–1634.

Index

A

B

C

H

I

J

L

M

N

O

P

Q

R

S

T

Y

Z